RADIATION SAFETY IN RADIATION ONCOLOGY

RADIATION SAFETY IN RADIATION ONCOLOGY

K. N. Govinda Rajan

CRC Press
Taylor & Francis Group
Boca Raton London New York

CRC Press is an imprint of the
Taylor & Francis Group, an **informa** business

CRC Press
Taylor & Francis Group
6000 Broken Sound Parkway NW, Suite 300
Boca Raton, FL 33487-2742

© 2018 by Taylor & Francis Group, LLC
CRC Press is an imprint of Taylor & Francis Group, an Informa business

No claim to original U.S. Government works

Printed on acid-free paper

International Standard Book Number-13: 978-1-4987-6224-3 (Hardback)

Library of Congress Cataloging-in-Publication Data

Names: Rajan, K. N. Govinda, author.
Title: Radiation safety in radiation oncology / K.N. Govinda Rajan.
Description: Boca Raton, FL: CRC Press, Taylor & Francis Group, [2017] |
Includes bibliographical references and index.
Identifiers: LCCN 2017008197| ISBN 9781498762243 (hardback; alk. paper) |
ISBN 1498762247 (hardback; alk. paper) | ISBN 9781315119656 (eBook) |
ISBN 131511965X (eBook) | ISBN 9781498762267 (eBook) | ISBN 1498762263
(eBook) | ISBN 9781498762274 (eBook) | ISBN 1498762271 (eBook) | ISBN
9781498762250 (eBook) | ISBN 1498762255 (eBook)
Subjects: LCSH: Cancer–Radiotherapy–Safety measures. | Medical physics.
Classification: LCC RC271.R3 R365 2017 | DDC 616.99/40642–dc23
LC record available at https://lccn.loc.gov/2017008197

Visit the Taylor & Francis Web site at
http://www.taylorandfrancis.com

and the CRC Press Web site at
http://www.crcpress.com

I dedicate this book to my parents Nagaraja Rao and Janhavi, wife Savitha,

son Jayesh, daughter-in-law Mridula Mahadevan, daughter Reshma, son-in-law Prasad,

grandchildren Arjun Rao and Rohan Rao, and the recently born granddaughter Zara Prasad.

Contents

Foreword

It is a great honor and pleasure to have the opportunity to perform an early review of *Radiation Safety in Radiation Oncology* authored by Govinda Rajan, PhD. Dr. Govinda Rajan is a world-renowned medical physicist from India with expertise in radiation protection. The writing of this textbook confirms his commitment and dedication to the teaching of radiation protection to the younger generation. This textbook as written is directed to the medical physicists specializing in radiation oncology. However, it has been written with general principles that can be used by those interested in radiation protection in the imaging and radiation oncology fields.

As an experienced medical physicist, this textbook is well written, covering the basic principles of radiation protection to the changes in modern radiation oncology. The general topics of radiation protection principles cover (a) the atomic and nuclear structures and their instabilities leading to the emission of radiation, (b) the different types and characteristics of the emitted radiation, (c) the interaction of radiation with matter, (d) the different types of radiation monitoring instruments, and (e) transport of radioactive materials. In addition to these radiation protection concepts and radiation management, this book also provides an understanding of the evolution of radiation protection to the adoption of radiation risks–associated concepts. This philosophical change has introduced new radiation protection terminologies. The first nine chapters update the principles of radiation protection, consistent with current regulation terminologies. The next three chapters deal with radiation protection in radiation oncology. Chapter 10 explores the shielding requirements and the safe operation of the radiation dose delivery machines for external beam radiotherapy. It also discusses new radiation therapy machines introduced during the last three decades. Chapter 11 deals with both low-dose-rate and high-dose-rate brachytherapy. In low-dose-rate brachytherapy, the chapter discusses the preparation, transportation, implantation, and removal of brachytherapy sources. Awareness of brachytherapy sources by the public in areas hosting patients is an important aspect of public safety. In high dose-rates, the major concerns are the operation, storage, and replacement of the radioactive sources. Dr. Govinda Rajan reviews a number of radiation incidents in the book's last chapter. While the message is clear that radiation accidents do occur, preventive measures have been evolving including the formation of comprehensive quality assurance programs, systems of double-checks, auditing processes, and security of radiation sources. In summary, this book provides up-to-date principles of radiation protection for modern radiation oncology.

The role of medical physicists in radiation medicine has been evolving primarily from radiation safety and equipment management to the inclusion of patient care. This inclusion of patient care has led to the specialization of medical physicists in one of the three principal areas of radiation medicine. Today, medical physicists in radiation oncology are highly involved in treatment planning, specific quality assurance, in-vivo dosimetry, verification of machine parameters, and treatment records of the individual patients. It is anticipated that these responsibilities will constitute the major workload of a typical medical physicist in addition to overseeing radiation protection and general quality assurance. This book will provide a good reference and an up-to-date resource on radiation protection principles for those practicing medical physicists.

Cheng B. Saw
Northeast Radiation Oncology Center
Dunmore, PA

Preface

This textbook is based on my experience in teaching radiation safety courses in the premier atomic energy research organization in India, the Bhabha Atomic Research Center, and in universities outside the research center. Radiation safety courses were taught to students of varied backgrounds, namely therapy X-ray technicians, medical physics students, and service engineers of radiotherapy equipment. We also periodically organized IAEA-sponsored radiation safety courses at our research center for the international students from the developing countries. Though this book is primarily intended as a text for the postgraduate medical physics or radiation oncology physics students, it will also be useful for practicing medical physicists and radiotherapy technologists, and therapy technology students.

The book is divided into 12 chapters. Chapter 1 is an introduction to atomic and nuclear physics. Chapter 2 is devoted to basic radiation physics. An elementary understanding of Chapter 1 and a deeper understanding of Chapter 2 are a prerequisite for the subject matter of this book. Chapter 2 deals mainly with the classification of radiations and their interactions with matter at microscopic and macroscopic levels. Since neutron monitoring is essential for radiation oncology centers operating >10 MV linacs, neutron interactions with matter have also been briefly discussed. The concept of kerma, dose arising out of radiation interaction with matter, is explained in some detail. The interactions cover only the energy range of interest in radiation oncology. Chapter 3 reviews the history of radiation protection, recognition of radiation risks, and the setting up of important radiation protection bodies to study the published research and recommendations in the field to recommend "acceptable radiation risks" and "dose limits" for staff to work with radiation in the field of medicine. Chapter 4 deals with the physical dosimetric quantities and the radiation protection quantities, and how the two quantities are related. Here, the units and standards of exposure and air kerma, and how they are disseminated to the users, are described in some detail. The radiation protection quantities, namely the organ doses and whole body dose, are defined in terms of equivalent dose and effective dose. In the absence of any primary standard to realize these quantities, operational quantities of ambient dose equivalent and personal dose equivalent were introduced in radiation protection to conservatively estimate the protection quantities. Since no primary standard could be established for the tissue equivalent phantom of hypothetical composition, in terms of which operational quantities are defined, the only way these quantities can be measured is by relating them to the measurable physical dosimetric quantities. Using these relationships, the operational quantities can be derived and the protection-level monitors can be calibrated in terms of operational quantities. In hindsight, it looks more reasonable to relate the quantities of exposure and air kerma directly to the effective dose and equivalent dose concepts instead of going through an intermediate step of operational quantities, which are also not measurable, like the protection quantities. This is the present thinking in the field and these concepts are explained in detail in this chapter. Chapter 5 deals with the system of radiation protection as developed by the International Commission of Radiological Protection (ICRP), which is the most general framework that is applicable not just in the field of medicine but in other equally important fields of application such as the nuclear industry or nuclear accident situations, to control man-made sources of radiation. This chapter discusses the natural radiation background exposures followed by the ICRP categorization of radiation exposures, and the concept of justification/optimization/dose constraint/dose limitation in optimizing radiation exposures. The chapter then covers basic regulations in radiation oncology, regulatory requirements for starting a radiation oncology facility, designation of areas and posting of areas around the treatment room facility, and the duties of the radiation safety officer to ensure the radiation safety of the installation. I would like to thank here Dr. Jean-Francois Lecomte of the French Institute for Radiological Protection and Nuclear Safety, for reviewing my write-up on the ICRP system of radiation protection. Dr. Jean-Francois Lecomte is the Secretary of ICRP Committee 4 that is responsible for developing principles, recommendations, and guidance on the protection of man against radiation exposure, and considering their practical application in all exposure situations. Chapter 6 deals with the principles of calibration of protection monitors. The chapter describes the setting up of calibration fields, their characterization in terms of ambient dose equivalent and personal dose equivalent, and calibrating the workplace monitors and personal monitors in terms of these quantities. Chapter 7 covers the principles of radiation detectors used in workplace monitoring, their operation characteristics, and their design and suitability for monitoring different ambient exposure situations. The main features of survey meters, their maintenance, and use for surveying purposes are also covered. Chapter 8 is in the same format as Chapter 7 but deals with personal monitors used for individual monitoring. Chapter 9 discusses the transport of radioactive materials. Radiation oncology uses radioactive sources and radioisotope machines for brachytherapy practice. There are standard regulations for procuring, possessing, packaging, transporting, and disposing of the radioactive sources. These aspects are described in this chapter. Chapters 10 and 11 deal with radiation protection in the practice of external beam therapy and brachytherapy, respectively. External beam therapy equipment and remote afterloading brachytherapy equipment, both involve treating patients from outside a shielded enclosure. Incorporation of safety in the design of the equipment, acceptance testing of equipment, treatment room design, shielding calculations, and regulatory approval requirements for initiating treatment are some of the common features of radiation therapy for these two kinds of equipment. All these aspects are described in the two chapters. In addition to conventional linac treatments, advanced treatment techniques such as IMRT, SBRT, and TBI are carried out in many centers. The chapter also briefly explains how they influence room shielding. Shielding concepts for other types of treatment delivery equipment (e.g., tomotherapy, cyber knife treatment, gamma knife treatment) and imaging equipment (e.g., simulator, CT simulator) are also mentioned. Time/distance/shielding concepts in the control of radiation exposures of the exposed individual and their relevance in external beam therapy and brachytherapy are also explained. For gaining more knowledge on shielding calculations, one must refer to excellent materials available on the net. Particularly, I encourage students to read the excellent talks presented by Melissa C. Martin, Peter J. Biggs, Jim Rodgers, and Rajat Kudchadker. Other good references are the International Atomic Energy Agency (IAEA) publication, Safety Report Series 47, *Radiation Protection in the Design of Radiotherapy Facilities* and the most popular textbook written by Patton H. McGinley, *Shielding Techniques for Radiation Oncology Facilities*. Principles of installation survey are also explained in Chapter 10. Chapter 12 lists the accidents that have occurred in external beam therapy and brachytherapy over the years, the causes, and consequences of the accidents. More details regarding the accidents can be found in the IAEA publication referenced in the Chapter 12. Every medical physicist must read this publication to avoid repeating the same mistakes that led to these accidents.

Many problems are worked out as examples in each chapter. Many figures have been included in each chapter to illustrate the concepts. At the end of each chapter there are some general references and a set of review questions.

I sincerely hope the book does well to justify the faith the publisher has placed in me. I welcome any kind of feedback from the readers, which will help me to improve the content of the book in the future. My contact email is kngrajan@gmail.com.

Govinda Rajan K.N.
PSG Institute of Medical Sciences and Research

Acknowledgments

I must thank all the medical physics experts working in the field of radiation oncology who have given presentations at the American Association of Physicists in Medicine meetings and elsewhere and in such great detail about each of the topics discussed in this book. I did not have time to read all their work in detail but whatever material I did read was very helpful to me while writing this book.

My special thanks to Dr. Tharmar Ganesh, senior and chief medical physicist, and his colleagues at Fortis Memorial Research Institute, Gurgaon, India, for clarifying my doubts on several occasions, and also helping me with relevant reports and sending me the pictures for my book.

I must thank CRC press for giving me an opportunity to write this book. Though the thought of writing such a book was always in my mind, I had not found time to commit myself to take up this work, which I knew would require considerable amount of time. When the publisher approached me with the proposal, I was in the United States and had some free time and so I readily agreed to do this. However, I had to come back for professional reasons, and soon after my return, I became busy with my teaching and academic work, which delayed my writing by many months. Somehow, I managed to give the book a shape and managed to complete it.

I must also thank the Dean of PSG Institute of Medical Sciences and Research at Coimbatore, India, for giving me the liberty to write this book during my office hours. Without his support, the book would have taken much longer to complete.

Author

Govinda Rajan K.N. retired as the Head of the Medical Physics and Safety Section at Bhabha Atomic Research Center (BARC), Mumbai, India, in the year 2006. Since then, he has been working as a professor of medical physics at PSG College of Technology and PSG Hospitals, Coimbatore, India. As the Head of Medical Physics at BARC, he was in charge of radiation safety and quality assurance (QA) of all medical institutions in the country. He has vast experience working and teaching in radiation oncology physics, radiotherapy calibrations, medical dosimetry, and medical radiation safety. He has initiated several programs in India in the fields of mammography, diagnostic radiology, and radiation therapy with the objective of optimizing diagnostic information to patient dose in diagnostic procedures and target dose to normal tissue dose in radiation therapy treatments. In the fields of brachytherapy and external beam therapy, he had developed quality audit (QA) procedures and audited several centers to ensure radiation safety and in-house QA.

Dr. Rajan has contributed significantly to medical radiation dosimetry at the international level. He has represented his country in several technical committee meetings at the IAEA, Vienna, Austria, relating to medical radiation dosimetry, medical radiation safety, and safety and security of radiation sources. His contributions have appeared in several IAEA publications relating to these fields. His *Medical Dosimetry Certification Study Guide* published by Medical Physics Publishing (MPP), Madison, Wisconsin, in 2004 was extremely well received by the practicing medical dosimetrists. A second edition of the guide was published by MPP last year.

Dr. Rajan now coordinates radiation safety research activities at the PSG Institute of Medical Sciences Research and Hospitals, Coimbatore, India, with funding from the Indian Atomic Energy Regulatory Board. He also directs the radiological physics activities at the Institute.

1

Basic Atomic and Nuclear Physics

1.1 Introduction

We all have a basic idea about the atom as resembling the planetary model. There is a small nucleus with dimensions of the order of 10^{-13} cm, with electrons swirling around the nucleus like the planets around the sun. The size of the atom is larger by about 10^5 times compared to nuclear dimensions. Though the quantum mechanical idea of an atom is not that simplistic, this classic picture explains many of the characteristics of the atom, and we will make use of this to understand the observed properties of the atom.

An atom is electrically neutral and hence the nuclear charge must be equal to the total electronic charge. The nucleus contains neutrons and protons carrying a charge of 0 and +1, respectively, expressing the charge in terms of e = 1.6×10^{-19} C. The charge of the electron is −1. The neutrality of the atom suggests that the number of electrons must be equal to the number of protons in the nucleus.

The mass of atoms is minute and so the usual units like grams or kilograms are not suitable units to express them. The unit of mass must be the mass of the atom itself, so that the atomic masses can make more sense.

1.2 A Short Introduction to Atomic Structure

Niels Bohr, in his atom model proposed in 1913, introduced the concept of quantization (or discreteness of matter at submicroscopic level), to derive the discrete energy levels of the hydrogen atom. This could successfully explain the experimentally observed spectral lines of hydrogen. The energy levels corresponded to the atomic shells named K, L, M, etc., K being the first shell of lowest energy, L the second shell of higher energy, and so on. The shells were characterized by a principal quantum number n which assumed integer values (n = 1 for K shell). The transitions between the atomic levels gave rise to the emission of discrete spectral lines in the visible or ultraviolet regions, or in the X-ray region, depending on the strength of binding and the energy level differences. An atom, in its ground state, is in its lowest energy state (i.e., occupies the orbitals closest to the nucleus) according to the rules governing their occupancy. The unoccupied energy levels are vacant in the ground state. When the electrons of an atom absorb energy, they go to the higher energy states. The electrons quickly (in about 10^{-8} seconds) return to the lower vacant states emitting the excess energy in the form of photons. The transitions occur because the electrons would like to be closest to the nucleus for maximum binding.

When the energy supplied is greater than the binding energy (BE) of the electron, the electron can be knocked out of the atom—this is called ionization—and the atom becomes an ion. The electron will exist in its lowest energy state if all the electrons occupy the first shell. However, Pauli's exclusion principle forbids all electrons to occupy the same state. According to this principle, no two electrons (or nucleons) in an atom can have identical quantum numbers.

NOTE: Since the electrons can exist in two spin states (spin up or spin down) each orbital can be occupied by not more than a pair of electrons in opposite spin states. The atomic particles have half integral spin = ½ (expressed in units of $\hbar$). These particles follow the Pauli exclusion principle. This principle is the most fundamental principle followed by atomic particles.

We will briefly see how the electrons are organized in an atom, as per the above principle.

In an atom, the atomic shells are divided into subshells and the subshells are further divided into orbitals. Quantum mechanics determines the configuration of an atom and we will not get into its details. We will only look at it in a qualitative way. The easiest way to visualize all the orbitals is as follows:

1. Shell quantum number = n (takes values 1, 2, 3, …).
2. Subshell quantum number = l = 0, 1, 2, … (n−1). The electron energy depends on n and l values.

For each l, orbital quantum number m_l = +l to –l = (2l + 1) orbitals; Δm_l = ±1.

This gives the number of orbitals in the subshell and can be thought of as different orientations of the orbitals. These orbitals are degenerate (meaning having the same energy) but in a magnetic field, the degeneracy will be removed and the energy levels split up to show a fine structure.

Number of electrons in each orbital = 2 (spin up and spin down).

In spectroscopy, the subshells are referred to as s, p, d, f, (referring to sharp, principal, diffuse, fundamental series, respectively). Beyond f, the letters g and h are used for the subshells.

Example 1.1

How many electrons are in the shell n = 2?

Subshells: l = 0 and 1 (s and p)

No. of electrons in s state = 2 (m_l = 0)

For the p subshell, there are three orbitals (m_l = 1, 0, −1).
The number of electrons in the p subshell = 3 × 2 = 6.
Therefore, shell two can contain eight electrons.
The electron configuration is written as follows: $1s^2$ refers to s state containing two electrons. If the first and second shells are full, the configuration will be $1s^2 2s^2 2p^6$. So, it is easy to write the electron configuration of atoms. The simple formula $2n^2$ gives the maximum number of electrons a shell of principal quantum number n can take.

1.3 The Filling of Sublevels

In the hydrogen atom model, the electron energy is governed only by n. In the case of multiple electron atoms, the subshells do not have the same energy, as can be seen from the energy level diagram. This is because of the mutual repulsive interaction between the electrons. The energy of the subshell gradually increases with increasing l value. However, the orbitals in a subshell have the same energy, as stated earlier (see Figure 1.1).

The unpaired electrons have less energy compared to paired electrons, so they tend to occupy unoccupied orbitals first before they start pairing. There is an overlapping of the orbital energy levels of shells for higher Z elements, since the atomic energy levels are not decided by just the n value, but by n + l values. Figure 1.2 shows the paired electrons in the orbitals and the overlapping of higher energy states.

The orderly filling of atomic levels takes place up to the first 18 elements, beyond which the overlapping occurs and an Aufbau diagram can be used as a simple rule to predict the filling of levels (see Figure 1.3).

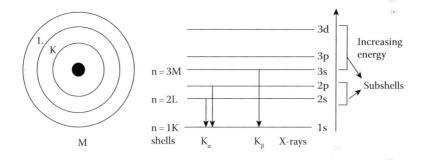

FIGURE 1.1
Atomic shells, energy levels, and transitions.

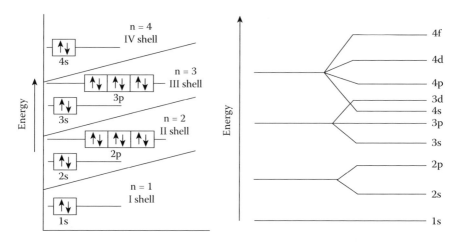

FIGURE 1.2
Energy levels in low Z and high Z atoms.

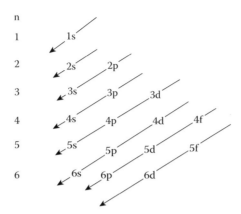

FIGURE 1.3
Aufbau diagram.

As per the above principle, the order of filling is 1s, 2s, 2p, 3s, 3p, 4s, 3d, 4p, 5s, 4d, ...
We will not discuss the order of filling further since the interactions inside the atoms, especially for the high Z atoms, are much more complex involving electromagnetic interactions between the nucleus and the electrons, and the mutual repulsive interactions between the electrons themselves. There is no simple rule to predict the order of filling, including the Aufbau principle, for all the elements.

1.4 A Short History of the Nucleus

The nucleus was discovered by Ernest Rutherford in 1911, from the concentrated scattering of alpha particles (details about this will be given later) by a central core of the atom, which was named the nucleus. Rutherford was not expecting to find a core inside an atom and was totally surprised by the alpha scattering behavior. He was expecting small deflections of the alpha in a sphere of uniform charge distribution. The large angles of scattering including back-scattering showed that there must be a concentrated charge core in the atom to produce such deflections.

Since the atom is neutral, it was known that there must be positive charges inside the atom neutralizing the electron charges. The discovery of the nucleus led to the idea that the positive charges must be confined to the nucleus. Later, Rutherford's experiments on transmutation of elements by alpha bombardment revealed a hydrogen nucleus in the final products.

$$^{14}N + \alpha \rightarrow {}^{17}O + p \text{ (hydrogen nucleus)}$$

The hydrogen nucleus was therefore identified as a fundamental particle of all atoms and was referred to as a proton.

The neutron had not been discovered in 1919 when the above experiments were conducted. It was known at that time that the atoms weighed roughly twice as much as the mass of the total number of protons, suggesting that there must be an equal number of protons inside the nucleus, pairing with electrons to cancel their charges. The discovery of the neutron is a long and interesting story but in essence, the discovery of matter waves and quantum mechanics led to the conclusion that electrons could not be part of the nucleus.

Further experiments by Chadwick in 1932 led to the conclusion that there were neutral particles inside the nucleus that accounted for the nuclear mass, and the particles were named neutrons. The nuclear constituents are now known by the common name of nucleons.

Further research has shown that protons and neutrons also have an internal structure, but in the energy ranges used in the radiations in medicine, for example in neutron therapy, proton therapy, electron therapy, or X- or gamma-ray therapy, the radiation interactions do not affect the integrity of the nucleons, because of the exceedingly large apparent binding energies of their internal constituents. This situation is very similar to that in chemistry where the nuclear structure did not matter for the energies involved in chemical reactions. All the elemental characteristics like the chemical, optical, thermal, and electrical properties are only concerned with the external valence electrons. No amount of heat or pressure in the laboratory can get the nucleons out of the nucleus. The atomic binding energies are of the order of electron volts (eV) to kilo electron volts (keV) at the maximum. The nuclear binding energies are in the million electron volts (MeV) range and therefore the nucleus can lose its integrity only when the particles used in radiation oncology and interacting with matter or with patient are in this range. This will be discussed later in this chapter.

1.5 Properties of the Nucleus

1.5.1 Nuclear Size

Various scattering experiments suggested that the nucleus must be approximately spherical with uniform density, except at the peripheral region (Figure 1.4). The radius of the nucleus varies as $R = R_0 A^{1/3}$, where A is the mass number of the nucleus and $R_0 \approx 1.2 \times 10^{-15}$ m or 1.2 fm. The nucleons are obviously smaller than the nucleus.

The nuclear radius equation shows that the nucleus gets larger and larger as A increases. The nuclear density is $\rho_n = 2.3 \times 10^{17}$ kg/m³. (Compare this density with density of elements where the atoms are pretty much closely packed.) The fact that the nuclear density is independent of the number of neutrons or protons in the nucleus suggests that the nuclear force between the nucleons is the same, independent of whether they are neutrons or protons.

Example 1.2

What is the radius of a ^{12}C nucleus?

$$R_C = R_0 A^{1/3} = 1.2 \times 10^{-15} \times (12)^{1/3} = 2.7473 \times 10^{-15} \text{ m.}$$

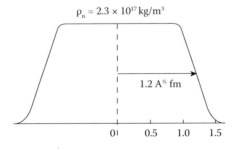

$\rho_n = 2.3 \times 10^{17}$ kg/m³

1.2 A^⅓ fm

0   0.5   1.0   1.5

FIGURE 1.4
Nuclear density variation from the nuclear center.

1.5.2 Nuclear Mass

We all know that the SI unit of mass (kg) was not appropriate for using with atoms and molecules, where the unit of mass must be of the order of atomic masses. So, a new unit was defined—atomic mass unit. The definition is:

Mass of one carbon atom, ^{12}C = 12 amu = 12 u (the symbol u is used to represent the amu).

Since the electron masses are negligible, the mass of a nucleon is roughly one twelfth the mass of a carbon atom, or 1 amu. In order to know the mass of nucleons in kg, we must be able to relate amu to kg. In chemistry, the term mole is used to represent molecular weight (or atomic weight) in grams. So, one mole of ^{12}C is 12 g. One mole of ^{1}H is 1 g and so on. One mole of any substance contains the Avogadro number (N_A) of atoms or molecules, or electrons, depending upon the substance chosen. Scientists figured this out from the way elements combined. For instance, Na and Cl combine in 1:1 ratio to form NaCl, by atom, but by mass they combine on a 23 g to 35.5 g ratio, which are their atomic masses. This indicated that if a mole is defined as molecular or atomic weight of the elements expressed in grams, they should have equal number of atoms or molecules. This number was evaluated as 6.22×10^{23} and named in honor of Avogadro.

Example 1.3

Express 1 amu in kg

6.22×10^{23} ^{12}C atoms weigh 12 g and each atom weighs 12 amu.

$6.22 \times 10^{23} \times 12$ amu = 12 g

1 amu = $1 \div (6.22 \times 10^{23})$ g = 0.16×10^{-23} g or 0.16×10^{-26} kg.

1.5.3 Isotopes

A nuclide of an element can have different number of neutrons, known as the isotopes of the element. Of course, they cannot have different number of protons since that would change the element into some other element. As to why isotopes exist, someone aptly answered that they exist because they can. If any configuration gives a stable atom, we can hope to find that atom since it is energetically possible. So, some elements have one, and some have two, and so on. For instance xenon has eight stable isotopes and tin has a maximum of 10 stable isotopes.

Given the mass of each isotope of an element and the abundances, we can calculate the average atomic mass and vice versa. If W_i represents the abundance (or weight) of i-th isotope of an element,

$$\Sigma \, W_i \, M_{x,i} = M_x$$

where x refers to the element and i refers to the i-th isotope, and M_x is the average mass of the atom.

Example 1.4

Nitrogen has two isotopes ^{14}N and ^{15}N. M_{N-14} = 14.003074 amu; M_{N-15} = 15.000108 amu.

W_{N-14} = 99.63%; W_{N-15} = 0.37%. Calculate the average atomic mass of nitrogen.

M_N = 0.9963 × 14.003074 + 0.0037 × 15.000108 = 14.006763 amu

Mass numbers are approximately the atomic mass, and so if only the mass numbers are given, we can still calculate the approximate atomic mass value. Here, the value would be 14.004.

Sometimes we may need to know the mass of atoms or nucleons in the usual SI unit of kg that is commonly used.

1.5.4 Binding Energies

Another important concept of importance is the binding energy (BE). The attraction between the nucleus and the electrons leads to the formation of atoms and the attraction between the nucleons leads to the formation of nucleus. Where is the energy of attraction or the BE energy coming from? If it is possible to weigh the mass of a nucleus, it will turn out to be less than the total mass of its constituents (the neutrons and the protons). This deficit in mass is known as the mass defect. Mass defect (ΔM) multiplied by c^2 (i.e., $\Delta M \times c^2$), where c is the velocity of light, gives the BE of the nucleus. This was the energy originally released when the nucleons came together to form the nucleus and this is the energy to be expended again to get back the original state of free nucleons, as shown in Figure 1.5.

> **Example 1.5**
>
> Calculate the BE per nucleon for carbon atom using the following data:
>
> $m_n = 1.008665$ amu; $m_p = 1.00782$ amu; 1 amu $= 1.66 \times 10^{-27}$ kg; $M_C = 12$ amu; 1 eV $= 1.6 \times 10^{-19}$ J
>
> Mass defect $= \Delta M_C = M_C - (6\ m_n + 6\ m_p)$
>
> $$= 12 - (6 \times 1.008665 + 6 \times 1.00782) = -0.099 \text{ amu}$$
>
> $$BE = \Delta M_C\ c^2 = 0.099 \times 1.66 \times 10^{-27} \times (3 \times 10^8 \text{ m/s})^2$$
>
> $$= 1.478 \times 10^{-11} \text{ J}/(1.6 \times 10^{-19}) \text{ J/eV} = 92.37 \text{ MeV}$$
>
> BE per nucleon $= 92.37/12 = 7.7$ MeV/nucleon ≈ 8 MeV/nucleon

NOTE: According to Einstein's theory of relativity, $E = mc^2$ and so any mass can be converted into equivalent energy. This does not mean we can just destroy an electron and convert it into some form of equivalent energy, but if any transformation can take place it will take place only when conforming to the above equation. For instance, when an electron meets a positron at rest, they will annihilate, and will be completely converted into energy in the form of gamma rays. According to Einstein's equation, the energy equivalent of an electron's rest mass = 511 keV. So, when the electron and positron are annihilated two gamma rays of energy 511 keV are created, as per the above equation. The mass defect is a classic case where part of the neutron and proton masses are utilized

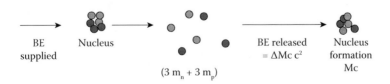

FIGURE 1.5
Concept of binding energy.

to bind them into a nucleus, with an attractive force. In order to get back their masses as free nucleons, that much energy must be supplied back to the system. These inter-conversions take place exactly as per Einstein's equation, as explained and illustrated above.

The average BE per nucleon for low Z elements is about 8 MeV. So, to eject a neutron out of the nucleus of these elements the average incident energy must be higher than 8 MeV. Thus a 10 MV X-ray beam used in radiotherapy is capable of knocking a neutron out of the nucleus since the reaction is energetically possible for photons of energies greater than 8 MeV, in the 10 MV spectrum. For the same reason, photons produced in a linac operating at 6 MV, photons emitted by a ^{60}Co source in a teletherapy machine, or a gamma knife unit, or photons from an ^{192}Ir high dose rate unit or ^{137}Cs other sources used in brachyther-apy cannot dislodge a neutron from the equipment-shielding components or from the wall, or from the body of a patient. Changing the nuclear configuration, by neutron ejection, often leads to nuclear instability and nuclear radioactivity. These aspects will be discussed elsewhere in the book.

1.5.5 Fundamental Forces of Nature

Table 1.1 compares the four fundamental forces that exist in nature and their characteris-tics. The nuclear force is quite complex; it is the strongest of the four fundamental forces in nature.

The characteristics of the forces are summarized below:

1. One thing in common for all the forces is that they are all exchange forces. In other words, the forces are transmitted by the field quanta. The range depends on the mass of the exchange particle. This is usually explained by the constant exchange of a ball between two boys. Heavier balls would bring them closer, and this exchange interaction would keep them bound. Similarly, the forces are mediated by the field quanta emitted and absorbed by the particles. They are called virtual quanta or particles since they are constantly emitted and absorbed.

2. The neutron and proton are composite particles made up of three quarks, the fun-damental particles of matter. There are two types of quarks, the *up quark* (u) and the *down quark* (d). Types u and d have charge of ⅔ e and ⅓ e, respectively. Since the proton charge is 1 e and the neutron charge is 0, uud forms a proton, and ddu

TABLE 1.1

Comparison of Four Fundamental Forces of Nature

Interacting Forces	Relative Strength	Force Range	Mediating Field Quanta	Mass/Charge of the Quanta
Nuclear (strong)	1	Short ($\approx$ 1 fm) Varies as r^2	Gluon Pion π^+, π^-, π^0	$0/0 \approx 130$ MeV
Electromagnetic	$\approx 10^{-2}$	Long ($\sim 1/r^2$)	Photon	$0/0$
Weak	$\approx 10^{-6}$	Short ($\approx 10^{-3}$ fm)	W^+, W^-, Z^0 bosons	$= 90$ GeV/+1, -1, 0
Gravitational	10^{-43}	Long ($\sim 1/r^2$)	Graviton	$0/0$

forms a neutron. There are other properties of quarks, which are not of relevance to our subject matter and so we will not discuss them in this book.

3. The strong force binds the quarks together and this force is mediated by gluons. Theory and experiments in elementary particle fields suggest that quarks and gluons cannot be extracted from nucleons, in the way that an electron can be extracted from the atom, or a nucleon from the nucleus by applying the BE. The force inside the nucleon varies as r^2, which means if energy is applied to pull a quark out of the nucleon, this energy increases with increasing distance, like a spring. There is no way a quark can be pulled out of a nucleon and the vast energy put in only goes to produce more quarks and antiquarks, producing more hadrons but not individual quarks.

4. The force that binds the nucleons together is the residual force of this nuclear force "spilling out" of the nucleons. It is this residual force that binds the nucleons in the nucleus. It becomes repulsive when the two nucleons are too close (say 0.7 fm), is maximally attractive around 1 fm, and becomes almost insignificant at around 2.5 fm. Figure 1.6 illustrates these concepts.

5. The repulsive core arises from the Pauli exclusion principle defined earlier. So, an attractive nuclear force cannot allow the nucleons to come closer than a certain distance. This ensures that the nuclear force cannot crush the nucleons together and the nucleus will have finite size.

6. The electromagnetic force is the force acting between moving charges and is responsible for binding the electrons to the nucleus, and the repulsive forces between the protons inside the nucleus. The force is mediated by photons with very long range. Therefore, many of the forces that are present in nature are electromagnetic in nature, acting at atomic and molecular levels.

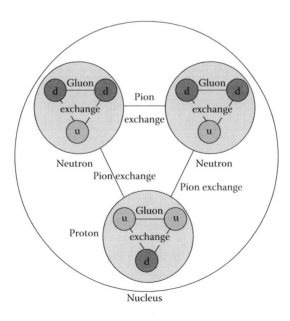

FIGURE 1.6
Mediation of strong force between nucleons and between quarks.

7. A weak force acts at extremely small distances (range ≈ 0.15 of the nucleon dimensions) and so the quanta mediating this force (W^+, W^-, and Z bosons) are the heaviest.

8. The free neutron decay and the transformation of neutrons and protons in radioactivity are attributed to the weak forces. We will discuss these aspects in the section on radioactivity. The transformations occur at quark level and hence the reason for discussing some aspects of the nucleon structure.

9. The gravitational forces are the weakest of the forces since they act on large masses like the sun and the planets. The gravitational force between people, for instance, is negligible.

1.5.6 Nuclear Energy Levels

Like the electronic energy levels, the nucleus also exhibits discrete energy levels, as shown in Figure 1.7.

The existence of nuclear energy levels suggests some kind of shell structure in the nucleus, similar to the electronic shell structure and indeed, there is a shell model of the nucleus but we will not explore this much further. The nuclide in the ground state has the least energy. All the excited states have higher energies and are vacant. When the nuclide is suitably excited (by absorbing energy) the nucleons move to the excited state. Since the nucleons stay at the lowest energy levels for stability, the excited states have an extremely small lifetime (of the order of 10^{-12} seconds) and any nuclide in the excited state usually releases the excess energy in some fashion to come back to the ground state. Sometimes the excited state could be metastable, or will have an extremely long lifetime amounting to minutes or years.

The fundamental difference between the electronic and nuclear energy levels is that the electronic energy levels are in the eV to keV energy range (emitting radiations in the visible, ultraviolet range to the X-ray range) while the nuclear energy levels are in the MeV range due to the stronger nuclear binding. That is why photons (or gamma rays) emitted

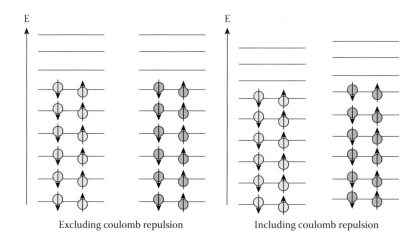

Excluding coulomb repulsion   Including coulomb repulsion

FIGURE 1.7
Nuclear energy levels.

by the nuclides are in this range. So, if we are interested in photons in the MeV range, these nuclides are the ones we have to use, as in the case of cobalt teletherapy.

Nuclear theory is not fully understood yet; so the different properties of the nuclides will be explained by different models (such as the shell model, or the liquid drop model).

1.5.7 Nuclear Stability and Radioactivity

There are two parameters that have a strong bearing on the nuclear stability: the neutron/proton ratio, and whether the nuclides are even–even, even–odd, or odd–odd. Table 1.2 shows that nuclides with even number of neutrons or protons, or even number of both, exhibit greater stability compared to nuclides having odd number of protons and neutrons. This shows that pairing of neutrons or protons bring in extra attraction and BE to the nuclides.

The first 92 elements in the periodic table, H to Pb, are stable elements (but for a few exceptions), and all the elements occurring beyond Pb are inherently unstable and give rise to the phenomenon of radioactivity (i.e., spontaneous emission of radiations).

Radioactivity was discovered by Becquerel while studying the properties of a uranium compound. Following the discovery of X-rays, he was trying to find out if uranium compound could emit some form of X-rays following the absorption of sunlight, but to his surprise he found the compound emitting some radiation (when it blackened a photographic plate), even without sunlight exposure. His experiments to deflect the radiations in a magnetic field showed one type deflecting to the right, one deflecting to the left and one not deflecting at all. He concluded that the emitted particles were carrying a positive charge, negative charge, and no charge, respectively. Not knowing the nature of the radiations, they were named as α, β, and γ radiations, by Rutherford. We now know that α is a helium nucleus, β is an electron, and γ is electromagnetic radiation. Madame Curie named the phenomenon radioactivity. When Curie extracted uranium from the ore and studied its activity, she found a lot more radioactivity and other radioactive elements in the ore. The subsequent studies revealed that all elements beyond lead were inherently unstable and were radioactive. The radioactivity emanates from the elements ^{238}U, ^{235}U, and ^{232}Th resulting in a radioactive chain that ends in stable ^{206}Pb, ^{207}Pb, and ^{208}Pb, respectively. All the uranium and thorium have not decayed by now because their half-lives are of the order of the age of the earth.

Later, it was discovered that radioactivity can be induced in any element by adding more protons or neutrons to the nucleus. This phenomenon is known as artificial radioactivity.

TABLE 1.2

Nuclear Stability and Number of Nucléons

No. of Protons	No. of Neutrons	Occurrence of Stable Nuclides	Stability
Even	Even	168	Most stable
Even	Odd	57	↓
Odd	Even	50	
Odd	Odd	4	Least stable

1.5.8 An Explanation for Radioactivity

By examining the stable elements that occur in nature (hydrogen to lead), it will be discovered that they have a certain n/p ratio (i.e., the neutron to proton ratio). It can also be seen that a given element can exist in more than one isotopic state, with slightly different neutron to proton (n/p) ratios. The fact that each element existed only in a few isotopic states in nature suggests that with any other n/p ratio, the elements are likely to be radioactive and turn toward stability by emitting radiations. In other words, if attempting to add some neutrons or protons to the existing nuclides—how to do that is a different story that will be explained shortly—this is likely to cause them to become unstable, and change their configuration by some means, to become a familiar stable nuclide that already exists in nature. This can be understood from the chart of the nuclides shown in Figure 1.8, which is a plot of all known nuclides as a function of proton number, Z, and neutron number, N.

All stable nuclei and known radioactive nuclei, both naturally occurring and manmade, are shown on this chart, along with their decay properties. The thick gray line represents the island of stability, enclosing all the naturally occurring nuclides. The nuclides appearing above this line are proton-rich nuclides and the ones appearing below are neutron-rich nuclides. They can be produced by irradiating the stable elements with neutrons (in a reactor) or protons (in an accelerator or a cyclotron) which will produce neutron- and proton-rich nuclides, respectively. These nuclides will then decay toward the island of stability by changing neutrons into protons or protons into neutrons, with the emission of β^- and β^+, respectively.

Figure 1.9 shows where a radioactive nuclide moves, on the chart of nuclides, when it emits various types of particles.

The common decay modes are α decay, β decay, EC (electron capture) decay, and isomeric transition (IT). There are also other rarer forms of decay like proton emission, double proton emission, neutron emission, etc., which can be learned from standard textbooks on nuclear physics. The common decay modes and the resulting nuclear transformations are summarized in Table 1.3.

As a neutron is heavier than a proton, it is energetically possible for a neutron to decay into a proton, in free state. The following example illustrates β decay of free neutron (with a half-life of about 12 minutes).

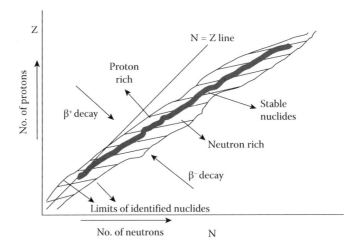

FIGURE 1.8
Chart of nuclides.

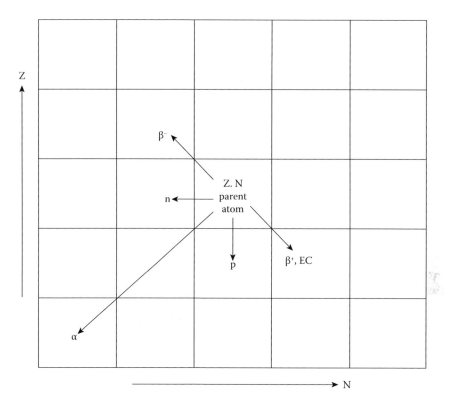

FIGURE 1.9
Influence of decay on the position of the nuclide on the chart of nuclides.

TABLE 1.3

Common Modes of Decay and Nuclear Transformations

Nature of Decay	Nuclear Transformation
Nucleón Emission	
α decay	$(A, Z) \rightarrow (A{-}4, Z{-}2)$
β Decay Modes	
Electron decay	$(A, Z) \rightarrow (A, Z{+}1)$
Positron decay	$(A, Z) \rightarrow (A, Z{-}1)$
Electron capture	$(A, Z) \rightarrow (A, Z{-}1)$
Nuclear Transitions in the Same Nucleus	
Isomeric transition	$(A, Z)^* \rightarrow (A, Z)$
Internal conversion	$(A, Z)^* \rightarrow (A, Z)$

$$^1_0\mathrm{n} \rightarrow {^1_1}\mathrm{p} + {^0_{-1}}\mathrm{e} + \bar{\nu}$$

This transformation also occurs during β^- decay of a radioactive nuclide.

It is energetically not possible for a free proton to decay into a free neutron since a proton has slightly lower mass compared to a neutron. (I think some theories predict other modes of decay for a free proton but so far it has not been observed, which suggests a half-life of at least 10^{36} years, in case it is found to be unstable. So, for all practical reasons, a proton can be considered stable). However, proton to neutron conversion does take place in radioactive nuclides via positron decay. The proton can find additional energy for this transformation internally, from the uncertainly relation $\Delta E\ \Delta t \approx h/2\pi$ governing such transitions.

$$^{1}_{1}p \rightarrow {}^{1}_{0}n\ +\ {}^{0}_{1}e + \nu$$

Though the decay energy is constant, both beta and positron exhibit a spectrum of energies. This is because of the existence of a third particle electron–antineutrino, and electron–neutrino which carry away the rest of the decay energies.

Like positron emission, EC also reduces the Z by 1. So these two are competing modes of decay. Electron capture leaves a shell vacancy so EC will lead to the emission of characteristic X-rays. Sometimes the excitation energy is directly transferred to an orbital electron which becomes ejected from the atom; these electrons are called Auger electrons.

A question that always comes to the mind of readers is regarding the electron and positron emission from the nucleus, and the nature of interaction that is responsible for the neutron to proton or proton to neutron transformations. Figure 1.10 explains this transformation.

The weak interactions discussed earlier are mediated by W^+ and W^- bosons. These particles have extremely low lifetimes and they decay into electron–antineutrino and positron–neutrino pairs. Since a proton loses its positive charge in positron decay, W^+ must be involved in positron decay and because a neutron loses -1 e to convert to a proton, W^- must be involved in this decay. In both cases, the total charge is conserved. Figure 1.10 illustrates the decay phenomenon. It can be seen that the neutron transformation involves a d quark changing into a u quark, and positron decay involves a u quark turning into a d quark, through the field quanta that mediate the weak force.

Readers might have noticed that the mass of the emitted W quanta is around 90 GeV (see the table given earlier) while the nucleon itself weighs only about 930 MeV. There are many such instances in nuclear physics, suggesting apparent violation of energy conservation laws. However, there is no violation of Heisenberg's uncertainty principle which is one of the cornerstones of modern physics. According to this principle, it is not possible to have exact knowledge of certain pairs of physical quantities. There will always be an uncertainty in the values of these quantities (say A and B) as per this principle given below:

$$\Delta A\ \Delta B \approx h/2\pi$$

Position and momentum is one pair of such quantities. Energy and time is another such pair of quantities ($\Delta E\ \Delta t \approx h/2\pi$). By applying this principle, we do not have to explain

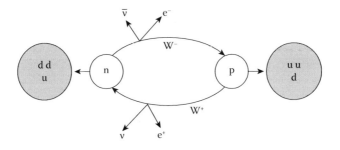

FIGURE 1.10
Neutron–proton transformations in beta and positron decay.

where the extra energy came from in the neutron to proton transformation, as long as this energy disappears within the time Δt, satisfying the above principle. The lifetime of W bosons is less than Δt derived from this uncertainty relation. Nature can fool us through the uncertainty principle!

In some decays, the excited state of the daughter nuclide happens to be a metastable state with a lifetime in minutes to days. Such nuclides behave like radioactive isotopes in their own right, and can be considered as pure gamma emitters. In some decays, instead of an IT, the excitation energy is directly transferred to the orbital electrons and they are knocked out of the atom. This is known as internal conversion (IC). IC is a competing mode of de-excitation mechanism for IT. When an IT is forbidden due to any conservation law, IC is an alternative mode of de-excitation for the nuclides. The decay of Mo^{99} is an excellent example of these transformations.

Mo^{99} decays by beta emission to two different excited states of the daughter nuclide Tc^{99} (see Figure 1.11). Here 80% of the decays lead to the metastable Tc^{99m}. Tc^{99m} exhibits two excited states, the 142 and 140 keV energy levels. The 142 keV energy level de-excites by IC mechanism to the ground state (2%), and to the 140 keV energy excited state (98%). This is an example of an alternative mode of decay. The 140 keV energy level de-excites to the ground state by IC (10%) or by IT (88%). This is an example of competing modes of decay.

Both IC and EC lead to shell vacancies, so in both cases characteristic X-rays and Auger electrons are emitted.

(Alpha decay involves a mass change of four and a charge change of two. The natural radioactive series involves several α decays before reaching the stable element. However, this is not particularly relevant in radiation oncology and hence will not be discussed further.)

Particle decay is a rule rather than an exception. All particles decay to lighter particles (except the fundamental particles) obeying some conservation laws. Radioactive decay also takes place conserving the following quantities:

1. Energy
2. Momentum (linear and angular)
3. Charge
4. Nucleon number

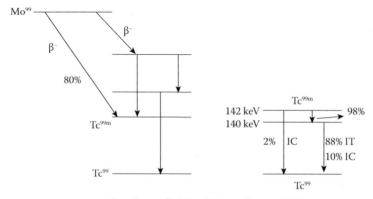

Beta decay of Mo^{99} and decay scheme of Tc^{99m}

FIGURE 1.11
Isomeric transitions.

One of the obvious rules of radioactive decay is that the mass of the products must be less than the mass of the decaying nuclide. Otherwise, the decay is energetically not possible.

Example 1.6

The atomic mass of a $_2^4$He atom is 4.0026 u. The mass of a hydrogen atom $_1^1$H is 1.0078 u. The mass of a neutron, m_n is 1.00870 u. Calculate the binding energy of the helium nucleus.

$$(1 \text{ amu} = 1 \text{ u} = 931.5 \text{ MeV.})$$

If nuclear mass is given, the problem is simpler. When atomic mass is given, the mass of electrons must also be accounted for. Two hydrogen atoms have two electrons. The He atom also has two electrons.

$$\text{Mass deficit of } _2^4\text{He} = 2 \times M_H + 2\, m_n - M_{He}$$

$$= 2 \times (1.0078) + 2 \times (1.0087) - 4.0026 = 0.03040 \text{ u}$$

$$\text{BE of He atom} = 0.03040 \text{ u} \times 931.5 = 28.3 \text{ MeV}$$

Example 1.7

^{218}Po$_{84}$ can decay by α or β emission. Calculate the energy released in the two decays using the following data:

$$M_{Po\text{-}218} = 218.008965 \text{ amu};\ M_{Pb\text{-}214} = 213.999798 \text{ amu};\ m_{At\text{-}218} = 218.00868 \text{ amu}$$

$$M_\alpha = 4.002602 \text{ amu};\ 1 \text{ amu} = 931.5 \text{ MeV}$$

The decay scheme is as follows:

$$_{84}^{218}\text{Po} \rightarrow\ _{82}^{214}\text{Pb} + \alpha$$

$$_{84}^{218}\text{Po} \rightarrow\ _{85}^{218}\text{At} + \beta^- + \bar\nu$$

α particle is the He nucleus. Here, we have taken the atomic mass of He into account for the two extra electrons in $_{84}^{218}$Po compared to $_{82}^{218}$Pb.

$$\Delta M_{Po,\alpha\ decay} = 218.008695 - (213.999798 + 4.002602) = 0.006565 \text{ amu}$$

$$\text{BE} = 0.006565 \times 931.5 = 6.12 \text{ MeV} = \text{released decay energy for } \alpha \text{ decay.}$$

Beta decay

We should not take electron mass into the equation since the $_{85}^{218}$At atom contains one extra electron. The equations will be written both in terms of nuclear mass (m) and atomic mass (M), assuming the rest mass of the neutrino as zero.

$$\Delta M_{Po,\beta\ decay} = M(_{84}^{218}\text{Po}) - M(_{85}^{218}\text{At})$$

$$\Delta m_{Po,\beta\ decay} = m(_{84}^{218}\text{Po}) - \{m(_{85}^{218}\text{At}) + m_\beta\}$$

Here, the first equation is used:

$$\Delta M_{Po,\beta\text{-}decay} = 218.008965 - 218.00868 = 0.000285 \text{ amu}$$

$$\text{BE} = 0.000285 \times 931.5 = 0.27 \text{ MeV} = \text{decay energy for } \beta \text{ decay.}$$

1.5.9 Radioactivity Decay Law

Radioactive decay is probabilistic. So, it cannot be said when a particular atom will decay, but the decay for a given population of atoms can be statistically predicted. With a population of N radioactive atoms, the number –dN decaying is

$$\alpha N$$

α time interval dt. So,

$$-dN = \lambda N \, dt$$

where the constant of proportionality is known as the decay constant, λ. From the above equation, λ can be defined as the fraction of the atoms decaying in unit time.

$$-[(dN/N)/dt] = \lambda \quad \text{or} \quad -(dN/dt) = N \lambda \quad \text{or} \quad -(dN/dt) = A = N\lambda$$

where A is the activity defined as the number of nuclides transforming in unit time.
Rearranging the above equation,

$$-(dN/N) = \lambda \, dt$$

Integrating and using the initial condition of $N = N_0$ at $t = 0$ (or $A = A_0$ at $t = 0$), the atoms remaining at time t can be shown to be

$$N(t) = N_0 \, e^{-\lambda t} \text{ (or } A = A_0 \, e^{-\lambda t} \text{)}$$

The above equation shows that radioactive decay is exponential and it will take infinite time to decay completely, but in a finite number of half-lives most of the atoms would have decayed.

The half-life $(T_{1/2})$ of a radioactive isotope is defined as the time taken for a given population of atoms (or given activity) to decay by half. Using this definition, in the above equation, it can easily be seen that

$$T_{1/2} = 0.693/\lambda$$

Radioactivity being a probabilistic event, the life of a radioactive atom can be anything between zero and infinity. However, an average or mean life of an atom (τ) can be defined by

$$\text{Mean life } \tau = \frac{\tau_{1} + \tau_2 + \tau_3 + \tau_4 + \tau_4 + \ldots\ldots \text{ of all atoms}}{\text{Total number of atoms}}$$

It can be shown that the mean life is related to the half-life by the following equation (see the Exercises section):

$$\tau = 1.44 \, T_{1/2} \quad \text{or} \quad 1/\lambda$$

If the source decay rate at $t = 0$ is assumed to be constant over the time τ, then all the radioactive atoms of the source (equal to the area of the dotted rectangle in Figure 1.12) would have decayed in this time interval, and this number would be given by $A_0\tau$ mCi hours. This is the area under the decay curve. The concept of mean life, therefore, can be utilized to determine the total radiation emitted by an implant when it is permanently left inside a patient.

The effective life is another important quantity. It takes into account the biological half-life (of elimination) of the radionuclide incorporated in the body. Apart from physical decay, a part of the radionuclide will also be excreted from the body. So, there is also

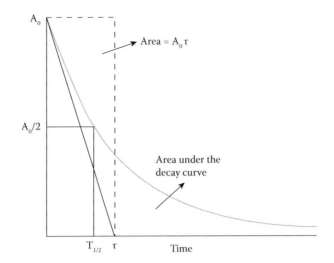

FIGURE 1.12
Concept of mean life.

a biological half-life ($T_{1/2,b}$) for a radioisotope when it is inside the body. For instance, $T_{1/2,p}$ for ^{131}I is 8 days, but its $T_{1/2,b}$ is 138 days. For ^{137}Cs, $T_{1/2,p}$ is about 30 years, but $T_{1/2,b}$ is only 70 days. It can be shown that

$$T_{eff} = \frac{T_{1/2,p} \times T_{1/2,b}}{(T_{1/2,p} + T_{1/2,b})}$$

The unit of activity is Curie (Ci) in conventional units, and Becquerel (Bq) in SI units.

- 1 Ci = 3.7×10^{10} nuclear transformations/sec.
- 1 Bq = 1 nuclear transformations/sec.

1.6 Secular and Transient Equilibrium

It is easy to understand why uranium still exists since the creation of the earth (its half-life is comparable to the age of the universe), but how come radon which has a half-life of a few days still survives? The answer to such questions lies in the equilibrium established between the parent and daughter radionuclides. It is physically easy to understand. Radon cannot decay faster than its creation, which is governed by radium decay. So, it has an apparent half-life ($T_{1/2,D}$) equal to radium's half-life ($T_{1/2,P}$). However, the two products must be together for this equilibrium. If Rn is separated (extracted) it will decay with its own half-life of 3.8 days.

There are two types of equilibrium: (1) secular equilibrium and (2) transient equilibrium.
For secular equilibrium, $T_{1/2,P} \gg T_{1/2,D}$ (reached in six to seven half-lives)
For transient equilibrium, $T_{1/2,P} > T_{1/2,D}$ (reached in four to five half-lives).
Figure 1.13 illustrates the equilibrium attained in the two cases.

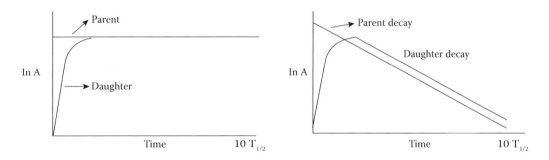

FIGURE 1.13
Concept of secular and transient equilibrium.

It can be seen from Figure 1.13 that during the first 10 half-lives of the daughter, there is practically no parent decay in secular equilibrium, but significant decay in transient equilibrium. A good example of secular equilibrium is the decay of Ra ($T_{1/2}$ = 1620 years) and Rn ($T_{1/2}$ = 3.8 days). In 10 half-lives of the daughter (38 days), there is practically no decay of Ra and so the line is almost parallel to the time axis. In about 6 × 3.8 days, Rn reaches equilibrium (maximum activity) and can be extracted every month for any use. Radon tubes were in high demand for brachytherapy during the 1940s to 1970s. In order to use Rn, a radon plant is necessary to extract Rn, the gaseous product from radium. If the extraction is earlier than this period, the activity would be less than the equilibrium or the maximum activity that can be obtained, since it is in the increasing portion of the curve. If it is not extracted, it will apparently have the same decay as the parent, indicated by the single line. A well-known example of transient equilibrium is the ^{99}Mo/^{99m}Tc generator used extensively in nuclear medicine. The half-lives of ^{99}Mo and ^{99m}Tc are 67 and 6 hours, respectively. During the 10 half-lives of the daughter, about 50% of the Mo would have decayed, as can be seen from the diagram, unlike the secular equilibrium where the parent decay is practically zero. Here, after reaching equilibrium, the decay of the daughter follows the decay of the parent—^{99m}Tc can be extracted every day for nuclear medicine use. If it is not extracted, it does not build up much further since it has reached equilibrium with the parent. If it is extracted earlier, there would be much less activity since it will be in the ascending portion of the curve.

1.6.1 Nuclear Fission

As seen earlier, nuclides beyond lead are unstable and reach stability by the phenomenon of radioactivity. Uranium is one of the few special elements, which by virtue of its atoms being large in size with a large number of protons and neutrons, is most unstable and needs only a little excitation energy to split typically into two medium size nuclides ^{141}Ba and ^{92}Kr. This phenomenon is known as nuclear fission. In this form of transformation, a thermal neutron entering a ^{235}U nucleus can give its BE to the nucleus to form a compound nucleus ^{236}U, which subsequently splits into two medium size nuclides. The fission process also emits an average of three neutrons per fission (see Figure 1.14).

The BE curve, given in Figure 1.15, shows where the fission energy comes from.

The BE per nucleon curve shows the high binding energy (HBE) and low binding energy (LBE) regions on the curve. When an LBE nuclide (i.e., very high Z and A) fissions into two HBE nuclides, the BE difference gives the BE per nucleon released in the fission process.

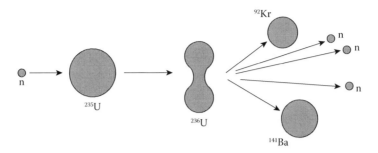

FIGURE 1.14
Fission of U^{235} nucleus.

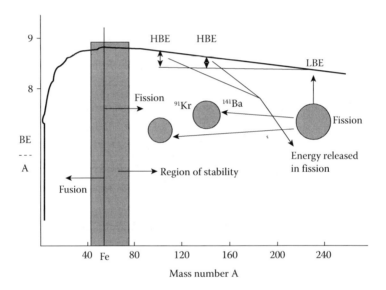

FIGURE 1.15
Nucleon binding energy curve.

Example 1.8

Calculate the energy released in the fissioning of a ^{235}U nuclide. BE per nucleon for ^{235}U is 7.6 MeV and for the fission product (LBE nuclide), 8.5 MeV/nucleon.

BE difference per nucleon = 7.6 − 8.5 = −0.9 MeV

BE released per fission = 235 × 0.9 ≈ 212 MeV

The distribution of energies is as follows: fission fragments kinetic energy = 175 MeV; fission gamma = 7 MeV; fission neutrons kinetic energy = 5 MeV; β/γ energies from decay = 13 MeV; neutrino energy = 10 MeV.

In a nuclear reactor a critical mass of fissionable material is used under controlled critical conditions (only one neutron must be available for fission at any time, making the nuclear reaction self-sustaining) to generate power. If the neutron available in fission is >1 and if the fission is not controlled it can lead to a nuclear explosion, the principle of nuclear bombs.

Nuclear reactor is a great source of neutrons and produces neutron-rich nuclides. If you want to induce instability in stable elements by neutron absorption and make them radioactive, all you have to do is to keep the stable elements inside a neutron reactor for a specific period, to absorb neutrons. The element becomes an artificial radioisotope and emits particles, just like in natural radioactivity, to readjust its n/p ratio until it reaches the region of stability, on the nuclide chart.

NOTE: In order to produce proton-rich radionuclides, we have to bombard the elements with charged particles in a cyclotron or in a particle accelerator. Neutron-rich nuclides convert excess neutrons into protons (undergoing beta decay) and proton-rich nuclides convert excess protons into neutrons (undergoing positron decay). Positron emitters are short-lived and are used extensively in nuclear medicine. However, a cyclotron is needed to produce positron emitters.

Spontaneous fission also occurs in these nuclei but it is much rarer and is explained by the quantum tunneling effect, but this does not concern our subject matter.

1.6.2 Nuclear Fusion

It can be seen from the BE curve that low BE (per nucleon) nuclides occur on either side of the curve. So, this suggests that BE can be released not only by the fissioning of a high Z nuclides, but also by the fusion of very low Z nuclides.

The bombarding nuclides must have sufficient kinetic energy to overcome the Coulomb repulsion, to fuse together. However, this topic is not of much relevance here and so will not be discussed further. The two references cited here [1,2] are the two most well known textbooks on Nuclear Physics.

References

1. R.D. Evans, *The Atomic Nucleus*, Krieger, Melbourne, FL, 1982.
2. J.M. Blatt and V.F. Weisskopf, *Theoretical Nuclear Physics*, Springer-Verlag, New York, 1979.

Review Questions

1. What are the allowed l values for a shell of principal quantum number 4?
2. What is the electron shell corresponding to n = 3?
3. How many orbitals are there in a subshell of quantum numbers n = 3 and l = 2?
4. How many electrons can occupy the state with n = 3, l = 2, and m_l = 0?
5. Which electron has higher ionization potential, n = 3 electron or n = 2 electron?
6. If there are four subshells, what is the principal quantum number of the shell?
7. Write the electron configuration for a carbon atom.
8. Calculate the number of electrons that the sublevels d, f, and g can hold.

9. What is the simple formula to calculate the maximum number of electrons in a shell? Using the formula calculate the maximum number of electrons for the shells with principal quantum numbers 2, 3, and 4.

10. What is the radius of a ^{238}U nucleus?

11. How will you determine the binding energy per nucleon for a nucleus?

12. Where do you find the subatomic particles the proton, the neutron, and the electron inside an atom?

13. Given the symbol $^{125}_{53}$I, identify the element, its atomic number, and the mass number.

14. The symbol of an atom is given as $^{19}_{9}$F. Identify the element, the number of protons, the number of electrons, and the number of neutrons in the atom.

15. What principle prevents not more than two electrons from occupying an orbital?

16. Express one eV in joules.

17. Express one amu in MeV.

18. What is the BE of the hydrogen nucleus?

19. Why do the nucleons inside the nucleus have less mass compared to free state?

20. Define mass defect and calculate the mass deficit for deuteron (nucleus of heavy hydrogen) using the following data:

    m_n = 1.008665 amu; m_p = 1.007825 amu; 1 amu = 931.5 MeV; M_D = 2.014102 amu.

21. How much energy will be released if all the atoms in just 1 g of iron (Z = 26) are separated into individual protons and neutrons? The following data are given:

    m_n = 1.008665 amu; m_p = 1.007825 amu; 1 amu = 931.5 MeV; M_{Fe-56} = 55.934942 amu;

    $$1 \text{ eV} = 1.6 \times 10^{-19} \text{ J}$$

22. Calculate the energy released (in MeV) in a free neutron decay, using the following data:

    $$m_n = 1.6747 \times 10^{-27} \text{ kg}; m_p = 1.6725 \times 10^{-27} \text{ kg}; m_e = 0.00091 \times 10^{-27} \text{ kg}$$

23. The two isotopes of boron are ^{10}B and ^{11}B. M_{B-10} = 10.0129370 u M_{B-11} = 11.0093054 u

    Abundances: W_{B-10} = 19.78%; W_{B-11} = 80.22%. Calculate the average atomic mass of boron.

24. $^{137}_{55}$Cs undergoes β^- decay to form $^{137}_{56}$Ba. 94% of the β^- decays lead to the excited state of Ba. The maximum energy of the emitted beta is 0.511 MeV. The atom returns to the ground state by emitting 0.662 MeV gammas. Draw the energy level diagram for this decay process.

25. For beta decay of a nuclide of given (Z,A) write the equation for the decay energy released in terms of atomic mass M and nuclear mass m.

26. For positron decay of a nuclide of given (Z,A) write the equation for the decay energy released in terms of atomic mass M and nuclear mass m.

27. For EC decay of a nuclide of given (Z,A) write the equation for the decay energy released in terms of atomic mass M and nuclear mass m.

28. ^{60}Co radionuclide decays to ^{60}Ni by beta decay. Calculate the decay energy from the following data:

$$M_{Co} = 59.933822 \text{ u}; \; M_{Ni} = 59.930791 \text{ u}$$

29. $^{18}_{9}$F is a positron emitter and is extensively used in positron emission tomography imaging. It decays to $^{18}_{8}$O. Calculate the decay energy from the following data:

$$M_F = 18.000937 \text{ u}; \quad M_O = 17.999160 \text{ u}$$

30. Describe the phenomenon of EC and IC.

31. Write an equation for IT and explain how the decay energy is shared by the gamma and the daughter nuclide.

32. Describe the phenomenon of radioactivity and explain the common modes of decay.

33. Why can't a neutron decay by positron emission, or a proton by an electron emission?

34. Why can't a free neutron decay into much lighter particles? Calculate the energy released (in MeV) in a free neutron decay, using the following data:

$$m_n = 1.6747 \times 10^{-27} \text{ kg}; \quad m_p = 1.6725 \times 10^{-27} \text{ kg}; \quad m_e = 0.00091 \times 10^{-27} \text{ kg}$$

35. What is the SI unit of activity? Express 10 Ci in SI units.

36. What conservation laws are obeyed in positron–electron annihilation at rest?

37. Derive the radioactive decay law and deduce an expression for half-life.

38. Explain the concept of decay constant. Explain what is meant by effective half-life and derive an expression for it.

39. I-131 sodium iodide has a $T_{1/2,b}$ of 24 days. What is $T_{1/2,eff}$?

40. A Tc-99m compound has a $T_{1/2,eff} = 1$ hour. What is $T_{1/2,b}$?

41. ^{82}Sr decays by EC to ^{82}Rb which in turn decays to ^{82}Kr by positron emission. State the nature of equilibrium reached and after how long maximum amount of ^{82}Rb can be extracted from the mixture. ($T_{1/2,Sr} = 25.4$ days and $T_{1/2,Rb} = 1.25$ minutes.)

42. ^{99}Mo decays by β^- to ^{99m}Tc which in turn decays to ^{99}Tc by gamma emission. State the nature of equilibrium reached and after how long the maximum amount of ^{82}Rb can be extracted from the mixture ($T_{1/2,Mo} = 67$ hours and $T_{1/2,Tc} = 6$ hours).

2

Basic Medical Radiation Physics

2.1 Introduction

Radiation therapy started with kilovoltage X-ray beams in the 1910s, and with time the beam energies gradually increased to give better penetration, better skin sparing, and higher beam output. Ra and ^{137}Cs units came into radiotherapy use in the kilovoltage X-ray era, mainly because of their higher energies. However, with the development of ^{60}Co teletherapy units by H. E. Johns in Canada in the 1950s, most of these treatment units were gradually replaced by ^{60}Co units and the existing kV X-ray units were reserved only for superficial treatments. Since it was impractical to increase the DC voltage of X-ray machines beyond a few hundred kV, linear accelerators (traveling-wave and side coupled standing-wave type) were developed for mega voltage therapy.

Today, almost all the treatment machines in hospitals in developed countries are linacs while in the developing countries, the majority of the treatment machines are still ^{60}Co units. In a linac, electrons are accelerated in the electric field of microwaves to the desired energies and are guided through a beam line to bombard a high Z target to produce an MV X-ray spectrum, which is suitably modified for the purposes of the treatment.

From the 1940s to the late 1960s there were many betatrons (more than 100 machines) operating in different parts of the world for electron beam therapy. These beams are ideal for treating skin and shallow depths, and avoiding any deep lying critical structures. Unlike the X-ray beams, they have definite range of penetration, decided by the energy of the electron beams. With the advent of linacs capable of operating in dual mode (both X-ray and electron mode), the latter quickly replaced all betatrons (and kV X-ray machines) for treating skin surface to moderate depths (up to 6 or 7 cm) by electron beams and any deeply seated tumors by X-ray beams. Electron beams are also used during intraoperative procedures when the tumor region is directly irradiated during a surgical procedure.

Fast neutron therapy studies started in the 1970s. The problem with a neutron beam is they carry no charge, so it is not possible to accelerate them in an electric field. So, there are three options for neutron therapy—use a nuclear reactor which is a source of neutrons, a radioactive neutron source (^{252}Cf), or use a particle accelerator again (like a cyclotron) and guide the beam through the beam line onto a suitable target to produce neutrons. Neutron beams were the first high linear energy transfer (LET) radiation used in radiation therapy. They are superior to photon beams because of higher biological effectiveness in killing tumor cells, and are particularly suitable for certain types of cancers.

Proton beams and carbon ion beams have high LET, limited range, and ideal dose distribution for radiation therapy. They are produced in cyclotrons and guided through the beam lines to the treatment rooms.

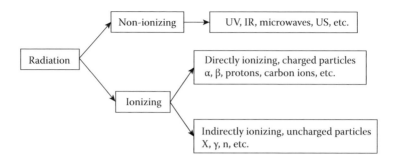

FIGURE 2.1
Classification of radiation.

The characterization of these radiation beams and their interactions leading to energy deposition in the medium forms the subject matter of this chapter.

2.2 Classification of Radiation

Radiation is classified as ionizing or non-ionizing [1,2]. Non-ionizing radiation cannot ionize and cannot generally produce a response in the system. The average energy required to produce an ion pair in air is around 34 eV. So, a few tens of keV can be assumed as the borderline between ionizing and non-ionizing radiations (i.e., beyond the ultraviolet frequency of electromagnetic radiation can be termed as ionizing). We further categorize ionizing radiation as directly ionizing (e.g., electrons, protons, and ions) and indirectly ionizing (X, γ, n, etc.). Figure 2.1 explains the classification of radiation.

2.3 Energy Deposition Concept

The radiation response of any medium (physical or biological) is due to the energy deposited in the system. So, to quantify radiation response it is essential to compute or measure the energy deposited in the medium. The energy deposition mechanisms for ionizing radiation are illustrated in Figure 2.2.

It can be seen from Figure 2.2 that for charged particles, the energy deposition is a one-step process of Columbic interactions with the atoms or electrons of the medium; these are directly ionizing particles (DIPs). For uncharged particles (e.g., photons and neutrons) energy deposition is essentially a two-step process, involving the generation of charged particles by the interactions these particles undergo with the atoms of the medium as a first step, and energy deposition in the medium through charged particle interactions as a second step. These uncharged particles are referred to as indirectly ionizing particles (IDIPs).

NOTE: The removal of the photons or neutrons from the beam in single interactions is responsible for the exponential attenuation of these beams. The large number of

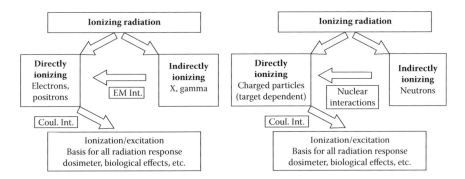

FIGURE 2.2
Energy deposition process for directly and IDIPs.

interactions experienced by each individual charged particle of a beam gradually losing energy, instead of being removed from the beam in single interactions, are responsible for characterizing the charged particles in terms of energy loss per unit length traversed in the medium. This is known as the stopping power and range, and will be explained later. When the charged particles also get lost from the beam in single interactions, as in the case of photons, the concept of stopping power and range will lose much of its meaning with charged particles as well. When such interactions predominate, as in the case of very high-energy charged particles in the GeV range, the charged particle interactions can also be characterized in terms of the concept of attenuation.

The energy deposition concept described refers to the physical media. In the case of biological systems, the primary goal is to quantify the biological harm caused to the system as a result of the physical energy deposition. This involves extending the physical dose concept to include an index of biological harm caused by the radiation type in question while interacting with the biological media, such as a radiation worker or the public. This topic will be further discussed in one of the following sections.

2.4 Types of Photon (X, γ) Interactions

Both radiation and matter being microscopically discrete, the interactions themselves constitute discrete individual events. In order to discuss the interactions on a macroscopic scale, radiation needs to be distinguished in terms of the field quantities (i.e., the number and energy of the particles incident per unit area of the absorber and the rates of these quantities) and the interactions in terms of attenuation coefficients (for IDIP beams) and stopping power (for DIP beams).

The first condition for the occurrence of any interaction is that it must be energetically possible. For instance, an 80 keV photon does not have enough energy to eject a K-shell electron in lead (because its binding energy is 88 keV, more than the photon energy). A 1 MeV photon cannot have a nuclear interaction since the average energy required to eject a nucleon out of a nucleus is around 7 to 8 MeV. So for any interaction to take place it must be energetically possible, and often that applies to more than one type of interaction. An important question that follows is, when more than one type of interaction is possible, which one will occur? The simple answer is that both interactions would occur but with different frequencies, depending on the interaction probabilities. We will discuss the interactions first and the probabilities later.

In the energy range of interest in radiation oncology, the only photon interactions of importance are:

1. Photoelectric effect
2. Compton effect
3. Pair production effect
4. Photo nuclear reactions

2.4.1 Photoelectric Effect

In the atomic photoelectric effect (PEE), a photon disappears and an electron is ejected from an atom (see Figure 2.3).

The electron carries away all of the energy of the absorbed photon, minus the shell binding energy (BE). (electron energy $E = h\nu - BE$). However, if the photon energy drops below the binding energy of a given shell, A PEE is not possible with those shell electrons. The photon can still interact with the electrons in the lower shells. However, when the photon energy is much larger than the BE, the probability of PEE falls as $(1/(h\nu)^3)$. So, this interaction is like a resonance effect. These characteristics and the well-known spikes at $h\nu = BE$ are reflected in the photoelectric cross section versus energy curve shown in Figure 2.4.

So, for a given photon energy, tightly bound electrons are more likely to undergo a PEE compared to loosely bound electrons. For instance, if the photon energy just exceeds the K-shell BE, these photons can eject both K and L shell electrons, but the probability of photoelectric (PE) interaction is much higher for K electrons (see Figure 2.5).

Since the electrons are bound to the atom, this is an atomic absorption process and the atom as a whole takes part in the interaction. So, the cross section for the PEE is an atomic cross section and varies as

$$\sigma_{pee} \sim [(Z^3/(h\nu)^3] \ cm^2/g$$

It is this Z^3 dependence that gives good contrast in imaging for variations in Z, since there is a vast difference in the absorption of radiation. (Figure 2.5 shows variation of σ_{pee} with energy for lead and water).

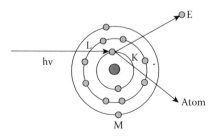

FIGURE 2.3
Concept of photoelectric effect.

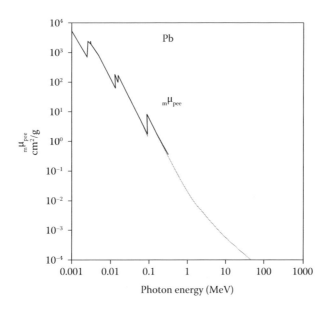

FIGURE 2.4
Variation of photoelectric attenuation with photon energy.

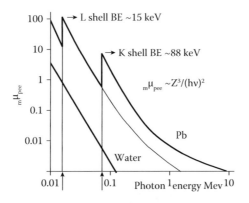

FIGURE 2.5
Variation of photoelectric attenuation with photon energy for Pb and H_2O.

Another point to remember is that the PEE gives rise to a shell vacancy. Any shell vacancy gives rise to characteristic X-rays as a result of electron transitions to lower levels to fill vacancies. The transition energy, instead of being emitted as photons, can knock out peripheral electrons. This is known as Auger electron emission.

The PEE has great significance in medical applications. The image contrast in diagnostic radiology is mainly due to the PEE. The Z dependence explains why bone gives such good contrast in the images. The densities of bone and soft tissue expressed as atomic numbers are approximately 13.8 and 7.4, respectively. So, bone absorbs more photons compared to surrounding tissue by a factor of $[(13.8)^3/(7.4)^3]$ which is roughly seven times. This is again the reason for introducing high Z contrast agents into the patient for imaging when there is no inherent contrast (e.g., Ba and I, which have higher Z and higher density compared to soft tissue). Also for radiation shielding, a fraction of a 1 mm thick Pb lining is enough for shielding diagnostic X-ray room doors since the K and L shell binding energies in Pb are much higher and a large fraction of the photons are lost (attenuated) in the photoelectric interactions. This interaction is of much less significance in radiation oncology.

2.4.2 Compton Effect

When the photon energy is much larger compared to the shell BEs, the photons tend to see the electrons as free electrons since the BE becomes negligible compared to the photon energy. So, there is no Z dependence of electronic cross section in the Compton effect. Since there are Z electrons in an atom, the atomic cross section varies as Z, that is, $_a\sigma_{ce} = Z\ _e\sigma_{ce}$. Compton attenuation only depends on the electron density of the materials (which is about the same for the low and medium Z materials and varies by a maximum of 20% from the low Z to the highest Z materials). So, a Compton scattering event is a direct collision of a photon with a free electron. It is called scattering event since the collision produces a scatter photon going at an angle θ and the electron recoiling at an angle ϕ with respect to the original photon direction, as shown in Figure 2.6.

The energy of the scatter photon and the recoil electrons depend on θ and ϕ and can be determined from the law of conservation of energy and momentum (see the review questions). The results are given below:

$$h\nu' = \frac{h\nu_0}{[1 + \alpha\,(1 - \cos\vartheta)]}; \quad \alpha = h\nu_0/(m_0 c^2)$$

$$E = h\nu_0\,\frac{\alpha\,(1 - \cos\vartheta)}{[1 + \alpha\,(1 - \cos\vartheta)]}$$

$$\cot\Phi = -(1 + \alpha)\,\tan(\vartheta/2)$$

The Compton effect decreases with increasing photon energy (see Figure 2.7).

As the photon energy increases, the collision becomes more energetic and the recoil electrons receive a larger fraction of the photon energy. Table 2.1 illustrates this important characteristic of the Compton effect. The electron rest mass energy is 0.511 MeV and this must be compared to the photon energy. As the photon energy decreases and nears this energy, the electron tends to appear more and more massive for the photon to cause a scatter. Similarly, when the photon energy is much larger than the rest mass energy of the electron, the electron appears too light for the photon to easily impart sufficient energy to it.

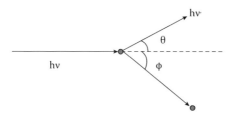

FIGURE 2.6
Concept of Compton effect (CE).

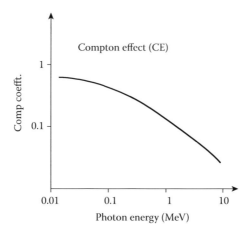

FIGURE 2.7
Variation of Compton cross section with energy.

TABLE 2.1

Scattered Photon Energy and Angle of Scattering in Compton Effect

Incident Photon Energy (MeV)	Photon Scattering Angle (degrees)	Scattered Photon Energy (MeV)	Scattered Electron Energy (MeV)	Scattered Electron Angle (degrees)
0.1	45.0	0.095	0.005	63.7
1.0	45.0	0.636	0.364	39.2
5.0	45.0	1.293	3.707	12.6
10.0	45.0	1.486	8.514	6.7

For a given photon energy, in a glancing collision the scattered photon travels in the same direction ($\theta \approx 0$) and the electron recoils with a maximum angle of 90° and minimum recoil energy. In a head-on collision, the scatter photon is back-scattered with minimum energy and the electron moves in the direction of the incident photon ($\phi = 0$) with maximum energy. The minimum back-scatter photon energy reaches a limit of 0.255 MeV, with increasing photon energy which means the maximum electron energy also reaches a limit of ($h\nu_0 - 0.255$) MeV. The scatter photon and electron energy equations given earlier can be used to derive the above statements.

Example 2.1

Using Figure 2.6, calculate the minimum energy of the scattered photon and the maximum energy of the Compton electron in a ^{137}Cs gamma collision. What are the scattering angles?

Energy of the ^{137}Cs gamma = 0.662 MeV

The photon transfers the maximum energy in a head-on collision resulting in the Compton photon back-scatter so $\theta = 180°$. Cos 180° = −1, with minimum energy.

$$(h\nu')_{min} = 0.662/(1 + 2\ \alpha); \alpha = 0.662/0.511 = 1.3$$

$$= 0.662/3.6 = 0.184\ \text{MeV}$$

The electron (maximum) energy = 0.662 − 0.184 = 0.478 MeV.
The photon and electron scattering angles are 180° and 90°, respectively.

Significance of CE

- This is the most important photon interaction in tissue for photon energies of 0.1–20 MeV, therefore it is the most important in radiotherapy. It is Z independent and so does not cause preferential bone absorption in the patients being treated.

- High-energy photons (≥1 MeV) are scattered predominantly in the forward direction, and the energy transfer is large. The scatter photons and scatter electrons (arising from the head of the treatment delivery equipment, known as beam contamination) affect the skin sparing advantage of megavoltage therapy and are likely to cause skin reactions if proper procedures are not followed to reduce the scatter reaching the patient. This point must be kept in mind while using beam modifiers close to the patient, in Co-60 teletherapy and in megavoltage therapy.

- The beam contamination also spoils the contrast of the portal images. The scatter can reduce the image contrast unless efforts are taken to reduce the scatter reaching the cassette (e.g., using a grid).

- The dose spreads laterally beyond the primary beam in a patient, due to the increasing ranges of the Compton electrons. In the 50 keV to 10 MeV energy range, much of the dose in tissue is delivered through the Compton effect (and at higher energies through pair production). This is an unwanted dose that the patient receives while undergoing radiotherapy treatments.

- The patient scatter and the primary wall scatter in the radiotherapy treatment rooms are two important components arising from Compton scatter, and special shielding calculations are done to shield the treatment room walls against scatter radiations.

2.4.3 Pair Production

The process of conversion of the photon energy into an electron-positron pair in a field of an atomic nucleus is called pair production (PP) (see Figure 2.8).

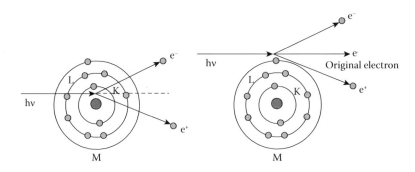

FIGURE 2.8
Concept of pair production.

Significant attenuation can take place through this process in the high Z room shielding materials, at higher photon energies used in radiation oncology.

Some of the important characteristics of the PP phenomenon are:

1. Quantum mechanical computations show that the PP actually occurs only in the vicinity of the nucleus and not inside the nucleus. This is confirmed by the fact that PP also occurs in the vicinity of an electron, which does not possess any internal structure.

2. As in the case of the PEE, it can be shown that the energy–momentum conservation laws do not allow a complete transformation of a photon into an electron–positron pair in vacuum and a third body, such as the nucleus, is necessary for momentum conservation. Since PP occurs in the vicinity of the nucleus, the PP cross section is usually defined at the nucleus or at the atomic level, as in the case of the PEE.

3. In the Compton effect, the scattering angle of the photon uniquely determines the recoil angle of the electron. However, in PP, the angle of emission of a positron does not uniquely fix the angle of emission of the electron because of a component of momentum carried away by the third body—the nucleus. Depending on the direction of recoil of the atom, the electron can go in any angle, for a given angle of emission of the positron. Hence, relations for the angles of the positron–electron pair cannot be written down exactly, as in the case of the Compton effect.

4. The positron produced in a PP process, unlike an electron, is an antiparticle, and hence is annihilated at the end of the range when it encounters an electron of the medium. Since a positron at rest does not have any momentum, the annihilation quanta are always emitted in opposite directions, so that the net momentum is zero. These annihilation quanta share the rest energies of the electron-positron pair (i.e., 0.511 MeV each). The positron also decays in flight, though with a much smaller probability compared to its annihilation probability at rest. In such a case, (a) the residual kinetic energy of the positron is also carried away by the annihilation quanta and (b) the two quanta are not emitted in opposite directions, since the position also has momentum.

5. Since the photon's rest mass is zero, the rest mass of the electron–positron pair must arise from the kinetic energy of the photon itself. Thus the kinetic energies attained by the electron–positron pair is given by $h\upsilon - 2m_0c^2$ or the threshold for PP is $2m_0c^2$ or 1.02 MeV. Below this energy, PP is not energetically possible.

6. It can be shown that the threshold for PP is given by $2m(m + M)/M$. Since M (nuclear mass) >> m (electron mass), this quantity is very nearly given by $2m$. However, when the PP occurs in the vicinity of the electron, the threshold for such an interaction becomes $\approx 4m$ or 2.04 MeV. Since here m = M, the recoiling electron receives appreciable kinetic energy, and hence the phenomenon of PP in the field of the electron is usually referred to as triplet production since three electrons are ejected from the atom.

7. $\sigma_{pp} = 0$ for $h\upsilon < 2m_0c^2$ and for $h\upsilon > 2m_0c^2$, σ_{pp} increases gradually with $h\upsilon$ until it starts leveling off at around 1000 m_0c^2.

8. $\sigma_{tp} \approx (1/Z) \sigma_{pp}$, which means that σ_{tp} is important (compared to PP) only for the low Z materials.

9. The triplet cross section per electron is independent of Z, and hence, the triplet cross section per atom is proportional to Z. Thus, the pp per atom is Z times

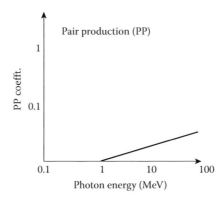

FIGURE 2.9
Variation of PP attenuation with energy.

stronger than the triplet cross section per atom. Triplet production, therefore, is of importance for the low Z materials.

The probability of the PP effect increases with increasing photon energy beyond the threshold energy as shown in Figure 2.9.

2.4.4 Photonuclear Interactions

Photons do not carry a charge, so they experience no Coulomb barrier in entering a nucleus. As the photon rest mass is zero, it can only deliver its kinetic energy to the nucleus on absorption. When this energy is greater than the BE of a nucleon in the nucleus, one or more nucleons may be emitted from the nuclei, depending on the energy of the photon and the nature of the nuclide (Figure 2.10).

The binding energy of the uppermost neutron in the nucleus is usually in the range of 5 MeV (^{13}C) to 20 MeV(^{4}He). Lower threshold energies exist for some nuclides (e.g., deuterium [2.23 MeV] and beryllium [1.67 MeV]). The cross section of the photonuclear effect is characterized as a giant dipole resonance and is most pronounced in high Z materials.

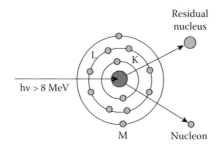

FIGURE 2.10
Concept of photonuclear interaction.

NOTE: There is another interaction that occurs at very low energies—the Rayleigh effect, or Rayleigh scattering. This is purely a coherent scattering effect and occurs when the photon energy is not enough to excite or ionize an atom (i.e., $h\nu < BE$). The bound electrons can at best oscillate in the electric field of the incident photon, and scatter the energy in all directions. This can be thought of as elastic scattering of photons by the atom as a whole, with the same frequency. The energy of X-ray photon is first completely absorbed and then re-emitted by the electrons of an atom. Since we are interested in energy deposition and biological harm, in radiation protection, this interaction is not of much consequence to us. Also, this occurs at very low energies, less than the excitation and ionization energies of the atoms.

These interactions occur on a macroscopic scale when a photon beam traverses a medium and cause the attenuation of the beam and energy transfer to the charged particles in the medium. When considering photon attenuation, the photonuclear effect is usually neglected, as PEE, CE, and PP have much higher interaction probabilities, but it plays an important role in the production of neutrons in radiotherapy linacs operating at >10 MV. These neutrons must be considered for reducing the neutron dose to the patient and also for reinforcing the linac treatment room doors to reduce the neutron dose in the control or supervised areas.

2.5 Characterizing Radiation in Terms of Field Quantities

A radiation field is characterized by the number of particles N, and the energy transported by these particles R (known as the radiant energy). The latter is also known as the energy fluence. Since the radiation is continuously incident on the medium (subject to statistical fluctuations in the emissions from radioactive sources, or pulsed nature of emissions by the X-ray machines), the rates of these quantities are required. The rate plays a critical role in causing damage to the tissues. Whether the tissues and organs receive the dose at low rates over a long period of time or at very high rates in a short period of time makes a difference as the time available for the cell repair mechanisms to operate can result in higher biological damage.

Particle fluence is defined by

$$\text{Particle fluence} = \Phi = dN/dA \text{ (unit: m}^{-2}\text{)}$$

where dN gives the number of particles crossing a sphere of cross-sectional area dA, irrespective of the energy or direction of these particles. The definition is illustrated in Figure 2.11.

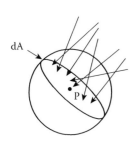

FIGURE 2.11
Definition of particle fluence = $\Phi = dN/dA$.

NOTE: Planar fluence can also be defined by taking into account only particles coming in one direction, but the energy deposition is a scalar quantity. It does not depend on the direction of the particle and is used in theoretical dosimetry, so will not be explored here.

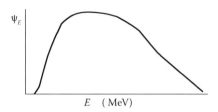

FIGURE 2.12
Typical linac spectrum.

Energy fluence is given by (number × energy)

$$\Psi = dR/dA \text{ (unit: J/m}^2)$$

where dR is the radiant energy incident on a sphere of cross-sectional area dA.

If each particle has an energy E (in a mono energetic beam),

$$\Psi = dR/dA = \{d(NE)/dA\} = \Phi E$$

Radiation is usually polyenergetic (e.g., linac spectrum) and the spectral distributions give an idea of the number of photons existing in the different energy ranges. To characterize a spectrum, a differential distribution Φ_E is needed, the number of particles with energy between E and E+dE.

$$\Phi_E (E) = d\Phi(E)/dE$$

Similarly, the differential energy spectrum is given by

$$\Psi_E = d\Psi(E)/dE = (d\Phi(E)/dE) \, E$$

In the same way the rate quantities (φ, ψ) can be defined as the number or the energy incident per unit area per unit time as $d\Phi_E/dt$ and $d\Psi_E/dt$. The units are $m^{-2} s^{-1}$ and $J \, m^{-2} s^{-1}$, respectively. The energy fluence rate is more commonly known as the beam intensity.

Figure 2.12 shows a typical linac photon spectrum, and illustrates the concept of differential fluence.

The energy bins dE, for any E, on the energy scale determine the differential energy fluence dΨ(E) on the Ψ_E scale. When the photon beam is incident on a medium each atom presents an area (Figure 2.13) that is proportional to the probability of each type of interaction that can take place.

2.6 Concept of Interaction Coefficients

Figure 2.14 illustrates the attenuation of a photon beam traversing a medium.

To characterize the attenuation of the beam, the (beam attenuating) interactions must be expressed on a macroscopic scale (i.e., not at electronic or atomic level but as per cm [or per g/cm^2] of beam traversal in the medium). If 'n' is the number of electrons per unit volume and ρ the material density, the total number of atomic interactions per unit volume or per gram is given by $n\sigma_t$ or $(n/\rho) \, \sigma_t$, respectively. The total interaction cross section per atom is a summation of all the different types of interactions that are energetically possible.

$$\sigma_t = \sigma_{PEE} + \sigma_{CE} + \sigma_{PP}$$

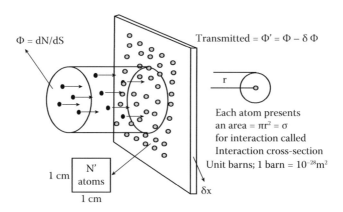

FIGURE 2.13
Concept of cross-section.

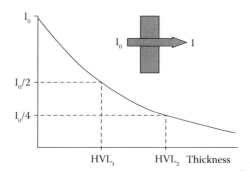

FIGURE 2.14
Attenuation of photon beams traversing a medium.

$n\sigma_t$ is called the linear attenuation coefficient, $_l\mu$, with dimensions of cm^{-1}. The attenuation coefficient can be defined as the fraction of the photons interacting in traversing unit distance in the medium, as shown below.

When a beam of photons passes through a thin absorber, under narrow beam conditions the attenuation coefficient can be deduced by measuring the transmitted fluence, as a function of absorber thickness and plotting a curve (see Figure 2.14).

The form of this curve can be determined as follows:

   Incident photon fluence = Φ

   Transmitted photon fluence = $\Phi - d\Phi$ in traversing an absorber thickness dx

   Therefore, number of photon interactions = $-d\Phi$

   Probability of interaction, $P = -d\Phi/\Phi$

   Fraction of photons interacting in unit distance of the medium traversed = $-[(d\Phi/\Phi)/dx] = _l\mu$

This equation can be compared to the equation we derived for the radioactivity law, in Chapter 1, using the following expression:

$$-(dN/N)/dt = \text{fraction of the atoms decaying per unit time} = \text{constant} = \lambda$$

This gave the radioactive decay equation. Using the same method of derivation, we would get the exponential law of attenuation (see the review questions or consult any standard textbook, e.g., Attic's book).

$$\Phi = \Phi_0\, e^{-i\mu x} \quad \text{or} \quad I = I_0\, e^{-i\mu x}$$

NOTE:

1. Since $_1\mu$ gives the attenuation per cm of the medium traversed, it represents the macroscopic attenuation coefficient. On the other hand, σ_1 represents the attenuation coefficient at the microscopic level, per atom level.

2. Total number of photons lost in all processes $= \Delta\,\Phi = \Phi_0 - \Phi = \Phi_0 - \Phi_0\, e^{-_1\mu\, 1}$. If the number of photons lost in a particular process (say process 1) is required, this is given by $\delta\Phi\, (_1\mu_1/_1\mu_t) = (_1\mu_1/_1\mu_t)\, \Phi_0\, e^{-_1\mu\, 1}$. The fraction of the total linear attenuation coefficient has to be used to obtain the fraction of photons lost in different interaction processes. Note the exponential includes all interactions, not just process 1, because the total number of interactions × the process 1 probability fraction gives the interactions due to process 1.

3. For thin absorber, $\delta\Phi \ll \Phi$, or $\delta\Phi/\Phi \ll 1$ and $\delta\Phi \approx \Phi_0\, (_1\mu\, 1)$.

   If $_1\mu = 0.02$, it means 2% of the photons is lost per cm of travel in the absorber. The transmitted fluence is $0.98\, \Phi_0$. This does not mean in traveling 50 cm, 100% of the photons will be attenuated. For larger thicknesses, the exponential equation must be used. The simplified equation applies only to thin absorbers. For thin absorbers, $\delta\Phi$ interacts due to process $1 \approx (_1\mu_1/_1\mu_t)\, _1\mu_t\, 1 = _1\mu_1\, 1$.

   By using the above approximate equation and the exponential equation to calculate to transmitted fluence for different thicknesses, the difference between the two will keep increasing because as the thickness increases, the approximate equation based on thin absorber assumption becomes more and more inaccurate.

4. By measuring the transmitted photon fluence (or the energy fluence, or a proportional quantity), only the total attenuation coefficient can be determined and they cannot be separated into their constituents. The attenuation due to individual processes can only be computed as mentioned above.

5. In the above discussion, $\Phi - d\Phi$ represents the transmitted fluence under narrow beam conditions. This means that the transmitted fluence must contain only the photons that have not interacted with the absorber and should not include any scatter photons. The attenuated coefficients tabulated by Hubbel [3] are all narrow beam attenuation coefficients. On the other hand, while doing shielding calculations for teletherapy or linac rooms, the largest beam incident on the wall is typically 120 cm × 120 cm (assuming the wall to be at a typical distance of about 3 m from the machine isocenter). This is typically broad beam geometry and considering the large wall thicknesses involved, the transmitted intensities will include a large buildup of scatter photons. So, for shielding calculations, the tenth-value layer (TVL) values given in National Council on Radiation Protection and Measurements (NCRP) reports refer to broad beam conditions obtained for typical linac treatment rooms.

6. The attenuation equation can also be written in terms of photon intensity I or energy fluence Ψ as shown below:

$$\Psi = \Psi_0\, e^{-_1\mu\, x}; \quad I = I_0\, e^{-_1\mu\, x}$$

7. The beams used in radiotherapy are usually polychromatic beams with typical spectra. When such a beam traverses a medium, usually the low-energy photons are preferentially absorbed and the beam hardness (and hence the attenuation coefficient) gradually changes with depth. This situation occurs in the case of low-energy X-ray beams. In the case of high-energy therapy photon beams, the low-energy components are already cut off by the thick flattening filter. However, some low-energy contamination photons (the Compton scattered photons) do accompany the beam incident on the phantom. Though these photons will be attenuated in the initial layers of the phantom (or patient), the beam penetrability does not monotonically increase with depth because of the attenuation of the higher energy photons by the PP processes and because of the softening of the spectrum by the Compton scattered photons produced in the phantom. The attenuation coefficients for these beams give effective attenuation coefficient values.

8. The term analogous to half-life (in the case of radioactivity) is half-value layer (HVL), the thickness that cuts the intensity (or the dose) to half of the incident intensity (dose). The term most commonly used in shielding calculations in radiation oncology is the tenth-value layer (TVL), the thickness that reduces the intensity (dose) to one tenth of the incident intensity (dose).

9. Keeping the radioactivity decay law in mind, we can write (see review questions)

$$HVL = 0.693/_1\mu$$

Since a TVL reduces the dose or the intensity by a factor of 10, 1 TVL is approximately 3.3 HVL.

Example 2.2

A parallel beam of 6 MV is incident on the wall of the linac treatment room. The incident kerma is 410×10^3 mGy/wk. The permitted weekly dose on the other side of the wall is 0.1 mSv/wk. What must be the wall thickness? TVL = 37 cm concrete.

Incident dose = 410×10^3 mGy.

Permitted dose at the exit side of the wall = 0.1 mSv/wk.

The reduction factor required = $410 \times 10^3 / 0.1 = 4.1 \times 10^6$.

6 TVL will reduce the dose by 10^6 times.

2 HVL will reduce the dose by a factor of 4.

So, 6 TVL + 2 HVL = $6 \times 37 + 2 \times 11.2 = 222 + 22.4 = 244.4$.

About 2.5-m thickness of concrete is required.

When dx is expressed in terms of g/cm^2, as ρdx, the attenuation coefficient μ_l also must be replaced by ($_l\mu/\rho$ or $_m\mu$) so the product becomes dimensionless. The mass attenuation coefficient $_m\mu$ is used so that density influence is taken care of.

$$_m\mu = \frac{\Delta\Phi}{\Phi} \cdot \frac{1}{\rho\Delta x} \, cm^2/g = \frac{\Phi_1 - \Phi_0}{\Phi_1} \cdot \frac{1}{\rho\Delta x}$$

$$I = I_0 \, e^{-\mu \, (x/\rho)} = I_0 \, e^{-_m\mu x}$$

Example 2.3

A photon beam is incident on a thin Al layer of dimensions 10 cm × 10 cm. Given the mass of the absorber as 50 g, calculate the absorber thickness in g/cm^2. Calculate its linear thickness in mm (given Al density $\rho = 2.7\ g/cm^3$).

Areal density = mass ÷ area = 54 ÷ 10 × 10 = 0.54 g/cm^2.

$$\rho x = 0.54\ g/cm^2;\ x = 0.54/2.7 = 0.2\ or\ 2\ mm$$

The contribution of various interaction processes to mass attenuation coefficient as a function energy is shown in Figure 2.15.

The mass attenuation coefficient curves for two typical materials of interest in dosimetry and teletherapy room shielding, water and Pb, are shown in Figure 2.16. It shows the contribution of the important interaction processes at different photon energies.

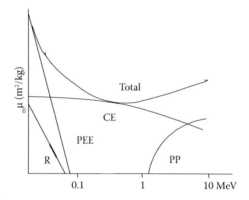

FIGURE 2.15
Contribution of various attenuation processes to $_m\mu$.

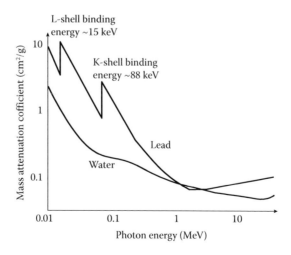

FIGURE 2.16
Mass attenuation coefficient of water and lead.

It is noticeable that as the probability of one process decreases, the probability of a competing process starts increasing. As the PEE probability decreases, the CE probability starts increasing and as this probability starts decreasing, the PP probability starts increasing.

2.7 Concept of Kerma

In radiation oncology, the primary interest is in the energy transferred to charged particles (by the IDIPs since it is this energy which is subsequently deposited in the medium causing tumor destruction or biological harm. So, the fraction of the photon energy converted to charged particle energy needs to be determined, at the point of interaction, in δm of mass. This energy per unit mass is known as kerma—kinetic energy released per unit mass.

Table 2.2 shows the photon interactions that involve transfer of energy to charged particles.

2.7.1 Fraction of Photon Energy Transferred to Electrons

In PEE,

Electron energy released = $h\nu_0$ – BE.

Fraction of photon energy converted to electron energy $f_{pe} = (h\nu_0 - BE)/h\nu_0 = [1 - (BE/h\nu_0)]$.

For PP, a minimum of 2 $m_0 c^2$ energy is required to create an electron–positron pair and the excess energy of the photon is shared by this pair as their kinetic energies.

So, in PP, fraction of photon energy converted to electron energy $f_{pp} = [1 - (2\ m_0\ c^2/h\nu_0)]$.

For triplet production, in addition to the energy required to create the electron–positron pair, BE must be supplied to eject the interacting electron from the atom.

In triple production, the fraction of photon energy converted to electron energy is given by

$$f_{tp} = [1 - (2\ m_0\ c^2 + BE)/(h\nu_0)]$$

TABLE 2.2

Photon Interactions That Involve Charged Particle Emission

Interaction	Charged Particles
Photoelectric	X
Compton	X
Pair production	
Triplet production	X
Rayleigh scatter	
Photonuclear:	
(γ,n)	
(γ,p)	X

CE produces a spectrum of electron energies and so the average electron energy or the average scatter photon energy per interaction must be determined. $\overline{E}_c = h\nu_0 - \overline{h\nu}$, where $\overline{E}_c$ and $\overline{h\nu}$ represent the average energies of scattered electrons and photons, respectively. $\overline{E}_c = h\nu_0 \, (_e\sigma_a/_e\sigma)$.

The cross section for nuclear interactions is neglected here since it is a complex function of the nuclides and the data are not well tabulated. Its contribution to photon dose is usually negligible (<0.1% of the photon dose).

The mass attenuation coefficient × fraction of the photon energy converted to charged particle energy gives the mass energy transfer coefficient. When we write the mass attenuation coefficient for each process in terms of the energy transferred and sum the coefficients, this produces the total mass energy transfer coefficient $(_m\mu)_{en}$.

$$(_m\mu)_{pee}(1 - BE/h\nu_0) + (_m\mu)_{ce}(\overline{E}_c/h\nu_0) + (_m\mu)_{pp}[1 - (2m_0c^2/h\nu_0)]$$
$$+ (_m\mu)_{tp}\{1 - [(2m_0c^2 + BE/h\nu_0]\}$$

This is known as the mass energy transfer coefficient, $(_m\mu_{tr})_t$; the suffix t is usually dropped while writing this coefficient. The expression $_m\mu_{tr}$ is a measure of the fraction of the photon energy converted to charged particle (CP) energy.

If the incident energy fluence Ψ of a monoenergetic beam is known, at the point of interaction in δm of mass, the kerma (kinetic energy released per unit mass) can be calculated as

$$K = \Psi \times (_m\mu_{tr})$$

This is a very important equation for computing kerma theoretically. In the case of a spectrum, the equation is written in terms of the differential energy fluence Ψ_E (i.e., energy fluence between E and E + ΔE).

$$K = \int \Psi_E \, _m\mu_{tr}(E)dE$$

To find K in J/kg, $(_m\mu_{tr})$ must be expressed in m^2/kg, and Ψ in J/m^2. If the photon energy is in MeV, then the conversion factor 1 MeV = 1.6×10^{13} J must be used.

It is important to emphasize that this is a point quantity. When a photon beam traverses a medium, medium kerma can be defined at every point along the beam path (see Figure 2.23). In other words, it represents a kerma at each point = $\delta E_{tr}/\delta m$ (in the limit $\delta m \rightarrow 0$ (or represents dE_{tr}/dm).

The definition also implies that it is a non-stochastic quantity, having a definite value at each point. This in turn implies that the photons in the fluence are quite large and the statistical fluctuations in the energy transfer are negligible, and do not show up compared to the precision of measurements. This means, if the measurements are repeated a reproducible value of kerma at each point will be obtained.

Students usually have some misunderstanding regarding this quantity from the units of J/kg. It is not the energy absorbed in 1 kg of mass. It is the energy absorbed in an infinitesimal mass, δm of mass at the point of measurement so that the unit becomes J/kg. (So a statement such as "energy absorbed in kg of mass" referring to kerma, is totally erroneous). Since Ψ varies with depth, in an absorber such as $\Psi(x) = \Psi_0 \, e^{-_m\mu \, x/\rho}$, K(x) also varies as $K_0 \, e^{-_m\mu \, x/\rho}$ with a well-defined value at each point along the depth.

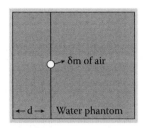

FIGURE 2.17
Air kerma K_a at depth d in water.

Kerma depends on the medium of interaction, since $_m\mu_{tr}$ is characteristics of the medium. So, when talking about kerma the medium should always be referred to, such as air kerma K_a, tissue kerma K_t, or water kerma K_w.

Air kerma can also be calculated in a medium other than air, for example in a graphite phantom or in a water phantom. There is some confusion among students regarding the meaning of expressions such as "tissue kerma in air" and "air kerma in air." Since kerma is a point quantity, it is not difficult to define it at any point, including inside a phantom. For instance, air kerma at a particular depth, in a water phantom, only means the energy released in an infinitesimal mass δm of air at the specified depth in the water medium. Figure 2.17 illustrates the meaning of this concept.

An infinitesimal mass δm of air in a graphite medium or in a water phantom is not a hypothetical situation. Air kerma can always be measured in a water phantom using an air kerma calibrated chamber in water method. In fact, this was the standard method of measurement in the days of cobalt-60 dosimetry. We will discuss these aspects in more detail in the next chapter.

If $\Psi_{E,W}$ represents the differential energy fluence in water at depth d, air kerma K_a at depth d in water is given by

$$[K_a(d)]_W = \int \Psi_{E,W}[_m\mu_{tr}(E)]_a dE$$

The important point to note is the use of the mass energy transfer coefficient of air, and not of water.

NOTE: There is also a quantity defined for the total energy released by the photons including the scatter photon energy (not just the charged particle energy that defined kerma). This is TERMA—total energy released per unit mass (including bremsstrahlung produced by the charged particle). It is computed from the following equation:

$$T = \int \Psi_{E,W}(_m\mu(E)) \, dE$$

using the attenuation coefficient in place of energy transfer coefficient.

Since the quantity of interest is the energy deposition in physical and biological systems, the quantity TERMA is not often required to be used in radiation oncology, but may be used in connection with scatter photon energy re-absorption.

Example 2.4

a. δm of graphite is exposed to a fluence rate of 2.8×10^7 photons/cm^2 s of energy 1.5 MeV. The total atomic attenuation coefficient for graphite, $_a\sigma_{t,gr}$ is 1.032 barn/cm^2. Calculate the mass attenuation coefficient for graphite energy.

b. Calculate the energy fluence rate, $d\Psi'/dt$, and the graphite kerma rate. in cGy/min (graphite mass energy transfer coefficient, $_m\mu_{tr,gr}$ is 0.0256 cm^2/g).

Graphite atoms/g $= 6.022 \times 10^{23}/12 = 0.5018 \times 10^{23}$ (a gram contains the Avogadro number of atoms)

$$_a\sigma_{t,gr} = 1.032 \text{ barn/cm}^2 = 1.032 \times 10^{-24} \text{ cm}^2$$

$$_m\mu_{gr} = (N_A/A) \, _a\sigma_{t,gr} = 0.5018 \times 10^{23} \times 1.032 \times 10^{-24} = 0.0518 \text{ cm}^2/\text{g}$$

$$_m\mu_{tr,gr} = 0.0256 \text{ cm}^2/\text{g}$$

To calculate the energy fluence rate:

$$\text{Fluence rate} = 2.8 \times 10^7 \text{ photons/cm}^2 \text{ s}$$

$$d\Psi'/dt = \text{fluence rate} \times 1.5 \times 1.6 \times 10^{-13} \text{ J/cm}^2 \text{ s } (1 \text{ MeV} = 1.6 \times 10^{13} \text{ J})$$

$$= 2.8 \times 10^7 \times 1.5 \times 1.6 \times 10^{-13} \text{ J/cm}^2 \text{ s}$$

$$= 6.72 \times 10^{-6} \text{ J/cm}^2 \text{ s}$$

Graphite kerma $= (d\Psi'/dt) \, _m\mu_{tr,gr} = 6.72 \times 10^{-2} \text{ (J/m}^2) \times 0.0256 \times 10 \text{ m}^2/\text{kg}$

$$= 6.72 \times 10^{-2} \times 0.256 \text{ J/kg} = 1.72 \times 10^{-2} \text{ Gy/s}$$

$$= 1.72 \times 10^{-2} \times 60 \text{ Gy/min} = 1.032 \text{ Gy/min or } 103.2 \text{ cGy/min}$$

If the energy released, δE_{tr}, can be measured at a point in δm of mass, kerma can be measured using the equation.

$$K = \delta E_{tr}/\delta m \text{ J/kg}$$

Figure 2.18 shows a hypothetical situation of three photons interacting in δm of mass, releasing electrons of kinetic energies E_1, E_2, and E_3, respectively.

$$K = (E_1 + E_2 + E_3)/\delta m = \sum_{i=1}^{3} E_i/\delta m \text{ J/kg}$$

But in reality, the fluence incident on δm will be very large (say about 10^7 or 10^8 photons/cm^2) and the corresponding interactions will also be large, but to measure kerma only requires the summing of all the electron energies released in δm (the measurement details will be dealt with in the next chapter).

It is difficult to measure kerma since the CPs usually escape the defined mass of measurement δm and what is measured in the defined volume will be less than kerma. In fact, the measurement becomes more complex because the mass δm is not isolated in space, and so there will be an inflow of charged particles from the surrounding medium (e.g., air).

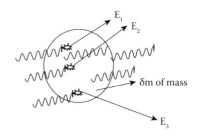

FIGURE 2.18
Concept of kerma.

2.8 Concept of Collision Kerma

All the energy transferred to the charged particles is not in turn transferred to the medium. Figure 2.19 illustrates the concept.

The figure (on the left) is the same as Figure 2.18 but here two electrons produce bremsstrahlung photons, one inside δm and one outside δm. Kerma is still the same as explained in the earlier paragraph. However, part of the electron energy goes as bremsstrahlung energy (ΔE_{2br} and ΔE_{3br}) and this energy generally escapes from the medium and cannot contribute to energy deposition in the medium. So, to relate to energy deposition in the medium, kerma is apportioned into collision kerma, K_c, and radiated kerma K_{rad}. If g represents the fraction of the electron energy converted into bremsstrahlung,

$$K_c = K(1 - g)$$

Similarly, the bremsstrahlung fraction can be removed from $_m\mu_{tr}\cdot _m\mu_{tr}$ $(1 - g)$ is then known as the mass energy absorption coefficient. The earlier equations then become

$$K_c = \Psi \times (_m\mu_{en}) = \Psi \times (_m\mu_{tr})(1 - g)$$

and for a photon spectrum

$$K_c = \int_E \Psi\ _m\mu_{en}(E)dE = \int_E \Psi\ _m\mu_{tr}(E)(1-g)\ dE$$

It is called collision kerma since it refers to collisions of charged particles that result in ionization and excitations of the atoms of the medium. Similarly radiated kerma refers to collisions of charged particles that result in bremsstrahlung production.

$$K = K_c \text{ and } K_{rad}$$

In Figure 2.19,

$$K = (E_1 + E_2 + E_3)/\delta m$$

$$K_c = \{(E_1 + E_2 + E_3) - (\Delta E_{2br} + \Delta E_{3br})\}/\delta m = \delta E_{tr,net}/\delta m$$

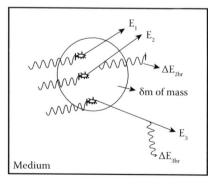
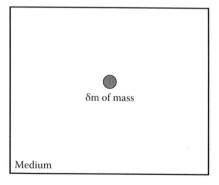

FIGURE 2.19

Concept of collision kerma (δm exaggerated on the left).

where $\delta E_{tr,net}$ is the net energy transferred to charged particles that result in energy deposition.

Example 2.5

Calculate graphite collision kerma rate (in cGy/min) in the above example assuming $_m\mu_{en,gr}$ is 0.02551 cm^2/g. See the extent of agreement if collision graphite kerma is calculated using g = 0.9966.

$$K_{C,gr} = 6.72 \times 10^{-2} \times 0.2551 \text{ J/kg} = 1.714 \times 10^{-2} \text{ Gy/s}$$

$$= 1.714 \times 10^{-2} \times 60 \text{ Gy/min} = 1.028 \text{ Gy/min, or } 102.8 \text{ cGy/min}$$

The relation $K_{C,gr} = K_{gr} (1 - g)$ can also be used; the bremsstrahlung fraction correction $(1 - g)$ is 0.9966.

$$K_{C,gr} = 103.2 \times 0.9966 = 102.8$$

It agrees with the earlier calculation as expected since $_m\mu_{en,gr}$ is derived from $_m\mu_{tr,gr}$ by correcting for the bremsstrahlung fraction.

2.9 Concept of Absorbed Dose

The quantity of interest in dosimetry is always the energy locally deposited in δm of mass, and not the amount of charged particle energy produced in δm. Often, δm is the mass of interest in dosimetry representing a dosimeter sensitive volume, the core of a calorimeter, or the cells in a biological system. The response in all these cases is due to the energy deposited in δm. Look at Figure 2.20 used in the definition of kerma again (with the figure slightly modified) to see what constitutes absorbed dose.

Part of the electron energies (ΔE_1, ΔE_2, and ΔE_3) are deposited in δm, through ionization and excitation events. Now what is the absorbed dose, D, or energy locally deposited in δm is straightforward.

$$D = (\Delta E_1 + \Delta E_2 + \Delta E_3)/\delta m = \sum_{i=1}^{3} \Delta E \delta m \text{ J/kg}$$

As mentioned earlier, in reality the interactions are very large in number and so we must add all the energies deposited in δm to determine the absorbed dose. It is again a point quantity with a well-defined value (i.e., non-stochastic) at each point in the medium traversed by the photon beam. How the dose equation can be written in terms of charged particle fluence will be examined in the next section.

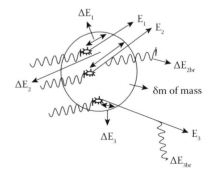

FIGURE 2.20
Concept of absorbed dose.

For experimental determination of an absorbed dose, dose D is defined as

$$D = \delta\varepsilon/\delta m \ /kg$$

The name of the SI unit of dose is the same as for kerma—Gray.

2.10 Relating Kerma and Dose

Here, we will first relate collision kerma to dose since only the collision part of kerma deals with energy deposition in the medium, the bremsstrahlung part escaping from the region. We can then easily relate K_C to kerma, using the equation given earlier, taking bremsstrahlung into account.

So far, we have discussed photon beam incident on δm of mass, but as mentioned earlier, it is also difficult to isolate a mass in the real situation. Photons leaving the source (a radioactive source or a teletherapy treatment machine), interact all along the beam path until they reach δm—after all, air is not free space where no interactions can take place. The electrons produced in these interactions (volumes marked around δm) get scattered at different angles and enter δm of mass. Figure 2.21 illustrates this situation.

So, there is always an envelope surrounding δm of mass that contributes to energy transport into δm.

This situation becomes handy for measuring collision kerma in δm. As per the definition of collision kerma, remembering our earlier figures, we only have to know the total energy $E_1 + E_2 + E_3$ and the bremsstrahlung fraction, produced in δm by the photon fluence, to determine K_C. But these electrons deposit only part of their energies, $\Delta E_1 + \Delta E_2 + \Delta E_3$, in δm and carry much of the energy outside δm. The inflow from the buildup layer can partially compensate for this outflow. If δm is surrounded by a buildup layer of thickness equal to the maximum range of the electrons, no electrons outside this envelope can reach δm; this thickness of the buildup layer is known as the charged particle equilibrium (CPE) thickness. Under CPE thickness, the outflow of CP energy is exactly compensated by the inflow of charged particle energy, and the local energy absorbed in δm per unit mass (absorbed dose) is equal to the net energy transferred to charged particles by the photon fluence in δm.

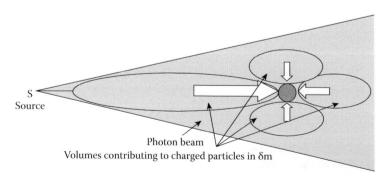

FIGURE 2.21
δm of mass receiving charged particles from all directions around δm.

Under CPE condition, radiation (charged particle) energy R flowing into δm of mass is exactly compensated by the CP energy flowing out.

$$(R_{in})_{CP} = (R_o)_{CP}$$

So the collision part of CP energy produced, K_c, in δm can be equated to dose since the electron energy flowing out is compensated by inflow of CPs.

$$[D]_{CPE} = K_C = \Psi_m \mu_{en}$$

The necessary and sufficient conditions for establishing CPE are:

1. The δm must be surrounded by the same material (or material of same density and atomic composition) of equilibrium thickness
2. Photon attenuation in the buildup thickness must be negligible

Conceptually, the second statement is important. A pertinent question to ask is whether exact CPE condition can be realized in δm because of the attenuation of photons in the buildup layer; it is possible to the extent the attenuation is negligible. If it is not negligible, the electron fluence produced in the δm by the photon fluence will be less than that produced in the buildup layer, and collision kerma and the dose curve will not be equal. That is the reason for the kerma and dose curve profiles not coinciding in Figure 2.23, beyond the equilibrium depth. A correction needs to be applied for ^{60}Co energies, because of finite attenuation of photons over electron ranges but this correction is too small (<0.5%) to consider except at primary standards dosimetry laboratories, where highest accuracy is required in measurements.

We will now derive a general expression for energy deposited in δm from the hypothetical example discussed earlier.

The net energy (i.e., deducting bremsstrahlung conversion) transferred to charged particles is given by

$$\delta E_{tr,net} = [(R_{in})_u - (R_o)_u] + [(R_{in})_{CP} - (R_o)_{CP}] = \delta\varepsilon$$

u refers to uncharged and here it refers to X-ray energy inflow and outflow, respectively. δε gives the energy deposited in δm.

In Figure 2.20 (assuming three Compton interactions),

$$(R_{in})_u = h\gamma_1 + h\gamma_2 + h\gamma_3$$

$$(R_o)_u = h\gamma_{sc1} + h\gamma_{sc2} + h\gamma_{sc3} + \Delta E_{2br}$$

$$\delta E_{tr,net} = E_1 + E_2 + E_3 - \Delta E_{2br}$$

The photons traversing δm without any interaction need not be taken into account, since the inflow is equal to the outflow.

Under CPE condition,

$$(R_{in})_{CP} = (R_o)_{CP}$$

In addition, the ΔE_{3br} loss should also be taken into account for δm even if it occurs outside δm, because CPE demands that incoming and outgoing CP energies are equal. So, if an outgoing CP produces a bremsstrahlung photon, the equivalent incoming CP also produces a bremsstrahlung photon, on average.

So, under CPE,

$$\delta E_{tr.net} = E_1 + E_2 + E_3 - (\Delta E_{2br} + \Delta E_{3br}) = \delta\varepsilon$$

$$[\delta E_{tr.net}/\delta m] = \delta\varepsilon/\delta m$$

or

$$K_c = [D]_{CPE}$$

NOTE: The expressions must include one more factor, ΣQ, if any of the energy goes as internal energy of the atom (causing, say, nuclear reactions). Here we assume this to be 0. Otherwise, we have to also include this term in the expressions for δE or $\delta\varepsilon$ above.

So, dose measured under CPE condition is numerically equal to the collision kerma. Correcting this for 'g' gives kerma.

In Figure 2.22, we have shown a thin shaded layer around δm, for establishing the CPE in δm. Though theoretically a maximum electron range for CPE condition is required, much of the electron fluence comes from the immediate layers surrounding δm and only a small percentage of contribution comes from the remaining layers. For instance, for a ^{60}Co beam the requirement may be about a 5 mm (500 mg/cm^2) envelope for CPE, but more than 90% of the contribution to δm comes from a thin layer of about 70 mg/cm^2. So, for practical purposes, thickness less than the theoretical thickness may also suffice.

The following three statements are important and should be learned.

1. All the energy transferred to the charged particles (mainly electrons and positrons) is not necessarily imparted to the medium at the same point.

2. While photons lose their energies at the point of interaction, electrons are always absorbed over their ranges (and not locally). In other words, charged particles (electrons) transport energy transferred from photons, away from the origin. The larger the photon energy, the larger is the energy of the electrons produced and the farther away they can carry the energy, in the process of depositing the energy.

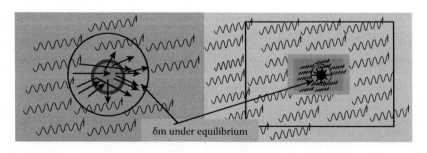

FIGURE 2.22
δm of mass under equilibrium in air (left) and in a medium (right).

3. Any system (physical or biological) responds to the energy locally deposited in δm and not to the energy released in δm. It is the energy absorbed that gives rise to a response in a dosimeter, or damage in a cell, or damage in tissues.

2.11 Dose and Kerma Profiles in a Phantom

Figure 2.23 shows the dose and kerma profile in a water phantom, irradiated by a photon beam. The profile is explained in the same manner as described earlier for the dose under CPE conditions. The phantom has two regions—the buildup region where there is no CPE, and a transient equilibrium region where CPE exists (with a minor correction for the finite attenuation of the photons in the buildup region).

There is a photon fluence in each layer right from the phantom surface, so there is an electron fluence produced in each layer. A small amount of this energy is locally absorbed (in δm) and the rest of the energy is carried forward and absorbed over their ranges. In the buildup region, the electron fluence slowly builds up in each layer, because the fluence flowing from the earlier layers adds to the fluence produced in the layer in question. Therefore, the dose also gradually increases with depth. Since the electron fluence beyond the electron ranges will not be able to contribute to a point of interest, there is an equilibrium fluence (and an equilibrium dose) reached at a certain d_{max} depth, after which the electron fluence decreases with the photon fluence (or D decreases with K_c) as shown in the figure. The region where the D and K_c curves are parallel is known as the transient equilibrium region. This is very similar to the transient equilibrium curve in radioactivity where the parent activity equals the daughter activity; in this case, K_c equals dose.

In Figure 2.23, δm of mass is marked at two depths, for $d < d_{TE}$ and for $d > d_{TE}$.

For $d < d_{TE}$, $D < K_c$ since it is in the fluence buildup region.

The dose slowly increases until we get to the region where $D = K_c$.

In fact, there is a small gap between the two curves and this gap increases with photon energy, as shown more clearly in Figure 2.23.

The gap seems to suggest that at any given depth, in the tissue equivalent (TE) region, D (d) > K_c (d) the energy deposited is more than the energy transferred, which appears as a contradiction. This happens because of the electron energy transport from its origin. D(d) is

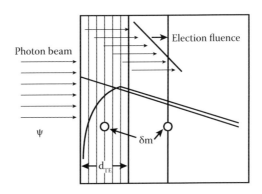

FIGURE 2.23
Dose building up before equaling collision kerma.

representative of K_c, not at d but at $(d - \bar{x})$, where the photon fluence is slightly higher creating higher electron fluence. From the curve it can be seen that the relation between K_c and D can be expressed in two different ways:

$$K_c(d - \bar{x}) = [D(d)]_{CPE}$$

or referring to same depth

$$K_c(d) = [D(d)]_{CPE} \, \beta \, (\beta > 1)$$

So, kerma is measured indirectly by measuring dose under CPE conditions, that is, by measuring $[D]_{CPE}$.

This correction β is difficult to evaluate accurately. The gap between the curves depends on the finite attenuation of photons over the electron ranges, which increases with photon energy. For kilovoltage X-ray beams, the electron ranges are very short and the photon attenuation over the ranges is negligible, and the two curves coincide. For ^{60}Co beams, the electron ranges are around 5 mm for unit density materials, and there is a small gap amounting to a correction of about 0.5%. For higher energy photon beams, the gap (and hence the correction) is larger but kerma is not measured for higher energy beams, but only the local absorbed dose. (This is one reason why kerma is not measured for higher energies).

J/kg is called Gray (Gy) in dosimetry, and an absorbed dose or air kerma of 10 Gy would mean an energy deposition or release of 10 J/kg. The older energy unit used was ergs/g and 100 ergs/g, the rad, which is no longer in use; 1 rad = 0.01 Gy.

A summary of attenuation coefficients is given in Table 2.3.

A summary of the photon interactions and their characteristics is given in Table 2.4.

TABLE 2.3

Units for Attenuation Coefficients and PL

Attenuation Coefficient	Symbol and Probability Per Unit Length	PL
Electronic attenuation coefficient	$_e\sigma$ m^2 or cm^2 $_m\mu \, N_0$	Can be related to PL ($\rho x. \, N_0$) kg/m^2 or g/cm^2
Atomic attenuation coefficient	$_a\sigma$ m^2 or cm^2 $Z \, _e\sigma$	Can be related to PL ($\rho x. \, N_0/Z$) kg/m^2 or g/cm^2
Linear attenuation coefficient	$_l\mu$ m^{-1} or cm^{-1}	x m or cm
Mass attenuation coefficient	$_m\mu \, (=_l\mu/\rho)$ m^2/kg or cm^2/g	ρx kg/m^2 or g/cm^2
Mass energy transfer coefficient	$_m\mu_{en}$ m^2/kg or cm^2/g	ρx kg/m^2 or g/cm^2
Mass energy absorption coefficient	$_m\mu_{ab}$ m^2/kg or cm^2/g	ρx kg/m^2 or g/cm^2

No. of atoms/g = (N_a/A); N_a = Avogadro no. = 6.022×10^{23}
No. of electrons/g = $N_a \, (Z/A)$; $\times 10^3$ gives per kg.

Note: PL, path length.

TABLE 2.4

A Summary of Photon Interactions

	Photon Interactions				
	Electronic Collisions			Nuclear Collisions	Nuclear Reaction
$hv < BE$	$hv \geq BE$	$hv \gg BE$	$hv > 2.04$ MeV	$hv > 1.02$ MeV	$hv > 10$ MeV
Rayleigh Scattering	Photoelectric effect	Compton effect	Triplet production	Pair production	Photo nuclear interactions
Interaction with the atom	Interaction with bound atom	Interaction with free electron	Pair production in the coulomb field of electron	Interaction with Nuclear Coulomb Field	(γ,n), (γ,p) (γ,m) type Interactions take place
$\sigma R \sim z^2$	$_a\sigma_{pe} \sim z^3/(hv)^3$	$_a\sigma_e \sim z$	$_a\sigma_{tp} \sim z$	$_a\sigma_{pp} \sim z^2$	$\sigma_{pn} \sim z$
<25 keV	Few keV – 100 keV	0.5–5 MeV	>5 MeV	>5 MeV	>10 MeV
No energy deposition	All these interactions produce energetic electrons which will lead to energy deposition in the medium.				n, charged particles

Note: BE, binding energy.

2.12 Interaction of Charged Particles with Matter

Charged particles carry electric charge, and so do the nuclei and electrons of atoms of the medium. So CPs undergo Columbic interactions with the electrons and nuclei of the medium. The CPs are classified as LCPs (light charge particles i.e., electrons or positrons) and HCPs (heavy charged particles, i.e., protons and carbon ions). Their masses being much greater than electrons and positrons, they are called HCPs. The large mass difference introduces subtle differences in their paths and in their energy loss mechanisms.

The charged particle interactions are also called collisions, which are classified as elastic and inelastic.

An elastic collision is one in which the total kinetic energy of the system is conserved, so the total kinetic energy before interaction is equal to the total kinetic energy of the products after the interactions. For instance, when the target particle in a collision simply recoils, the interaction is elastic and the energy lost by the charged particle appears as the recoil energy of the target particle. Such elastic collisions usually take place when the incident particle has no means of transferring energy to the nuclide (e.g., inadequate energy to excite or ionize or disintegrate the target atom or when certain forbidden transitions prevent energy absorption by the nuclide).

An inelastic collision, on the other hand, is one in which some part of the kinetic energy of the charged particle is used up to change the internal state of the target atom, which, therefore, does not appear as the kinetic energy of the resultant products.

The following CP interactions occur in a medium:

1. Elastic and inelastic collisions with medium electrons.
2. Elastic and inelastic collisions with medium nuclei.

The nature of interaction depends on the energy and the distance of approach of the charged particles with respect to the atomic dimensions. The closest distance of approach is known as the impact parameter, as shown in Figure 2.24.

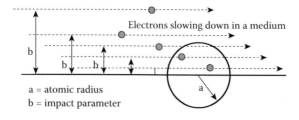

FIGURE 2.24
Impact parameter and nature of collisions.

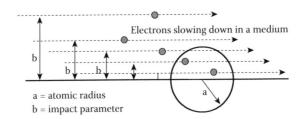

FIGURE 2.25
Electron collisions with an atom.

2.12.1 Coulomb Electronic Collisions

If the Coulomb force, F, acts for a time δt on the electron of the medium, it transfers a momentum p = F δt. This results in a transfer of energy. The amount of energy transfer for a given CP depends on the impact parameter (Figure 2.25).

For b >> a, the energy transferred is very small and the collisions are referred to as soft collisions. These are distant encounters. The energy transfers in such collisions are very small (i.e., less than the BE of the medium electrons) but can cause excitations or nearly elastic collisions.

The collisions are, therefore, mostly elastic in nature. However, no energy is imparted to the medium and the electron does not suffer much deflection due to the large mass of the atom.

For b > atomic dimensions, the interaction is with the atoms of the medium (i.e., with the bound electrons of the medium) causing ionization and excitation of the medium atoms. The energy losses of the charged particle in such interactions are just enough to cause ionizations, or in other words, the ionized electrons carry very little kinetic energies with them (a few electron volts).

These collisions are usually termed distant collisions (or soft collisions) and involve small energy losses. Because of the large impact parameter compared to the size of the atom, the finer structure of the interacting particles (e.g., the spin of the charged particle or the nuclear/Coulomb field of the nucleus) does not influence the interaction in this type of collision.

For b atomic dimensions, the charged particle can detect the finer details of the atom and can interact with individual electrons. The collisions corresponding to these impact parameters are called hard or close collisions and involve considerable energy transfer to the medium electrons. These electrons, known as rays, play an important role in ionization dosimetry. They have sufficient energy to produce further ionization and excitations of the medium, as in the primary particle.

The mechanism of ionization and excitation events are illustrated in Figure 2.26.

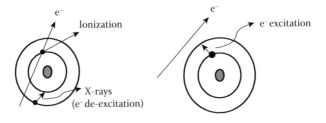

FIGURE 2.26
Inelastic collisions causing ionization/excitations of medium atoms.

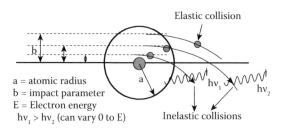

FIGURE 2.27
Elastic and inelastic nuclear (Columbic) collisions.

2.12.2 Coulomb (Nuclear) Collisions

For b << atomic dimensions, the interactions of the charged particles with the nuclei of the atoms become more probable and assume significance. The interactions are Coulomb elastic collisions, resulting in the scattering of electrons from their main paths, and Coulomb inelastic collisions, resulting in scattering and some energy loss in the production of bremsstrahlung photons. These interactions are illustrated in Figure 2.27.

2.12.3 Nuclear Collisions in Nuclear Force Field

HCPs can also engage in nuclear interactions which give rise to the release of nuclear particles (alpha, neutron, proton, etc.). Their contribution to the dose deposited in the medium is small in the energy range of therapeutic interest. Table 2.5 summarizes the CP interactions.

2.13 Concept of Stopping Power

Stopping power characterizes the ability of a material to stop charged particles. While discussing photon energy transfer coefficients, it was shown that not all the electron energy is transferred to the medium and a part of the electron energy is radiated out as bremsstrahlung. Here we have to differentiate between energy lost to the medium, and the energy radiated out. These two quantities are characterized by collision stopping power and radiative stopping power. The sum of the two is known as SP or total stopping power.

2.13.1 Concept of Collision Stopping Power

The continuous slowing down of CPs by electronic collisions that cause ionization/excitations in the medium gives rise to a definite energy loss of CP per unit path length traversed in the medium (−dE/dx). It is known as linear collision stopping power, S_c, with units of MeV/cm

$$S_c(E) = -\overline{dE}/dx$$

Since the interactions are statistical, all the electrons of a given energy will not lose exactly the same amount of energy in traveling unit path length. So, stopping power gives the mean energy loss. After traversing a distance dx, the mean energy loss is actually $\overline{dE}$ and the electrons of initial energy E (shown as a single line) exhibit a spectrum of energies (energy straggling), while exiting from the absorber, as shown in Figure 2.28.

TABLE 2.5

Types of CP Interactions

	CP Interactions in a Medium				
	Electronic Collisions		**Nuclear Collisions**		
	Elastic	**Inelastic**	**Elastic**	**Inelastic**	
	Coulomb Field	**Coulomb Field**	**Coulomb Field**	**Coulomb Field**	**Nuclear Force Field**
Scattering		Excitation Ionization Byproduct: Characteristic Byproduct: δ rays X-rays	CP deflection or scattering	Bremsstrahlung	Nuclear reactions
		CP slows down by losing energy to the medium. Energy loss per collision <100 eV. Followed by emission of Characteristic X-rays and Auger electrons.	Beam angular spread Important for Electrons; HCP tracks are straight	CP slows down by radiating out energy Important for energies $\approx M_0c^2$	Nuclear excitation Nucleón emission
No energy Loss for CP		All these interactions produce energetic electrons which will lead to energy deposition in the medium.			n, charged particles

Note: CP, charged particle; HCP, heavy charged particle.

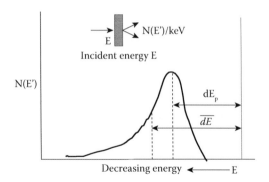

FIGURE 2.28
Energy straggling.

Mass collision stopping power $= -\overline{dE}/\rho dx = {}_mS_c(E)$ (MeV cm^2/g) gives the energy loss per g/cm^2 of the material.

2.13.2 Concept of Radiative Stopping Power

The energy radiated per unit path length (in Columbic nuclear deflections) leads to the concept of radiative stopping power, ${}_mS_r(E)$.

$$ {}_mS_r(E) \propto Z^2\, E $$

The total mass stopping power $= {}_mS_t(E) = {}_mS_c(E) + {}_mS_r(E)$

2.14 Stopping Power Expression for Heavy Charged Particles

The energy lost by the CPs while traversing a medium along their paths is characterized by the concept of stopping power which gives the average energy lost by the CP, in unit path length distance traveled. When the path length is expressed in cm it is referred to as linear stopping power, and when expressed in g/cm^2 it is known as mass stopping power. The SP mainly depends on the particle velocity and its charge, and it varies as $(1/v^2)$ as the stopping power increases at the near end of the range of the CPs.

For heavy charged particles, Bethe and Bloch derived a non-relativistic quantum mechanical expression for calculating the mass collision SP in different materials.

$$ dE/\rho dx = [4\pi(e^2/4\pi\epsilon_0)^2\, N_A]\,(Z/A)\,(z^2/m_e\,v^2)\,[\ln(2m_e v^2/I)] $$

where e is the elementary charge, v the velocity of the HCP, N_A the Avogadro's number, ϵ_0 the permittivity of free space, A the atomic number of the target, Z the atomic number of the target material, z the atomic number of the HCP, m_e the electron rest mass, and I the mean ionization excitation potential.

For relativistic velocities, Bethe derived the following expression for the HCPS:

$$ -\overline{dE}/\rho dx = [4\pi(e^2/4\pi\epsilon_0)^2 N_A](Z/A)\,(z^2/m_e v^2)\{\ln(2m_e c^2/I) + \ln[\beta^2/(1-\beta^2)] - \beta^2\} $$

where c = velocity of light and $\beta = v/c$.

Two corrections were later incorporated into Bethe's expression for the mass collision stopping power, shell correction CKz, and density correction or polarization correction δ. These will be discussed next.

2.14.1 Shell Correction CKz

When the velocity of the CP is in the order of the velocity of the electrons in orbits then most electrons do not easily absorb energy reducing the SP. So, a correction to this effect must be applied in the SP equation. Since the K-shell electrons have the highest velocities, when the CP slows down, the first to be affected are these electrons. The correction is the same for all CPs of the same velocity, and depends on the medium since K-shell velocities are characteristic of the medium. Shell correction is important at low kinetic energies and in high Z materials.

2.14.2 Density Correction or Polarization Correction δ

The CP polarizes the atoms in its vicinity while moving through the material. In a dense material, the polarization of the medium by the approaching particle screens its electric field from distant atoms thus reducing the SP. Therefore, a δ correction needs to appear in the SP equation. This effect increases with the increasing energy of the CP. Density effect flattens the dE/dx curve at higher energies. It makes a difference when a SP ratio of solid to air is computed, as in water/air SP ratios used in ionization chamber dosimetry. For the HCP beam energies used in charged particle therapy, the effect is negligible since the energies used in therapy are $<M_0c^2$, and the particles are non-relativistic. Since $m_e c^2$ is only about 0.5 MeV this correction is important in photon and electron dosimetry where electron energies are in the MeV and tens of MeV region, and are relativistic.

$$-\overline{dE}/\rho dx = [4\pi(e^2/4\pi\varepsilon_0)^2 N_A](Z/A)\,(z^2/m_e v^2)\{\ln(2m_e c^2/I) + \ln[\beta^2/(1-\beta^2)] - \beta^2 - C/Z - \delta\}$$

The SP depends on the charge and the velocity of the CP.

2.15 Stopping Power Expression for Electrons

In the case of electrons, due to their small mass, they usually attain relativistic velocities in the MeV region. Bremsstrahlung loss becomes significant and the slowing down of electrons in the medium is governed by both losses, so total SP must be considered.

For electrons, the following equation ICRU 35 [4] has been derived for calculating the collision energy loss

$$-\overline{dE}/dx = K\,(Z/A)\,(1/\beta^2)\left\{\ln\left[\tau^2(\tau+2)/2(I/m_e c^2)\right] + F(\tau) - \delta\right\} \text{MeV cm}^2/\text{g}$$

where K = constant = 0.1535, $\tau = E/m_e c^2$, and $F(\tau) = 1 - \beta^2 + [\tau^2/8 - (2\tau+1)\ln 2]/(\tau+1)^2$.

2.15.1 Angular Stopping Power

As a result of their low mass, electrons are easily scattered by the electrons and nuclei of the atoms of the medium. The large angle scatterings are much rarer and the majority of the collisions cause small deflections which add up, causing an angular spread in the beam. An initially near parallel beam gradually spreads laterally (Figure 2.29), and at the emerging surface the electrons emerge in different orientations with respect to the original direction. When the number of scatterings is greater than 20 or 30, the phenomenon is multiple scattering and a statistical theory can be applied to evaluate the mean deflections. Multiple scattering broadens a pencil beam and causes the emergent beam to exhibit angle straggling. Due to the statistical nature of interactions, particles of identical energy will not have identical cumulative angle of scattering, but will show a distribution of angles about a mean, called angle straggling. The electrons coming out of the absorber, after multiple scattering, are usually distributed in a Gaussian distribution, as shown. This phenomenon is made use of in radiation oncology to spread a pencil beam into a broad beam of finite field size to treat large-size tumors. Since the scattering angle is a random variable, the net scattering angle can be zero and hence the square of the angles are added and the root of the mean square is taken as the cumulative angular deflection.

The mass scattering power is defined as the increase in mean square angle of scattering per unit mass thickness (ρdx).

$$_mS_{sc} = (d\bar{\theta}^2/\rho dx) \text{ radian}^2 \text{ cm}^2/\text{g} \text{ or radian}^2 \text{ m}^2 \text{ kg}^{-1}$$

As the thickness of the medium through which the electron beam travels increases, the slowed down electrons become more and more scattered and disoriented. At this stage the mean square scattering angle attains a constant value of about 33°. This occurs near the depth of dose maximum, in a phantom, for the clinical electron beams.

HCPs scatter less because of their larger mass, and spread much less compared to electron beams.

The rate of collision energy loss is greater for low atomic number Z absorbers than for high Z absorbers, because atoms in high Z materials are too tightly bound to inner shells to absorb small amounts of energy. Material with more electrons per cm^3 have more electrons to interact and will have higher stopping power. Faster CPs have less time to interact and transfer momentum and will lose less energy per unit length. As they slow down, they will have more time to impart momentum and will show increasing energy loss per unit length.

The approximate electron energy E (total energy) at which the collision and radiative SPs are approximately equal in a material is given by the formula

$$_mS_r/{}_mS_c \approx ZE/800$$

where E is in MeV.

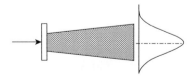

FIGURE 2.29
Lateral spread of electrons emerging from a thin absorber.

Example 2.6

Calculate the electron kinetic energy at which the collision and radiative stopping powers will be approximately equal in Pb.

Solution

Using the above formula, when the two SPs are equal, the ratio must be 1.

$$ZE/800 = 1 \text{ or } E = 800/82 \approx 9.8 \text{ MeV:}$$

$$E_k = 9.8 - 0.5 = 9.3 \text{ MeV}$$

2.15.2 Stopping Power Energy Dependence

Figures 2.30 and 2.31 show the trend in the typical stopping power curves for electrons and heavy charged particles in water medium. The shape of the curve remains the same for HCPs too but is shifted toward the right on the energy axis, because for the same kinetic energy, the velocity of HCPs is much less due to the heavier mass.

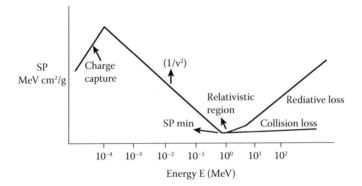

FIGURE 2.30
Stopping power energy dependence for electrons.

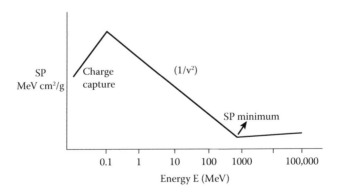

FIGURE 2.31
Stopping power energy dependence for HCPs (protons).

The SP curve has four prominent portions:

1. Region 1—charge capture region
2. Region 2—$(1/v^2)$ region
3. Region 3—relativistic effects region
4. Region 4—radiative region

SPs fall in Region 1 is due to the very slowly moving CPs that capture charges from the medium, reducing its dE/dx. SP fall in Region 2 is due to the dominance of the $(1/v^2)$ term in the SP equation. The SP then reaches a minimum, which occurs around 1–2 MeV for electron beams. The CP is minimum ionizing at this point. Beyond the minimum, the curve slowly starts rising again due to the relativistic terms in the SP equation becoming significant. This is due to the increase in the relativistic mass of the CP and occurs in the MEV region itself.

The SP drastically increases in Region 4 due to bremsstrahlung losses and becomes predominant at CP energies $> M_0c^2$. So, for HCPs (protons and heavy ions), this is insignificant in the energy ranges of interest in radiation oncology, but is significant for the electron energy range of interest. For example, for a 15 MeV electron beam traversing a water phantom, radiative SP contributes nearly 13% to the total SP. For the same beam penetrating a Pb absorber, radiative SP overtakes collision SP and contributes more than 50% to the total SP. A 1 MeV electron beam loses nearly 11% to bremsstrahlung in Pb at 1 MeV). So electron SP has two components, the collision part that causes ionization and excitation of the medium electrons, and the radiative part that usually escapes from the region of interest. This is where it becomes important to distinguish between the energy lost by the CP and the energy lost to the medium. While talking of energy deposition in the medium, the concern is only with the collision part and when talking of the range of CP, the concern is with the total energy loss by the CP and the total SP must be taken for computing the continuous slowing down approximation (CSDA) range. Since the density (or polarization) correction reduces the collision SP, the collision SP curve starts flattening out in the relativistic energy region, as shown in Figure 2.30.

The radiative part is not seen for the HCPs even beyond thousands of MeV and is not of concern in charged particle therapy.

2.16 Radiative Stopping Power and Shielding for β Sources

Shielding of beta ray sources is an important topic in radiation safety. The above explanation of electron SP clearly suggests that a suitable shielding material for beta radioactive sources is a low Z material like perspex or aluminum and not a lead container. Aluminum is better than perspex which is likely to crack with time due to radiation damage. Assuming the most approximate formula used in therapeutic electron beams, the beam loses approximately 2 MeV per cm thickness in a unit density material, a 1 MeV beta will require a less than 5 mm thick perspex container and if the source is of high activity, a lead lining may be required outside the aluminum container to attenuate the bremsstrahlung to safe levels. There are other empirical formulas in the literature for determining the range of beta particles, but the above formula applied to the maximum beta particle energy (= decay energy) of the beta source is a very simple way to estimate shielding thickness for beta sources.

To determine the fraction of the beta energy that would be converted to bremsstrahlung in any absorber, the following approximate formula may be used.

$$f_{br} = 3.5 \times 10^{-4} \, Z \, E$$

Example 2.7

a. A one curie ^{90}Sr source is stored in a lead container. What fraction of the beta energy (in percentage) would be converted to bremsstrahlung?

b. Calculate the photon fluence at a distance of 30 cm from the source (placed in air).

Given: ^{90}Sr half-life 28.8 years; beta decay energy = 0.546 MeV.

Daughter product ^{90}Y half-life = 64 hours and beta decay energy = 2.28 MeV.

The two radionuclides are in secular equilibrium and so the decay of ^{90}Y is governed by the decay of ^{90}Sr.

Beta decay involves a spectrum of electron energies varying from zero to the maximum of decay energy. The average energy of the beta spectrum is approximately one-third of the maximum energy. The bremsstrahlung also exhibits a spectrum. For a conservative estimate from radiation protection point of view, the maximum electron energy for all the photons can be assumed. This will give a conservative estimate with a margin of safety in the calculations. The bremsstrahlung from the container wall can be assumed to be coming from the source in all directions. This will also simplify calculations (Figure 2.32).

Fraction of bremsstrahlung emitted by an electron:

$$f_{br} = 3.5 \times 10^{-4} \, Z \, E = 3.5 \times 10^{-4} \times 82 \times 2.28 = 654 \times 10^{-4} \text{ or } 6.5\%$$

Total number of electrons emitted per second for a 1 Ci source = 3.7×10^{7}

Total photon energy emitted/curie·sec = $3.7 \times 10^{10} \times f_{br} = 3.7 \times 10^{10} \times 654 \times 10^{-4}$ MeV/s

$$= 2.4 \times 10^{9} \text{ MeV/s}$$

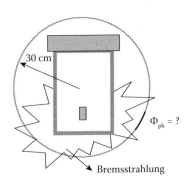

FIGURE 2.32
Bremsstrahlung emission from lead container for a beta source.

Photon energy fluence at 30 cm $= 2.4 \times 10^9 \div 4 \pi \, 30^2$ MeV/cm^2·s

$$= 2.1 \times 10^5 \text{ MeV/cm}^2 \cdot \text{s}$$

Assuming each photon has the maximum electron energy, $E_{ph} = 2.28$ MeV

$$(d\Phi/dt) \text{ at } 30 \text{ cm} = 2.1 \times 10^5 \div 2.28 = 9.2 \times 10^6 \text{ photons/cm}^2 \cdot \text{s}$$

2.17 Collision Stopping Power and Shielding of α Sources

Alpha sources do not pose any radiation safety problem. Because of their relatively high ionization loss per unit path length ($_mS$ α z^2 and alphas are doubly charged He ions) their ranges are very small and are stopped in a few centimeters, even in air medium.

The following simple formulas are easy to remember and would give a conservative estimate of the range of alpha particles in air. More accurate range formulas are available in the literature, but for shielding purposes a conservative and approximate formula would suffice).

$$R_{air} \text{ (cm)} = 0.6 \, E_\alpha \text{ MeV} < 4 \text{ MeV}$$

$$R_{air} \text{ (cm)} = E_\alpha \text{ MeV } 4\text{–}8 \text{ MeV}$$

For any other absorber, the range is inversely proportional to density.

$$R_z \, \rho_z = R_{air} \, \rho_{air}$$

The alpha energies in the naturally occurring radioisotopes lie in the range 3–7 MeV. The most energetic alphas will travel only a few tens of mm in air. The outer layer of skin, approximately 7 mg/cm^2 in thickness, will absorb alpha particles up to 7–8 MeV. Since 7 mg/cm^2 forms the dead layer of the skin, alpha sources pose no external hazard, therefore they do not pose any shielding problem either. The energy of a beat particle for 7 mg/cm^2 range is around 70 keV, so beta sources with energies exceeding 70 keV can damage the skin and the eye lens of a radiation worker; personal protection may be required.

Let us work out an example to calculate the shielding thickness for the maximum alpha energy of 7 MeV. This thickness will be sufficient to shield all the naturally occurring alpha sources as their energies are <7 MeV.

Example 2.8

1. Determine the range in air of an alpha source emitting 7 MeV alphas. Can ordinary writing paper stop these alphas? (Paper density is 1 g/cm^3 and its approximate thickness is 0.1 mm). Will it constitute any external hazard to the users?
2. If the source is stored in an Al container, what will be its shielding thickness?

Solution
The range of 7 MeV alpha in air = 7 cm = $7 \times 1.293 \times 10^{-3} = 9 \times 10^{-3}$ g/cm^2

$$\text{Range in paper} = R_p \, \rho_p = 9 \times 10^{-3} \text{ g/cm}^2 \text{ or } 9 \text{ mg/cm}^2$$

$$R_p = 9 \times 10^{-3} \div 1 \text{ cm} = 0.009 \text{ cm or just} < 0.1 \text{ mm}$$

The most energetic alphas from radioactive isotopes can be easily stopped by a thin sheet of paper.

As for constituting any external hazard, the alphas must be capable of penetrating the skin. The thickness of the dead layer of the skin is 7 mg/cm^2. So, most of the alphas will be stopped by the dead layer of the skin and will scarcely constitute an external hazard. If the source is stored in an Al container, the shielding thickness required = $9 \times 10^{-3}/2.7$ or a thin foil of 0.03 mm will be enough to stop the alphas. Bremsstrahlung production is virtually zero for HCPs.

2.18 Concept of RSP and LET

Dose is a macroscopic quantity and is single valued (non-stochastic). In other words, if a dose measurement is repeated in medical dosimetry, the same value should be expected (within the precision of measurement). This means the statistical fluctuations in the interaction and energy deposition events in the measuring volume are negligible, compared to the precision of measurements. However, the volumes of interest in radiobiology are very small when studying the biological effects of radiation on biological structures. Here, the interest is in the biological effects caused by the tracks of CPs in its close vicinity, say a cell nucleus with dimensions of the order of micrometers, or a DNA, or a chromatin structure with dimensions in nanometers. At this level, the pattern of energy deposition along the CP tracks and the secondary electron (δ-ray) tracks, or the density of ionization along these tracks, plays an important role in determining the biological effects of radiation, in addition to the dose delivered to the volume of interest. For the same dose, radiation that deposits more energy along a unit path length causes more lethal damage to cells compared to radiations that deposit less energy per unit path length. Therefore, local energy transfer along the path of the CP assumes great significance in radiobiology and radiation protection.

For local energy deposition along the tracks, the collision stopping power discussed earlier is not appropriate since it counts all the hard collision electron energies as locally absorbed. This would overestimate the energy deposited around the track since many δ-ray particles would deposit energy outside the volume of interest. Depending on the biological system, the energy deposition in a cylindrical volume of certain radius, say r, around the CP track is of interest, as shown in Figure 2.33.

The energy locally deposited along the track depends on which δ-ray particles have ranges, on average, $\leq$ r. If this range r corresponds to an electron energy of Δ keV, then all δ-ray electrons of energy $\leq \Delta$ must be included in the stopping power and all δ-ray energies $> \Delta$ must be added to the electron spectrum, at the appropriate energy band. The SP that depends on Δ is known as restricted SP (RSP) and is denoted by $_m S(E,\Delta)$. To distinguish this from the RSP, the collision SP defined earlier is usually referred to as unrestricted SP. In radiobiology, RSP is referred to as linear energy transfer (LET) and is denoted by LET$_\Delta$. As Δ value increases, the value of RSP also increases since the volume of interest increases and larger energy δ-rays are now included in the RSP. When $\Delta = E/2$, $_m S_c(E) = _m S$ (E, $\Delta = E/2$).

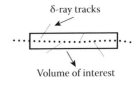

δ-ray tracks

Volume of interest

FIGURE 2.33
Defining RSP or LET for the volume of interest.

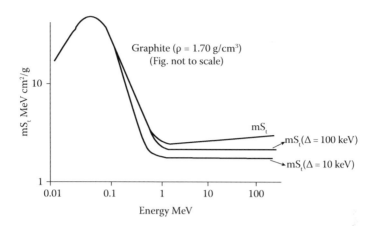

FIGURE 2.34
Unrestricted and restricted SP curves for graphite.

Figure 2.34 shows the electron RSP and unrestricted SP curves for graphite [1].

LET is a measure of how closely spaced the ionizations are along the track (specific ionization) and the term is frequently used in radiobiology in place of restricted stopping power, when dealing with energy deposition in microscopic volumes. For a given energy, electron velocities are much larger compared to HCPs, and the ionizations produced by electrons are far more widely spaced along the track compared to HCPs. Electrons, X-rays, and gamma rays (that produce electrons) are classified as low LET radiations, and the CPs and the radiations that produce heavy charged particles (e.g., neutrons) are classified as high LET radiations.

Table 2.6 gives the LET of various radiations.

When high LET radiations interact in a biological system (e.g., irradiated colony of cells), the HCPs can cause more than one ionization event within the span of a DNA molecule, resulting in double strand breaks (DSB) in the DNA. With low LET radiation interactions the ionizations are far more widely spaced compared to the dimensions of DNA where single strand breaks (SSBs) are more likely to occur. Cell DNA is the main target for the biological effects of ionizing radiation. Radiation-caused DNA lesions play a central role in understanding cell damage and carcinogenesis. Most of the SSBs induced by low LET ionizing radiation can be repaired. In contrast, DSBs caused by high LET ionizing radiation are responsible for mutations and carcinogenesis. This has lot of relevance in radiation protection where the same doses (in mGy or μGy) received by a radiation worker from a low LET radiation and a high LET radiation are not biologically equivalent. So, for radiation protection purposes, it is necessary to define a biologically equivalent dose concept to compare radiation risks.

2.19 Relative Biological Effectiveness

As explained above, LET is a measure of the biological effectiveness of radiation in causing damage to body cells and tissues. In general, the biological effectiveness of a radiation

TABLE 2.6

LET Values for Different Radiations and Energies

	Typical LET Values keV/um
High LET	
Proton 10 MeV	4.7
Neutron 14 MeV	12
Heavy CPs	>100
Low LET	
X-rays (e⁻) 250 kVp	2
^{60}Co γ-rays (1.2 MeV)	0.3
Electrons 100 keV to 1 Mev	0.3–0.2

Note: CP, charged particle; LET, linear energy transfer.

increases with its LET up to a value of LET = 100 keV/μm, and above this value starts to decrease due to overkill (i.e., energy deposited is much more than what is required for the cell kill). Relative biological effectiveness (RBE) is defined as

$$RBE = D_R / D_X$$

where D_R is the reference dose of (normalizing) radiation and D_X the dose of X-radiation in question that causes some biological damage. The reference radiation is usually 250 kVp X-radiation or ^{60}Co radiation. Early studies with low-energy X-ray beams and beta radiations had shown they were biologically equivalent for all cell types and so 250 kVp X-rays was chosen as a reference to compare the biological effectiveness of other radiations that came into use much later. RBE is 1 for X-rays and electrons since both are LLET radiations.

For HLET radiations, RBE is not so uniquely defined and depends a little on the tissue types involved and the end points chosen in biological studies. However, for radiation protection purposes where very high accuracies are not required, only a conservative value for the RBE of any HLET radiation is required to define a biologically equivalent dose concept. A biologically effective dose for an absorbed D_T in tissue was defined as RBE × D_T. This quantity was earlier referred to as an RBE dose. It is now known as an equivalent dose and for radiation protection purposes, RBE has been replaced by Q factors.

2.20 Range, Range Straggling, and Empirical Range Equation

Since CPs produce ionizations and excitations in the medium causing physical and biological effects, how far the CPs will penetrate matter is an important question in radiation biology and radiation protection. Unlike photons or neutrons that are exponentially atte-nuated and can penetrate any thickness, CPs have a finite penetration depth—a range—in any medium. The range is not always the track length or path length of the CPs. Path length is the total length of the actual particle track; range is the projected path length along the original direction, in the absorber. Figures 2.35 and 2.36 illustrate the concept of path length and range for electrons and heavier charged particles.

It can be seen that the particle tracks are a series of straight lines between scatterings. The two parameters differ because of multiple Coulomb scattering, which deflects the particles from their original direction of travel. The deflections depend on the mass of the CP. HCPs have much straighter paths and their ranges and path lengths are very nearly the same. On the other hand, electrons are easily deflected at larger angles and can even be back-scattered. There is significant difference between path length and range, which can differ by a factor of 2 in the case of electron beams.

Many electrons can have much shallower ranges because they can lose as much as half of their energy in a collision, and can also become scattered out of the beam. HCPs lose only a small fraction of their energy in a scattering collision, and are not much deflected from their paths. So, range concept is not well defined for electron beams, compared to HCP beams. The fall in transmitted number of electrons and protons with penetration depth is shown in Figure 2.37.

The electron beam profile clearly shows that many electrons are lost at shallower depths.

However, if we can compute a range for electron beams, neglecting large scatterings and δ-ray production (i.e., assuming that CPs continuously lose energy in small decrements, known as a continuously slowing down model), then this parameter can serve as maximum range of electron beams and will help us to know the maximum penetration of electron beams. Both SP and range are inversely related since if the particle has high dE/dx, it will lose its energy faster and will come to a stop in a shorter distance. If continuous slowing

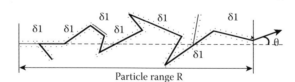

FIGURE 2.35
Electron track with delta ray tracks and ionization along the path.

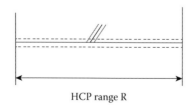

FIGURE 2.36
Heavy charge particle track and ionization along the track.

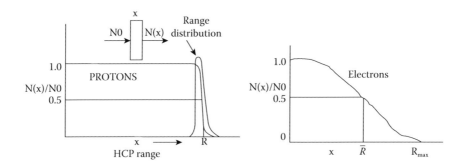

FIGURE 2.37
Transmission of protons and electrons as a function of penetration depth.

down without large energy loses in single collisions is assumed, the range from the SP can be calculated using the following relation:

$$dx = dE/\overline{dE}/dx = dE/S_t$$

$$\text{Range } R_{csda} = R \int_0^{} dx = \int_{E0}^{0} dE/(d\overline{E}/dx)t$$

The unit of range depends on the unit of dE/dx. If SP is in MeV cm^2/g, then R is in g/cm^2.

Range only depends on the total energy lost by the charged particles, and not whether the energy is locally absorbed or radiated, because once the CP loses energy and is not able to cause ionization or excitation, they come to the end of their ranges. Therefore, the total SP should be used in the range equation.

Due to the statistical nature of interactions and hence the energy loss, CPs of identical energy incident on an absorber will not have exactly the same ranges. Instead, there will be a statistical distribution in the ranges of the CPs centered around a mean value. This phenomenon is known as range straggling. The distribution is approximately Gaussian in form and its mean is known as the mean range (see Figure 2.37).

The range straggling is much less for HCPs compared to LCPs because they are scattered much less. The continuous slowing down model (CSDM) applies more accurately to HCPs (e.g., the maximum energy a 5 MeV α can transfer to an electron in a collision is just 2.5 keV making the electron slowing down almost continuous) and so the range is much better defined for HCPs. CSDM is not a good approximation for electrons as electrons can lose a large amount of energy in single collisions (because of equal mass) and many may have much shorter ranges. This explains the shape of the two curves in Figure 2.37.

There is a need for simple empirical equations to quickly determine the range of CPs in materials. Many such equations are available in the literature for various types of CPs. For electron beams, of energy in the range of interest in radiation oncology (for most of the materials) the SP is approximately constant and equal to 2 MeV/(g/cm^2). Using this in the range equation gives

$$R_{el} = E_0/2 \text{ g/cm}^2$$

This equation is used as a rule of thumb in electron beam therapy.

Similar empirical relationships exist for other CPs as well. For proton beams, for instance, the CSDA range is approximately given by the following [5]

$$\sqrt{R_P} = A_{eff} \ E^p g/cm^2$$

where A_{eff} is the effective atomic mass of the material and the constant p is approximately 1.75. No charged particle dose is delivered to the medium beyond the range of CPs.

Example 2.9

What is the range (in cm) of a 20 MeV electron beam in a water phantom?

Using the range formula, R = 20/2 = 10 g/cm². Since water is a unit density material, the electron beam range in a water phantom is about 10 cm.

Example 2.10

What is the lead thickness you would use to shield part of the patient's skin from a 12 MeV incident electron beam? (Assume Pb density to be approximately 10 g/cm³.)

For water the electron range is approximately 6 cm. Since the Pb density is about 10 times higher compared to water, the electron range in Pb would be 10 times lower—about 0.6 mm. This is the Pb thickness to be used to shield the patient's skin.

2.21 Absorbed Dose from Charged Particle Interactions

We have seen in the earlier section that the energy transferred to charged particles and the fraction of the energy transferred to charged particles that is eventually deposited in the medium are computed using the following equations:

$$K = \Psi(E) \ [_m\mu_{tr} \ (E)] \ dE$$

$$K_c = \Psi(E) \ [_m\mu_{en} \ (E)] \ dE$$

for a monoenergetic photon beam of energy E.

For a CP beam, $\Phi(E)$ gives the CP fluence for the beam of energy E. The stopping power $(\overline{dE}/dx)_t$ gives the energy lost by the CPs (equivalent to $_m\mu_{tr}$ (E)), and $(\overline{dE}/dx)_c$ gives the energy given to the medium (equivalent to $_m\mu_{en}$ (E)). The absorbed dose equation for the CPs will be

$$D = \Phi(E) \ _mS_c(E)$$

For a CP spectrum, the dose will be

$$D = \int \Phi_E(E) \ _mS_c(E) \ dE$$

Even for a monoenergetic CP beam like an electron beam penetrating a medium, there is a slowing down spectrum from E to 0, and so the integral equation is the most general equation for absorbed dose computation. Since the energy absorption also depends on the medium stopping characteristics, both the electron spectrum and SP are characteristic of the medium and we must write the equation as

$$D_{med} = \int [\Phi_E(E)]_{med} \{_mS_c(E)\}_{med} \, dE$$

Defining mean collision stopping power as

$$m\bar{S}_c = \int [\Phi_E(E)]_{med} \{_mS_c(E)\}_{med} \, dE \Big/ \int [\Phi_E(E)]_{med} \, dE$$

$$D_{med} = \Phi_{med} \, {}_m\bar{S}_c$$

Example 2.11

1. A 10 MeV electron beam is incident on a mass Δm, having dimensions which can be covered by an electron of energy 10 keV. Calculate the absorbed dose in Δm in cGy. The following data are given: electron fluence $= 8 \times 10^9/cm^2$. $_mS_c(\Delta=10keV) = 1.44$ MeV cm^2/g (Figure 2.38).

2. How much would have been the overestimate of dose if the unrestricted SP ($_mS_c = 1.99$) had been used?

3. Why was the total stopping power not used in this problem?

$$D = \Phi \, {}_mS_c(\Delta=10keV) = 8 \times 10^9 \times 1.44 \text{ MeV/g}$$

$$= 8 \times 10^9 \times 1.44 \times 1.6 \times 10^{-13} \times 10^3 \text{ J/kg}$$

$$= 1.84 \text{ J/kg} = 1.84 \text{ Gy or } 184 \text{ cGy.}$$

$$_mS_c(\Delta = 10keV)/_mS_c = 1.44 \, / \, 1.99 = 0.72.$$

We would have overestimated the dose by nearly 30% if we had made use of the unrestricted SP.

Bremsstrahlung loss will be radiated out and so should not be accounted for in the local energy absorption. So, we made use of restricted collision SP.

A few lines regarding the concept of converted energy per unit mass (CEMA) is of relevance here. The secondary electrons are the δ-rays produced by primary electrons, and have sufficient energy to form their own tracks and transport energy away from

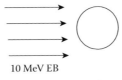

10 MeV EB

Δm of mass

(Diameter = range of 10 keV electron)

FIGURE 2.38
Local energy absorption in Δm of mass.

the origin. For this reason, for small volumes with dimensions less than the δ-ray ranges, the above equation would not give the dose locally absorbed, since part of the energy would have been carried out of the volume by the δ-rays. If the dose in δm is measured, a lesser value would be obtained and not what the above equation predicts. Instead of designating the above quantity as absorbed dose, the above quantity is designated as CEMA—converted energy per unit mass. If the δ-particles outflow are compensated for by providing a delta ray buildup layer around δm, then there would be a δ-particles inflow to compensate for the δ-particles outflow. The dose measured under δ-ray equilibrium (δ_{eq}) would then be what would be predicted from the above equation.

$$CEMA = (D)_{\delta eq}$$

The reason why this quantity is not heard about often is because the δ-ray ranges are very small and while taking measurements, say in a phantom, or in a dosimeter, there is always a thin buildup layer available, in practical situations for measuring the dose. When a high-energy electron beam is incident on a water phantom, as in electron beam therapy, in the initial layers there would not be equilibrium dose, but beyond the δ-particle range, what is measured is always the dose under δ-particles equilibrium. As in the case of photon beams, therefore, electron beam depth dose also exhibits a buildup region and one of the reasons for this buildup is the finite range of δ-particles. The other reason is the multiple scattering of electrons in the medium increasing the electron fluence.

2.22 Depth Dose Profile for Charged Particle Beams

A depth dose profile shows how the CP dose varies with depth, with the dose curve normalized to some standard depth (e.g., d_{max} depth).

To know the CP dose with respect to depth, we must know how S_c and Φ change with depth. There are some basic differences in the interactions of HCPs and electron beams while traversing a medium, which influence the shape of their depth dose profiles. These differences are:

1. HCPs lose energy very much like electrons in excitation/ionization interactions but the energy loss in each collision (because of their larger mass) is a very small fraction of their energy unlike for the electrons. So, the slowing down process is much smoother and continuous compared to the electrons.

2. HCPs are deflected less in Coulomb nuclear collisions compared to electrons and so have a much straighter path length, and are not lost from the beam.

3. The radiative collision losses are negligible, unlike for electrons.

4. HCPs are capable of experiencing nuclear forces and so elastic and inelastic scatterings due to the nuclear force field can occur with the nuclei of the atoms of the medium. An electron, being a lepton, does not exhibit this type of interaction at all. The nuclear interactions are much more complex compared to Coulomb interactions but have a much lower probability of occurrence compared to electromagnetic interactions.

2.22.1 Depth Dose Distribution of Heavy Charged Particle Beams

For radiation therapy treatment [6], dose distributions in a tissue medium are of particular interest. Water is nearly tissue equivalent, so beam data are established for a water phantom and are used for tumor dose calculations with appropriate patient inhomogeneity corrections.

HCPs, as already mentioned, travel straight until the end of their ranges so the depth dose profile is largely governed by the collision SP profile with respect to depth, as can be seen from Figures 2.39 and 2.40, for a typical 200 MeV proton beam. At low energies the LET drastically increases ($\alpha\ 1/v^2$), giving rise to a peak in the LET profile near the

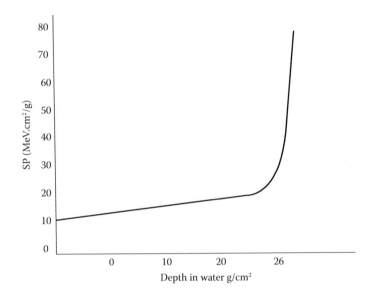

FIGURE 2.39
Variation of LET with depth for 200 MeV proton beam in water phantom.

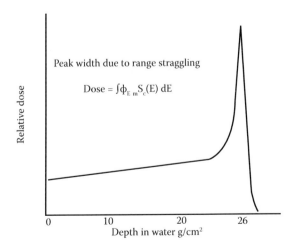

FIGURE 2.40
200 MeV proton beam relative depth dose in water.

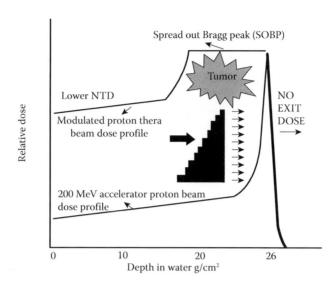

FIGURE 2.41
200 MeV proton beam relative depth dose in water, for sharp BP and SOBP beams.

end of the range. The resulting depth dose profile, known as the Bragg peak curve, therefore exhibits a peak also at the end of the CP range, as shown in Figures 2.39 and 2.40.

The BP is due to the increased LET for lower energies that appear near the end of the range. This shape is ideal for radiation therapy since the dose is low in the normal tissues in the proximal regions, and rises sharply in the peak region. If the peak depth is adjusted to be the tumor depth, the tumor would receive a very high dose.

For heavy ions such as carbon ions, nuclear interactions can cause a breakup of the HCP nuclei producing lower Z fragments (e.g., carbon ion disintegrates into three alphas) which have longer ranges. This produces a tail to the Bragg curve, which is not evident in the proton plot shown.

However, the Bragg peak is too narrow to accommodate the finite size tumor volumes for delivering a uniform dose. If this beam is passed through absorbers of varying thicknesses, the emerging beams will have energies of E, (E-ΔE), (E-2ΔE), (E-3ΔE) … with ranges of R, (R-ΔR), (R-2ΔR), (R-3ΔR) … so there will be a series of adjoining Bragg peaks and the resultant dose distribution would exhibit a spread out Bragg peak (SOBP), which can accommodate the tumor for uniform irradiation, as shown in Figure 2.41 for a typical 200 MeV proton beam traversing a water medium. The figure also shows the series of absorbers coming in the beam path to produce component BPs and the resultant SOBP.

From the known depth of the tumor, the beam energy can be adjusted to produce the SOBP at the required depth to cover the tumor volume for uniform dose delivery.

2.22.2 Depth Dose Distribution of Electron Beams

The electron beam dose distribution typically looks as it is shown in Figure 2.42.

The shape of the curve can be explained as follows:

1. The electron scattering and the δ-ray buildup lead to fluence buildup, and hence the dose buildup at the shallow depths.

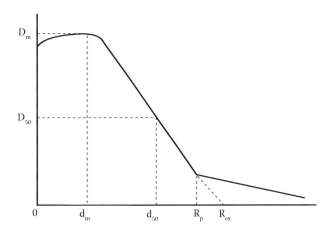

FIGURE 2.42
Typical electron beam depth dose profile.

2. The electrons lost to the beam due to scatterings wash out the BP and show a decrease in fluence and dose with respect to the depth.

3. The electron beam range is characterized by a range called the practical range (R_p).

4. The bremsstrahlung production by electrons gives rise to the tail region of the curve.

5. There is no exit dose beyond R_p for the electron beams, and any critical organ in this region does not receive any electron dose.

The slope of the dose fall is as a result of many electrons having shallower ranges due to the loss of a larger fraction of the electron energy in scatterings, and losing all their energies at shallower depths. This reduces the fluence with depth giving rise to the shape depicted in the figure. The range of the electron beam is, therefore, characterized by the practical range R_p.

The curve also shows the bremsstrahlung background (tail portion of the curve) in the water phantom due to radiative stopping power losses. Much of this bremsstrahlung comes from electron interactions in the head of the linac, mainly from the collimator surfaces. For higher energies there is greater probability of radiative collisions giving rise to larger tail in the dose profiles. Yet, the electron beam does not penetrate beyond its practical range and so the deeper structures in the patient do not receive electron dose, unlike the photon beam which shows a finite exit dose.

2.22.3 Depth Dose Distribution of Photon Beams

The photon absorbed dose and water collision kerma curve profiles are as shown in Figure 2.23. The shape of the photon dose distribution is the result of exponential attenuation of photon beams and the establishment of electronic equilibrium beyond the transient equilibrium depths. The electron buildup in the shallow depths is due to their production in every layer of the medium, giving rise to the dose buildup in the shallow regions.

Figures 2.43 and 2.44 compares the dose profiles of a photon beam, an electron beam, and the heavy charge particle beam (HCPB).

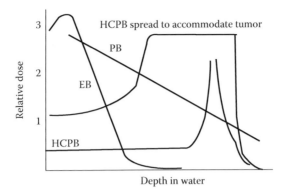

FIGURE 2.43
Comparison of photon and CP dose distributions.

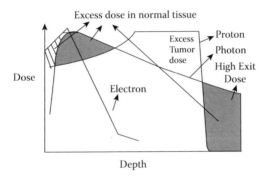

FIGURE 2.44
HCP delivering less dose to normal tissues and excess dose to tumor tissue.

The HCPB dose distribution is vastly superior to the photon beam therapy because of

1. Less dose in the entry portal
2. More dose at the tumor
3. No dose beyond the range
4. Higher biological effectiveness compared to photon or electron beams due to the HLET.
5. Higher Bragg peak dose-to-entrance-dose ratio

More details on HCPB therapy can be found in the book referenced here [7].

2.23 Neutron Production in Boron Capture Therapy

Boron neutron capture therapy is a treatment modality for treating primary and metastatic brain tumors. Here, boron compounds are locally absorbed in the tumor cells and the brain

is externally irradiated using a thermal neutron beam from a nuclear reactor. Two requirements for successful boron capture therapy (BNCT) are the availability of a thermal
neutron source of sufficient intensity, and selective absorption of ^{10}B in the tumor cells
(i.e., high tumor uptake and low normal tissue uptake).

Thermal neutron capture in boron nuclides gives rise to the following nuclear reaction:

$$n_{th} + {}^{10}B \rightarrow {}^{11}B \rightarrow {}^{7}Li + {}^{4}He + 2.79 \text{ MeV}$$

The ^{10}B has a very high absorption cross section for thermal neutrons. The reaction products have very short range and will be absorbed in the close vicinity of the point of production. They have high RBE due to their high LET; the thermal neutrons have
insufficient energy to damage normal tissues.

Thermal neurons (HVL in tissue about 1.5 cm) have a very limited depth of penetration
in tissue and can treat only few centimeters in depth. Epithermal neutrons (1 eV to about
10 keV) have also been used in BNCT for this purpose. They show skin sparing and thermalize before reaching the tumor depth, and so can be used for deeper lying tumors.

2.24 Neutron Production in Fast Neutron Therapy

Fast neutron therapy involves using neutron energies >14 MeV for cancer treatment. The
nuclear reactions normally used for the production of fast neutrons are the following:

$$p + {}^{7}Be \rightarrow {}^{7}B + n$$

$$^{2}H + {}^{9}Be \rightarrow {}^{10}B + n + \gamma$$

$$^{2}H + {}^{3}H \rightarrow {}^{4}He + n \text{ (14 MeV)}$$

Protons or deuterons are accelerated in a cyclotron and are directed to a Be target to produce the fast neutrons. Protons or deuterons must be accelerated to $\geq$50 MeV to produce
neutron beams with penetration comparable to megavoltage x comparable to megavoltage
X-rays.

In the case of a 14 MeV neutron generator, deuterons are accelerated to about 150 keV
before striking a tritium target. The interaction produces 14 MeV neutrons.

For effective treatment, the charged particle beam line is transported to an isocentric
therapy head before striking the Be target at the focus of the head.

The neutrons are biologically three times as effective as LLET radiation and hence a full
course of treatment can be delivered in one-third fractions compared to photons, electrons,
or protons.

2.25 Neutron Production in Linacs in Megavoltage X-Ray Therapy

Modern medical linear accelerators are used in two distinct operation modes: electron
mode and photon mode. In electron mode a primary pencil electron beam, spread into a
treatment field using a scattering foil, is used for treatment. In photon mode, the electron

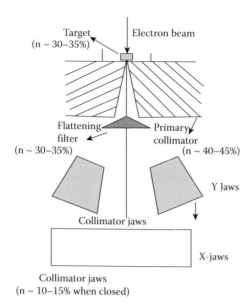

FIGURE 2.45
Source of neutron production from a typical linac.

beam impinges on a high Z target (instead of a scattering foil) and produces a photon beam which is collimated to produce treatment field sizes for patient treatment. The machine can produce several electron energy beams. In the photon mode, the machine usually offers two photon energies, a 6 MV beam and a 10 or 15 MV beam for treatment.

A typical linac head is illustrated in Figure 2.45, showing only the components that contribute to the neutron production.

The target is a tungsten disk. A 3-mm diameter electron beam of desired energy impinges on the target producing the linac X-ray photon spectrum. Because of the high momentum of the electron beam, the X-ray intensity peaks along the electron beam direction. A flattening filter is placed in the path of the beam to flatten the beam. The beam is collimated by the primary collimator into a circular beam. The beam is then shaped by the X and Y collimator jaws to produce beams of various field sizes, according to the purposes of treatment.

When the linac is operating at energies higher than 10 MV, the photon beam is capable of producing neutrons by (γ,n) reactions with the nuclei, which are called photoneutrons. The interactions are photodisintegrations or more correctly referred to as photonuclear reactions. The primary photon beam and the Compton scattered secondary photons interact with the components of the linac head producing these neutrons. The residual nuclide, following the emission of neutron, is in a stable or radioactive state. If the product nuclide is radioactive it emits short-lived β^+ and γ before attaining stability, as illustrated in Figure 2.46.

Since the BE of a nucleon is about 10–15 MeV for low Z materials and about 7 or 8 MeV for high Z materials, high-energy X-ray therapy units operating at >10 MV are capable of producing photonuclear interactions of the type (γ,n), (γ,p), $(\gamma,2n)$, and (γ,α). Because of the Coulomb barrier, emission of a proton or a charged particle becomes less probable than neutron emission. The most important interaction is the one producing neutrons. High Z nuclides exhibit high cross section for (γ,n) interactions, making the linac head an important source of neutrons. This gives rise to a neutron dose to the patient.

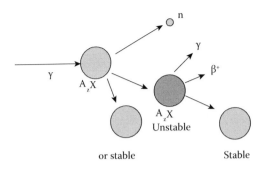

FIGURE 2.46
Concept of photodisintegration.

Also, the linac head components become radioactive (with short half-lives), so there would be a small gamma dose of very short duration near the collimator of the linac. Neutrons are not a serious issue for 10 and 15 MV beams, and this is one of the reasons for many treatment centers not going beyond 15 MV for patient treatment, though many centers still use 18 MV units in radiotherapy.

The neutron contributions from the various components of a Siemens Primus linac operating at 15 MV, as determined from Monte Carlo computations, is given in Table 2.7 [8].

The neutron production from the Varian linac head differs somewhat from Primus, depending on the linac head design.

Neutrons are also produced through $(\gamma, 2n)$ and (γ, pn) reactions but the yield is lower than for the (γ, n) reactions.

TABLE 2.7

Sources of Neutron Production from Siemens Primus Head (MC Computation)

Location	Contribution (%)
Primary collimator	54.85
MLC and jaws	26.72
Target	10.08
Target slide	5.64
Flattening filter	1.74
Bending magnet	0.61
Steel block	0.13
Steel and lead shield	0.11
x-low collimator	0.07
Steel skeleton	0.03
Absorber	0.01
Steel plate	0.003
Electronics	0.001
Other	0.006

Note: MC, monte carlo; MLC, multileaf collimator.

TABLE 2.8

Neutron Production Threshold for Different Nuclides

Isotope	Energy Threshold for Neutron Production (*MeV*)
^{1}H	2.23
^{12}C	18.72
^{16}O	15.66
^{27}Al	13.06
^{56}Fe	11.2
^{63}Cu	10.85
^{65}Cu	*9.91*
^{183}W	6.19
^{184}W	7.41
^{207}Pb	6.74
^{208}Pb	7.41

The photonuclear reactions are threshold reactions requiring minimum photon energy to cause the reaction, depending on the material involved. The threshold energy is decided by the neutron separation energy for the nuclide. Above the threshold energy, the neutron yield depends on the cross section for neutron production. The typical materials used in the linac head are Cu, W, Au, and Fe. Table 2.8 shows the neutron separation energies for some typical materials.

In general, heavy nuclides have lower threshold energies for photoneutron production, lower resonance peak energies, and higher photoneutron yields. For instance, from Table 2.8, it can be seen that the neutron production threshold is higher for Fe (i.e., neutron yield will be less) compared to Pb. So, for >10 MV X-ray beams, steel is a better choice as a shielding material compared to Pb from a neutron production point of view.

The nuclides of interest in a linac treatment room are ^{1}H, ^{16}O, ^{12}C, and ^{14}N in the patient's body, ^{16}O and ^{14}N in the air in the treatment room, ^{184}W, ^{208}Pb, and ^{27}Al in the linac treatment head and treatment accessories, and ^{1}H, ^{16}O, ^{28}Si, ^{40}Ca, and ^{12}C in the concrete walls. ^{10}B and ^{11}B are often used for maze or door lining to shield against neutrons.

The cross section for neutron production has a threshold of around 7–8 MeV, has a maximum value at 3–7 MeV above the threshold energy, and then decreases with higher photon energy. The curve shows peaks at photon energies that correspond to some excitation levels in the nuclei. This increases the probability of photonuclear reaction. The neutron yield has been found to saturate around 30 MV, so linacs operating at higher MV are not going to worsen the neutron contamination situation.

Example 2.12

Calculate the threshold energy (or neutron separation energy) for the photonuclear reaction ^{184}W(γ,n)W^{183}. Given: m(^{184}W) = 183.950931 u; m(^{183}W) = 182.950223 u; m_n = 1.008664 u;

$$\text{Rest mass of } \gamma = 0$$

Total mass on RHS [m(n) + m(^{183}W)] = 183.958887

Total mass on LHS [m(^{184}W) + 0] = 183.950931

Total mass on the LHS < Total mass on the RHS

This reaction cannot proceed unless the γ brings in a kinetic energy equal to the difference in mass.

Difference in mass = 183.950931 − 183. 9588887 = −0.007956 u

Energy deficit = −0.007956 × 931.h (MeV/u) = −7.41 MeV

The negative sign shows it is an endothermic reaction and the gamma must have a minimum energy of 7.41 MeV for the photonuclear reaction to proceed. This is also the neutron separation energy to get it out of ^{184}W.

Example 2.13

Neutrons are produced in a D-T generator using the following nuclear reaction:

$$^{2}H + {}^{3}H \rightarrow {}^{4}He + n$$

Calculate the energy of the neutrons produced. The following data are given:

m(^{3}H) = 3.016049 u; m(^{2}H) = 2.014101 u; m(^{4}H) = 4.002602 u; m_n = 1.008664 u;

Total mass on LHS = 3.016049 + 2.014101 = 5.030150

Total mass on RHS = 4.002602 + 1.008664 = 5.011266

Total mass on the LHS > Total mass on the RHS

So, this is an exothermic reaction.

Q value = mass difference × 931.5 MeV/u = 0.018884 × 931 = 17.6 MeV

The energy will be shared by the products. The neutron energy is given by 17.6 × 4/5 = 14.1 MeV.

In principle, this reaction can proceed without any energy input, but there is a Coulomb barrier and so in practice energy must be provided to overcome the barrier for the reaction to take place.

Example 2.14

Calculate the minimum energy to be supplied to the deuteron in the above case for the reaction to take place. Given: $(1/4\pi\epsilon_0) = 9 \times 10^9$; nuclear radius = $r_0 A^{1/3}$; $r_0 = 1.6 \times 10^{-15}$. The potential energy at the point of contact between the two nuclides is given by (Figure 2.47)

$$V(r) = k_0 (Z_1 e . Z_2 e)/r$$

where the constant $k_0 = (1/4\pi\epsilon_0) = 9 \times 10^9$

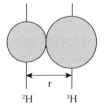

FIGURE 2.47
^{2}H and ^{3}H in contact.

$$V(r) = 9 \times 10^9 \times 1 \times 1 \times (1.6 \times 10^{-19})^2/[1.\ 6 \times 10^{-15} \times (3^{1/3} = 2^{1/3})]\ J$$

$$= [9 \times 10^{-14}/1.6 \times (2.694)]\ J \times [1/1.6 \times 10^{-13}]\ J/MeV$$

$$= 9 \times 10^{-14} \times 10^{13}/2.694\ MeV = 3.340 \times 10^{-1}\ MeV = 334\ keV$$

The deuterons must be accelerated to a few hundred kV voltage to strike the tritium nuclei in the target. The reactions may occur even at lower energies due to a tunneling effect.

2.26 Mechanism of Neutron Production and Neutron Activation

There are two groups of neutrons produced in photonuclear reactions—evaporation neutrons and direct neutrons. Evaporation of neutrons occurs when the absorption of a photon leads to the formation of a compound nucleus. The characteristic of a compound nucleus is that it forgets its history of formation and its subsequent behavior only depends on the energy gained by the nucleus. This energy is distributed to all the nucleons and one or more neutrons manage to gain sufficient energy to escape from the compound nucleus more or less isotropically. The emission resembles the evaporation of molecules from a liquid, hence the name. The evaporation neutrons have a Maxwellian energy distribution with an average energy of 1–2 MeV and contribute nearly 80% of the neutron fluence.

The production of direct neutrons involves a direct interaction between the photon and one of the surface neutrons of the nucleus. These neutrons are of higher energy compared to evaporation neutrons, because the photon transfers all its energy to the ejected neutrons, instead of sharing it with all the nucleons in the nucleus. This interaction usually occurs in glancing collisions. Direct emission neutrons have average energies of several MeVs and are ejected toward the patient and not isotropically emitted, as in the case of evaporation neutrons. The contribution of direct emission neutron is around 10–20% [9]. So for all practical purposes, the linac head can be considered as a neutron source of isotropic emission superimposed with higher energy direct neutrons directed toward the patient. From the photon dose and spectral information, the neutron spectral information and yield can be computed and also the neutron doses and risks. The yield depends on the photonuclear cross section.

2.26.1 Linac-Produced Photoneutron Spectrum

It is important to know the neutron spectrum and yield at the linac isocenter, for the following reasons:

1. It is important from the radiation protection point of view since many of the neutron interactions (e.g., activation of the linac head components) are energy dependent.
2. RBE is higher for neutrons compared to X-rays and electrons, and is also neutron energy dependent. So, radiation dose limits for occupational workers (staff in radiation oncology departments) are much more restrictive for neutron doses compared to X-ray doses.

3. These details are required for optimizing the neutron dose per Gy of the photon dose at the isocenter, so that the secondary cancer risk in the patient can be reduced to acceptable levels and the occupational risk due to neutrons from the treatment beam can be at ALARA (as low as reasonably achievable) levels.

In a photonuclear collision interaction, if the energy of the photon (allowing for recoil loss and neutron separation energy) is less than any of the excited states on a nuclide, the interaction can only be elastic scattering. When the energy exceeds the excited states of the nuclide, then a spectrum of neutrons will be emitted, the energy being decided by the excited state of the residual nucleus. The highest excited state corresponds to the emission of least neutron energy.

$$E_n = E_\gamma - S_n - E' - E_R$$

where E_n = neutron energy, E_γ = incident photon energy, S_n = neutron separation energy, E' = excitation energy of daughter nucleus, and E_R = the energy lost in nuclear recoil. For $E' = 0$, the neutron has maximum energy. For light nuclei, with only few appropriate excited states, there are only a few discrete neutron energies. On the contrary, for heavy nuclei with several exited states for decay, the discrete neutron spectrum becomes almost continuous.

The shape of the linac-produced neutron spectrum resembles a fission spectrum as shown in Figures 2.48 and 2.49 [8].

One popular method of estimating the neutron spectrum is to expose a set of foils (having different energy threshold for nuclear reaction) to the neutron fluence. From the activation of the foils the energies and from the activities of the foil the fluence at different energies can be determined.

2.26.2 Photoneutron Spectrum at Linac Isocenter

Because of multiple scattering of neutrons in air and in the linac room walls, floor, and ceiling, the spectrum incident on the patient is much more degraded compared to the fluence spectrum emerging from the linac head. Figure 2.50 [8] compares the differential neutron spectra per Gy at the location of neutron production (circles) and at the isocenter (squares).

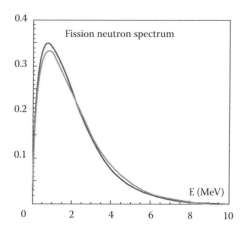

FIGURE 2.48
Fission neutron spectrum for U-235 and Pu-2339.

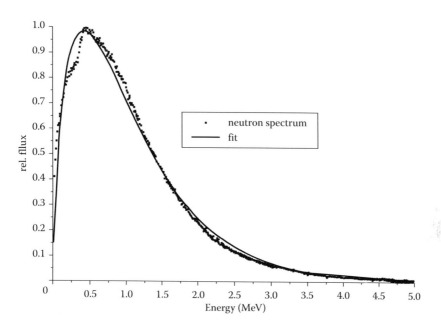

FIGURE 2.49
Normalized neutron source spectrum from a linac (Siemens Primus).

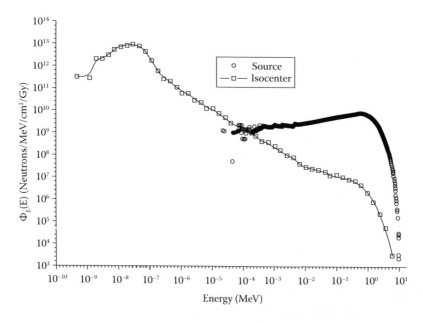

FIGURE 2.50
Differential neutron spectra per Gy at the location of neutron production (circles) and at the isocenter (squares).

The source does not produce neutrons with energies below 10 eV. All thermal neutrons at the isocenter come from scattering reactions throughout the treatment room. The mean neutron energy was calculated for source neutrons 1.06 MeV and at the isocenter 0.458 MeV. The most probable energy of source neutrons was E = 450 keV, and the maximum neutron energy found was E_{max} = 8.7 MeV.

2.26.3 Neutron Production in Electron Mode

Since the primary electrons have energies equal to the maximum energy of the linac photon spectrum, they are also capable of neutron production through (e,n) and (e,e'n) reactions. Both are Coulomb interactions between the electron and the nuclear charge field (not nuclear force field). The first one is an inelastic scattering event transferring excitation energy to the nucleus. The second interaction is mediated through virtual photon absorption by the nucleus and emitting a neutron. The Coulomb field of interaction between the electron and the nucleus is through the mechanism of virtual photon exchange between the two, the quanta of the Coulomb force field. The electron is scattered with reduced energy and the energy difference going to the virtual photon which is absorbed by the nucleus. The neutron yield however is smaller by two orders of magnitude compared to the photonuclear process because of the much lower beam currents in the electron mode of operation. So, the neutron yield in the electron mode is always neglected.

There is another, smaller component of photoneutron production compared to the linac head. The treatment room walls, the floor, and the air volume through which the beam passes in the treatment room and the patient body are also sources of neutrons from photon interactions.

2.26.4 Neutron Activation of Linac Head and Air in the Linac Room

Two important nuclear interactions occurring in the materials in a linac room are:

1. Neutron production i.e., photo disintegration: $\gamma + {}^{A}X \rightarrow {}^{A-1}X + n$
2. Neutron capture: $n + {}^{A}X \rightarrow {}^{A+1}X$

The product nuclide, following neutron production or neutron capture, may either end up as stable nuclide or as unstable nuclide in which case the product nuclide would be radioactive and undergo a β^{-} or β^{+}, γ decay with short half-life. For instance ${}^{180}W$, ${}^{182}W$, ${}^{183}W$, ${}^{184}W$, and ${}^{186}W$ are all stable nuclides. If ${}^{183}W$ absorbs a neutron (neutron capture) to become ${}^{184}W$ or if ${}^{184}W$ or ${}^{183}W$ loses a neutron (through γ absorption) to become ${}^{183}W$-183 or ${}^{182}W$, respectively, the residual nuclides are stable rather than radioactive. On the other hand, if ${}^{182}W$ loses a neutron to become ${}^{181}W$, or ${}^{186}W$ captures a neutron to become ${}^{187}W$, then the resulting nuclide is β^{-} or β^{+} radioactive. This means that immediately following a treatment exposure, the linac head is slightly radioactive and there is a small gamma dose to the technician if they are too close to the head.

The treatment room air also becomes activated through ${}^{14}N(\gamma,n){}^{13}N$ and ${}^{16}O(\gamma,n){}^{15}O$ reactions releasing ozone. In the electron mode, more ozone will be produced.

Ozone is very toxic even at a few ppm level of concentration (NCRP 1977 recommends a maximum permissible concentration of 0.1 ppm). Most of the staff can detect ozone at this concentration by its odor. More details on the reaction products and estimation of ozone concentration may be found in V.P. Singh et al.'s book [10].

2.27 Neutron Production in Particle Therapy

Particle therapy (PT) is external beam radiotherapy that uses positively charged particles such as protons and carbon ions, instead of photons or electrons. Neutrons are also produced in PT installations. Since the mean BE of nucleon is around 8 MeV and the BE falls in the low and high Z regions, protons (and heavier CPs) used in PT have enough energy to interact with the materials in the beam line and eject nuclear particles, including neutrons, from the target nuclei. For instance, the proton energy range of therapeutic interest is around 60–300 MeV. The low-energy proton (<10 MeV) interactions lead to compound nucleus formation followed by the evaporation of nuclear particles, including protons. At high energies, the proton interacts with individual nucleons creating a cascade of particles [7]. The neutron energies can go as high as the incident proton energies in this type of reaction.

2.28 Neutron Interactions

A basic knowledge of neutron interactions is necessary in radiation oncology since neutrons present a shielding problem and the functioning of neutron detectors is based on neutron interactions with the detector materials.

Neutron interactions in matter are quite different from those of photons or CPs. This is because a neutron does not carry a charge, so electromagnetic interactions are absent. The neutron can experience nuclear force and hence can interact readily with the nucleus when they are close to the nucleus. The absence of charge makes the Coulomb barrier of the nucleus zero. This facilitates easy entry of neutrons into the nucleus, irrespective of its energy. In some respects, neutron interactions with nuclei are similar to photon interactions with atoms. Photons excite the atom or knock electrons out of the atom through various interaction mechanisms. Similarly, neutrons excite the nuclei, or knock nucleons out of the nuclei by various interaction mechanisms. However, the neutron interactions are very much nucleus specific. Two isotopes of the same element can have vastly different probabilities for neutron interaction. As in the case of photons, the neutron interaction probabilities are also expressed in terms of cross sections, measured in barns.

As neutron interactions are highly energy dependent and the neutron energies span several decades, the neutrons in different energy ranges are classified as fast, slow, epithermal, and thermal, as shown in Table 2.9.

Neutrons interacting with atomic nuclei are either scattered or absorbed. The scattering can be elastic (no neutron energy transferred to the inner states of the target nucleus) or

TABLE 2.9

Classification of Neutrons According to Energies

Neutron Classification	Energies
Thermal	≈ 0.025 eV
Epithermal	≈ 0.5 eV–50 keV
Slow	≈ 1 keV–100 keV
Fast	≈ 100 keV–10 MeV

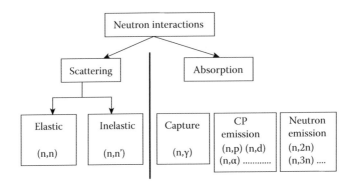

FIGURE 2.51
Neutron interaction mechanisms.

inelastic (part of the neutron energy goes as excitation energy of the target nucleus). In both types of interactions, a neutron reappears after the interaction. The inelastic scattering generally occurs at high neutron energies, with high Z nuclides. In absorption reactions, the nucleus forms a compound nucleus and gets rid of its excess energy by emitting various particles (γs, CPs, or neutrons). Figure 2.51 illustrates the concept.

2.28.1 Elastic Scattering

The elastic scattering is typically billiard ball type of collisions that were mentioned in connection with Coulomb nuclear scattering of electrons, or Compton scattering of photons. This is the main mechanism for the slowing down of neutrons in matter. This reaction channel is always open and can occur even when the neutron energy is near zero. When the neutron energy is not sufficient to excite the nucleus or eject particles out of the nucleus this is the only interaction that is energetically possible (Figure 2.52).

$$n + A \rightarrow n + A$$

The energy lost by the neutron is only due to the recoil of the nuclide. The fraction of energy transferred to the atom is given by $E_n \times [m_n/(M + m_n)]$, where m_n and M are the mass of the neutron and the nucleus, respectively. This can be approximated to $E_n \times \{1/(A+1)\}$. The equation shows that low Z nuclides are good at slowing down or thermalizing media for neutrons, and the best ones are hydrogenous materials (A = 1) which are invariably used for neutron shielding purposes. The energy lost by neutron in collisions with high Z nuclides are much smaller ($\alpha \, 1/A$). Figure 2.53 illustrates the neutron moderation concept.

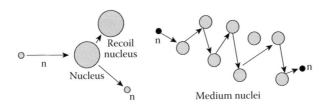

FIGURE 2.52
Neutron elastic scattering concept and slowing down in the medium.

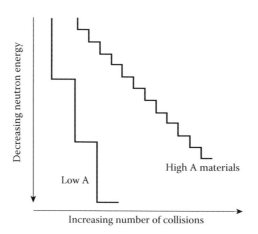

FIGURE 2.53
Neutron moderation in low Z and high Z materials.

2.28.2 Inelastic Scattering

The inelastic scattering of neutrons leading to the excitation of the target nucleus is illustrated in Figure 2.54.

$$n + A \rightarrow n' + A^*$$

The de-excitation of the nucleus occurs through γ emission. This reaction is responsible for the gamma fluence that always accompanies neutron fluence.

2.28.3 Neutron Capture

Neutrons are thermalized following elastic and inelastic scatterings, and are readily absorbed through thermal capture reactions. In this interaction, the neutron with very little energy (i.e., the thermalized neutron) is captured by the target nucleus, which then decays by the emission of a gamma ray, as shown in Figure 2.55. If the residual nucleus is unstable, then it is radioactive and emits β, γ radiations.

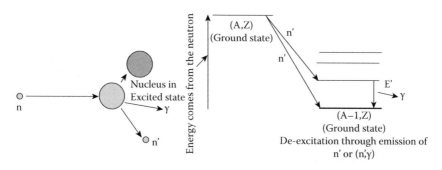

FIGURE 2.54
Concept of inelastic scattering and nuclear excitation/de-excitation.

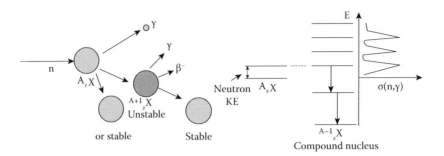

FIGURE 2.55
Concept of thermal neutron capture.

An important example of thermal neutron capture that is of relevance in neutron shielding of linac rooms is the reaction ^{1}H$(n,\gamma)^2$ that occurs in hydrogenous materials (e.g., the walls and the polyethylene enforced doors) of a linac room, for slowing down the neutrons. The thermal neutron capture results in deuterium formation and the emission of a prompt gamma ray. The capture gammas are in the MeV range and require further shielding considerations in a radiation oncology department.

The neutron shielding concept is illustrated in Figure 2.56.

The borated polyethylene reinforcement of the linac room door moderates and thermalizes the fast neutrons. The thermal neutron capture gammas produced are absorbed by the lead lining provided.

2.28.4 Other Neutron Interactions

The other types of interactions involve compound nucleus formation following the absorption of the neutron. The compound nucleus forgets its origin and the unstable nuclide returns to stability by emitting charged particles, or multiple neutrons, or fission fragments (that are not of concern here).

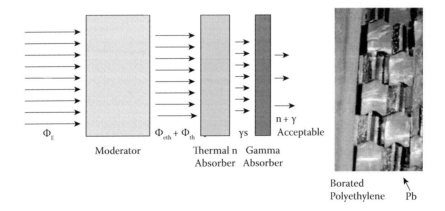

FIGURE 2.56
Neutron shielding concept.

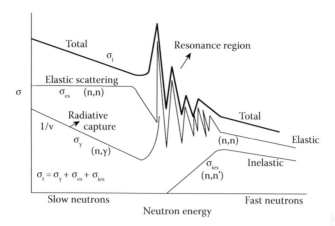

FIGURE 2.57
Individual interactions and their energy dependence.

$$n + A \rightarrow C^* \rightarrow b + B$$

The types of interactions occurring with the target nucleus depend on the neutron energy, the internal structure of the target nucleus, and the cross section for the reaction. Figure 2.57 and 2.58 illustrate the different types of possible interactions with a target nucleus. Excitation of the target nucleus (inelastic scattering) occurs in the resonance region where energy levels are available for excitation.

Table 2.10 gives a comparison of the basic photon and neutron interaction mechanisms and the corresponding coefficients.

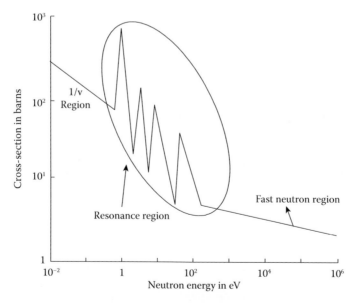

FIGURE 2.58
Total interaction probability as a function of neutron energy for a typical nuclide.

TABLE 2.10

Neutron/Photon Interaction Coefficients—A Comparison

	Interaction Cross-Sections		
Photons	**x-Section**	**Neutrons**	**x-Section**
PEE	σ_{pee}	(n, n)	σ_{es}
CE	σ_{ce}	(n, n')	σ_{ies}
PP	σ_{pp}	(n, 2n)...	σ_{nes}
		(n, p)...	σ_{cp}
		(n, γ)...	σ_{cap}
Total Cross-section	σ_t		σ_t

Note: Attenuation coefficient $\mu = (N_A/A)\sigma_t$ $\mu = (N_A/A)\sigma_t$. CE, compton effect; PEE, photoelectric effect; PP, pair production.

2.29 Energetics of Nuclear Reactions

One can easily determine if it is energetically possible for a nuclear reaction to occur, from the Q value of the reaction. The Q value is the mass difference in energy between the initial and final states of a nuclear reaction. Let us consider a general equation where a projectile bombards a nucleus A. The reaction products are b and B.

$$a + A \rightarrow b + B$$

$$Q = [(M_A + m_a) - (M_B + m_b)]c^2$$

Q > 0 (positive): Initial state has more energy than the final state. The reaction is energetically possible. The incoming particle need not bring in any extra energy for the reaction to proceed, although a CP must bring in energy to overcome the Coulomb barrier, which is zero for neutrons.

Q < 0 (negative): The neutron must bring in enough energy (>Q) for the reaction to take place, and also to supply the target recoil energy.

Since the nucleus will recoil in a collision reaction, E_{th}, the threshold energy for the reaction to occur is greater than Q.

$$E_{th} = Q [(M_A + m_a) / M_A]$$

$(E_{th} - Q)$ is the recoil energy of the target nucleus.

For elastic scattering, the Q value is immaterial.

2.30 Neutron Absorbed Dose

In the case of photons, it is easier to measure the energy absorbed in a medium rather than the photon fluence. On the other hand, it is easier to measure the neutron fluence of neutron fields. Hence the dose formalism is slightly different for neutron dosimetry. In the earlier section, the following equation was made use of for determining the photon kerma. The same equation applies to neutrons as well.

$$K = \int_E \Psi \, (_m\mu_{tr}(E) \, dE) = \int_E \Phi \, [E(_m\mu_{tr}(E)] \, dE$$

The term $[E \, (_m\mu_{tr} \, (E)]$ is the kerma factor, $k_E(E)$. For a monoenergetic neutron beam of energy E,

$$K = \Phi_E \, [E \, (_m\mu_{tr} \, (E))] \, dE \text{ or } k(E) = K/\Phi_E$$

So, the kerma factor can be defined as kerma per unit fluence. Knowing the neutron spectrum, kerma per unit fluence can be calculated for each bin and the total kerma can be computed; the kerma factor depends on the material. Much of the neutron dose, for the energy range of interest in radiation oncology, occurs from neutron–proton interactions and the energy lost by the recoil protons in the medium. A smaller fraction of the dose is contributed by other nuclei.

The kerma factor for tissue, k_T is given by

$$k_T(E) = \Sigma w_i k(E)_i$$

where w_i is the weighting factor for the i-th element in the tissue.

A typical kerma factor for water is shown in Figure 2.59.

Neutron kerma factors for tissue substitute materials are available in the literature, and can be used for determining the neutron dose in tissue substitute materials [11]. Most of the elements show a similar trend with a minimum kerma factor at a few eVs to few tens of eVs, and for fast neutrons the kerma factor increases with energy. Figure 2.60 shows the kerma factor for some tissue elements [11].

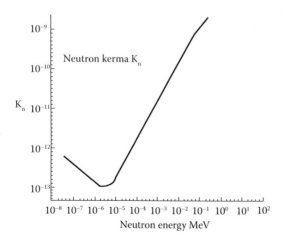

FIGURE 2.59
Kerma factor for water.

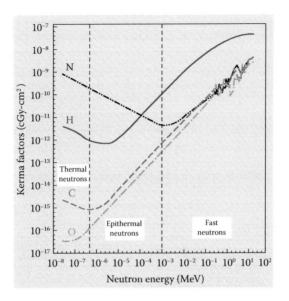

FIGURE 2.60
Kerma factors for tissue elements.

2.31 Neutron Dose Equivalent

The ranges of protons produced in (n,p) reactions are much smaller (a few micrometers) compared to electrons (produced by photons) and the ranges of heavier recoil nuclei are still smaller. So, unlike photons, collision kerma and dose are essentially the same for neutrons. Also, in the case of protons and heavier recoil nuclei, or charged particles produced in the interactions, bremsstrahlung produced in these charged particle interactions is practically zero (particle energy << rest mass energy), so kerma and collision kerma are equal. Hence, as in the case of kV X-rays, kerma is equal to dose. This is not true for (n,γ) interactions because gammas have a large mean free path and may escape from the region of interest without depositing much energy in the medium.

The dose equivalent (DE) for neutrons is given by

$$H_n = \Sigma_E \, D(E) \, Q_n(E)$$

$$= \Sigma_E \, \Phi(E) \, k(E) \, Q_n(E)$$

where Q_n is the quality factor of the neutrons of energy E (i.e., Q of the LET of the charged particles produced by neutrons of energy E). The Q(L) versus LET relationship is given in the International Commission of Radiological Protection (ICRP) documents and will be discussed in another chapter. If the neutron exhibits a spectrum, the average Q value of the neutrons is given by

$$\overline{Q}_n = \sum_E \Phi(E) \, Q_n(E) \Big/ \sum_E \Phi(E)$$

In megavoltage photon therapy, there is a negligible neutron DE component (about 0.1–0.2% of the X-ray dose in Gy at the isocenter) in the treatment beam, which delivers a small dose to the tumor. The neutron contamination in the room also contributes some dose to the patient apart from the tumor dose contribution. Because of the higher RBE of neutrons, the stray neutron dose received by the patient receives some attention compared to the stray X-ray dose, from the patient cancer risk point of view.

A neutron field is always accompanied by a gamma field due to the (n,γ) reactions, and so neutron dosimetry is essentially a mixed-field dosimetry and D_n and also D_γ need to be determined.

References

1. E.B. Podgorsak (Editor), *Radiation Oncology Physics: A Handbook for Teachers and Students*, International Atomic Energy Agency (IAEA), Vienna, Austria, 2005.
2. F.H. Attix, *Introduction to Radiological Physics and Radiation Dosimetry*, Wiley-VCH Verlag GmbH, KGaA, Weinheim, Germany, 1986.
3. J.H. Hubbel, *Tables of X-Ray Mass Attenuation Coefficients and Mass Energy-Absorption Coefficients from 1 keV to 20 MeV for Elements Z = 1 to 92 and 48 Additional Substances of Dosimetric Interest*, NISTIR 5632, National Institute of Standards & Technology (NIST), Gaithersburg, MD, 1995.
4. ICRU, *Electron Beams with Energies between 1 and 50 MeV*, ICRU Report 35, International Commission on Radiation Units and Measurement, Bethesda, MD, 1984.
5. S.N. Boon, *Dosimetry and Quality Control of Scanning Proton Beams*, Drukwerk: Ponsen & Looijen BV, Wageningen, The Netherlands, Chapter 2, p. 33, 1968.
6. H. Paganetti (Editor), *Proton Therapy Physics*, CRC Press, Boca Raton, FL, 2012.
7. J. Becker, Simulation of neutron production at a medical linear accelerator, Diploma Thesis, Institute of Experimental Physics, University of Hamburg, Hamburg, Germany, 2007.
8. IAEA, 1979, *Radiological Safety Aspects of the Operation of Electron Linear Accelerators*, International Atomic Energy Agency, Technical Reports Series (No. 188), Vienna, Austria.
9. P.H. Mcginley, *Shielding Techniques for Radiation Oncology Facilities*, Medical Physics, Madison, WI, 2002.
10. V.P. Singh et al., Neutron Kerma factors and water equivalence of some tissues, Applied Radiation and Isotopes, 103, 115–119.
11. H.R. Viga Carrillo and E. Manzanares-Acuna, Calculation of Neutron Kerma in Tissues, *Proceedings of International Joint Meeting*, Cancun, Mexico, 2004.

Review Questions

1. What are the radiation beams used for cancer treatment of patients?
2. Differentiate between DIPs and IDIPs in their energy deposition mechanisms.
3. Describe the three main photon interactions and the significance of Compton interactions in radiation oncology.
4. What is the relationship between particle fluence and energy fluence?
5. Knowing the photon or the neutron particle spectrum, how can the photon kerma or the neutron kerma be computed?
6. How are neurons produced in a linac treatment facility?
7. What is the relationship between collision kerma and absorbed dose in a medium traversed by photons and neutrons?

8. Name two important types of CP interactions that occur in a medium traversed by CPs and describe these two interactions.

9. Define impact parameter and the interactions that occur when the impact parameter is large, of the order, and small compared to the atomic dimensions.

10. Differentiate between stopping power, collision stopping power, and restricted stopping power.

11. What factor is of importance when providing shielding for a beta ray source?

12. Why is biological damage related to the LET of the radiation? Give two examples of low LET and high LET radiations.

13. Define path length and range. Why is path length much larger than the range for the electrons?

14. Multiplication of two quantities gives the collision kerma for the photons or absorbed dose for the charged particles. What are these two quantities?

15. Explain why the depth dose profile of a proton beam increases near the end of the range.

16. What is a Bragg peak? How can spread out Bragg peak be produced?

17. Why does the electron beam profile not show a well-defined range but instead shows a tail in the dose distribution?

18. Describe the advantages of HCP beams over the electron or photon beams in the treatment of cancer.

19. What is boron capture therapy?

20. How are neutrons produced for fast neutron therapy?

21. What components of the linac head give rise to neutron production? List the main components in order of importance.

22. Is there a minimum photon energy for neutron production and if so, why?

23. What reaction is responsible for the production of neutrons in the linac head?

24. Describe the mechanism of neutron production.

25. Does the linac produce neutrons in the electron mode?

26. How does the linac head become activated and how much dose will the technician receive due to this?

27. What is the significance of the activation of the air in the linac room?

28. How are neutrons classified in terms of their energies?

29. A neutron of a given energy is elastically scattered by a hydrogen atom and a carbon atom. Which interaction neutron loses more energy?

30. What is the principle of shielding linac room doors for neutrons?

3

Evolution of Radiation Protection and Radiation Risk Concepts

3.1 Introduction

It is difficult to trace the evolution of radiation protection chronologically with any authenticity, but at the same time a basic knowledge of this evolution will give a balanced perspective on the radiation dose and the accompanying risks. A few general references available on the topic [1–4] and the significant ICRP publications since its inception [5–10] give a basic idea on the evolution of radiation protection and radiation risk, since the discovery of X-rays and radioactivity. This chapter is a short summary of the evolution of radiation protection and the risk concepts.

Two discoveries that changed the face of medicine are (1) the discovery of X-rays by Wilhelm Conrad Röntgen in the year 1985 and (2) the discovery of radium in natural radioactivity by Madame Curie in 1898. The use of these radiations, in the field of diagnosis and therapy, started almost immediately but gradually radiation injuries also started manifesting in the patients and the physicians.

Marie Curie, who worked extensively with radioactive materials, contracted leukemia, apparently induced by inhalation and ingestion of radioactivity. Her daughter too, in the later years, became afflicted by the disease. The notebooks of Marie Curie are still so radioactive that they cannot be handled. Cataracts, one of the effects of radiation exposure, contributed to her failing vision. Marie Curie died in 1934 of pernicious anemia, most likely an effect of the radioactivity ingested and external radiation received from radium. This is almost identical to the diagnosis given for the first reported radiation-related death of a radium dial painter, in 1925: "rapidly progressing anemia of the pernicious type." Marie's eldest daughter and collaborator, Irene Curie, died in her mid-fifties of leukemia. They lived longer because of protracted radiation exposure. Pierre Curie too, with whom Madame Curie worked, suffered from bone cancer—Ra poisoning—but died in an accident.

Henry Becquerel, discoverer of radioactivity, died only 12 years after his discovery of radioactivity, at the age of 54. Although his cause of death was unspecified, he had developed serious burns on his skin, likely from the handling of radioactive materials. Röntgen died in 1923 of carcinoma of the bowel. It is thought that his carcinoma was not a result of his work with ionizing radiation because his investigations were for only a short time and he was one of the few pioneers in the field who used protective lead shields routinely.

The first radiograph was taken by Röntgen himself of his wife's hand showing clearly the internal structure, thus suggesting the potential use of X-rays in diagnosis. The same year physicians started using X-rays for diagnosis and cancer cure. The first medical radiograph was taken in 1896, demonstrating a fracture in the hand. The damaging effects of ionizing radiation were not known to many of the pioneers working with radiation or radioactive materials.

3.2 Invention of Fluoroscopy and Increase in Staff Injuries and Deaths

In 1896, Thomas Alva Edison invented a fluoroscope for direct viewing (Figure 3.1).

The first invention was a handheld fluoroscope to investigate the abnormalities inside the patient. This was later modified so it could be worn by the physician like a mask, so his hands were released for carrying out any procedure on the patient. In the beginning, nobody believed or even suspected that some invisible radiation could cause any harm to anyone. If you see any of these earlier pictures—do search for them on the internet—taken in the 1910s you will find that the tubes had no protective housing, the physician had no lead aprons, no eye protection or hand protection. The physician was looking directly into the transmitted X-rays. Obviously, the physician was susceptible to hand injury, eye injury, lung cancer and leukemia, the biological effects of the radiation that we will learn about in another chapter of this book.

C.M. Dally, Edison's assistant, who made all the X-ray tubes and fluoroscopes for Edison, was the first X-ray martyr. Mr. Dally tested his X-ray equipment on his own hands to ensure the tube was producing X-rays and at sufficient intensity. Eventually Dally lost both arms to malignant ulceration. He died a painful death in 1904, at the age of 39, and is now remembered as the first martyr to radiation (Figure 3.2). This led Thomas Edison to stop working on X-rays, fearing the radiation was too dangerous, but physicians continued to experiment with the device for the diagnosis of diseases in patients.

By 1897, barely two years since the discovery of X-rays, about 70 cases of skin damage were reported. By 1902, hundreds of cases of X-ray injuries were documented. Surgery was often needed to repair the damage. Many radiologists had to get their hands amputated

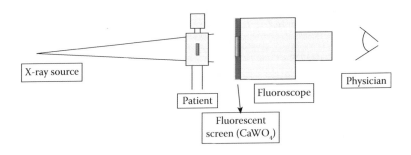

FIGURE 3.1
Birth of fluoroscopy.

C. M. DALLY DIES A MARTYR TO SCIENCE

Was Burned While Experimenting with X-rays.

WORKED WITH T. A. EDISON

For Seven Years He Constantly Underwent Operations, Finally Losing His Arms.

Special to The New York Times.

FIGURE 3.2
Dally's death as reported in the *New York Times*.

and ultimately succumbed to cancer. We are not sure how many of the X-ray tubes in the market, in that era, were producing X-rays of sufficient intensity to cause harm and it is quite possible some physicians escaped injuries due to the weak output of the X-ray machines. Gradually, two groups came into existence, the believers and the nonbelievers; one group saying, "Come on, can't you see the evidence around you?" and the other group countering, "Come on, how can invisible radiation hurt you?" Unfortunately, the majority believed radiation could not hurt, with the result that while the physicians became more responsible and started thinking about radiation protection, indiscriminate use of radiation exponentially increased in society. This must have caused harm to unknown numbers of people but there is no documentation of it. X-ray studios opened up in many cities for public exhibitions and taking "bone portraits" of healthy individuals. X-ray glasses for one dollar a pair, shoe fitting fluoroscopes for purchasing correct size shoes and so on had come into the market.

3.3 Radium Hazards

The frivolous use of radiation was not confined to X-rays. Radium too joined the list, as a wonder drug for general well-being and was used in many products from drinking water, to paste, to chocolates (Figure 3.3). The case of radium dial painters is well known. Young girls were employed to paint watch dials with self-luminous paint (containing radium) in a factory in New Jersey, around the 1920s. The women used to lick the tip of the brushes to make them a fine point and in the process ingested significant amounts of radium. This led to the so-called radium poisoning causing anemia, bone fractures, necrosis of the jaw (that came to be known as radium jaw), and cancer. Most of them died young and others suffered for a long time before their death. Radium (like calcium) concentrated in bones, causing leukemia, and in teeth, leading to their disintegration. A monument to the X-ray and radium martyrs of all nations, erected by the German X-ray Society in the garden of the St. Georg Hospital, Hamburg, in 1936, contains 159 names of the victims of overexposure.

Radium in beauty
products (1930s)

The physiological
action of
radio-active subs

Radium water,
Radium bread

Radium chocolate sold in
Germany 1931–1936

Radium paste

FIGURE 3.3
Indiscriminate use of radiation in society.

3.4 A Pioneer in Radiation Protection

Wise men always lived in this world. The first radiation protection advice came as early as 1986. Just a year after Röntgen's discovery of X-rays, the American engineer Wolfram Fuchs (1896) gave the following recommendations for radiation protection, advice far ahead of its times.

> **The Concept of Time/Distance/Shielding (1896)**
>
> - Make the exposure as short as possible—**Time**
> - Do not stand within 12 inches (30 cm) of the X-ray tube—**Distance**
> - Coat the skin with Vaseline (petroleum jelly) and leave an extra layer on the most exposed area—**Shielding**

Dr. William Herbert Rollins (1852–1929) is considered the father of radiation protection. He was one of the brightest to graduate from Harvard medical school in medicine and also dentistry. He is compared with Edison in his intellect, and he invented many useful devices in his medical career. He did pioneering work in oral radiology using X-rays. He developed the intraoral cassette and oral fluoroscope for this purpose, within one year of the discovery of X-rays, and also published details of his work.

Rollins suffered a severe burn on one of his hands after it was exposed to X-rays in 1898, which prompted him to study the biological effects of X-rays. He demonstrated experimentally the X-ray lethality to mammals and mammalian fetuses, following which he expressed concern about the use of X-rays in pelvic examinations of pregnant women.

- He redesigned the X-ray machine with more safety features, and introduced a fluoroscope with lead glass backing.
- He devised a wooden box—known as a Rollins box—lined with layers of lead paint with only one opening for the emission of an X-ray beam (X-ray protective housing).

- He introduced collimating diaphragms to narrow the beam to the region of interest.
- He introduced lead glass goggles (which were a full centimeter thick) for fluoroscopists as protection against cataracts; Edison himself suffered from vision impairment.
- He also invented and advocated the use of lead shields and lead aprons for physicians.
- He developed a series of guidelines like the use of radiation-absorbing glasses and lead aprons, standing at a minimum distance from an X-ray tube from the target to the film, the use of filters to remove unwanted radiation, using lead as a barrier against radiation exposure.

But he did not aggressively promote his ideas.

In 1921, the British X-ray and Radium Protection Committee was created with the objective of studying various ways of reducing radiation exposure to the physicians and the operators of X-ray equipment, but no progress in that direction could be made in the absence of radiobiology knowledge, and a means of quantifying radiation exposure.

It was soon realized that to control doses to patients or staff, there was a need to quantify the output of X-ray equipment. Though X-rays were being used by the physicians they had no idea how much radiation the X-ray tubes were producing and how much the physicians were receiving. There were no dose quantities, units, or measurement methods existing in the 1910s and early 1920s. All that was known was, prolonged exposure to radiation could cause skin erythema. So, skin erythema became the first unit of dose. Erythema dose (ErD) was taken to be the dose that caused skin reddening. (Following the introduction of the Röntgen unit, ErD was estimated to be about 600 R.)

3.5 Tolerance Dose Concept (to Avoid Skin Injuries)

In 1924 came the introduction of a dose quantity, the *tolerance dose*. Arthur Mutscheller, a German physicist who moved to New York in the 1920s to work for an X-ray equipment manufacturer, studied the physicians who had not exhibited any radiation effects during long periods in their professional career. He estimated their occupational doses, over a period of a month, as 1/10 of an erythema dose.

This could be considered a "safe" dose, since they did not exhibit any ill effects. Sievert, who was working on similar lines in Sweden, also arrived at the same conclusion in 1924. This concept of safe or tolerable dose was adopted from the field of toxicology where quantities that could be tolerated for toxic substances was assessed. Adding a safety factor of 10, Mutscheller defined the first tolerance dose for the radiation workers as

TD: 1/100 of Erythema dose/month

= 60 mSv/month = 2 mSv/day

Or 500 mSv/y (1924–1934)

Assuming five days/week and 50 weeks/year.

Mutscheller concluded that any X-ray worker not receiving one-hundredths of erythema dose, over a 30-day period, was "entirely safe." He then went on to establish shielding principles for X-ray work, by calculating the shielding thickness required for reducing the unshielded dose to less than the tolerance dose, by referring to the tables that gave the attenuation fraction for lead, for various X-ray wavelengths. That is how shielding calculations are done even today, the tolerance dose being replaced by an acceptable or design value dose. He realized that absolute safety was not possible and that safety must be balanced against the cost of construction. Again, the same principle is followed to this day. It took nearly a decade before the concept became the basis of safety standards. He presented his ideas in the 1924 meeting of the American Röntgen society and published his work the following year. There was just one problem that remained unresolved—a pressing need for a physical dose quantity.

3.6 Precursor to ICRP

The first International Congress of Radiology (ICR) was held in London in 1925. Though the issues of quantifying radiation was discussed, no decisions on measuring radiation or its effects were taken in the first congress, but a significant achievement of the first congress was to conceive an International Commission of Radiation Units and Measurements (ICRU) to address the issue. (It was originally known as the International X-ray Unit Committee and later renamed as the International Committee for Radiological Units before taking up the present name). It officially came into being in the second ICR held in Stockholm, in 1928. ICRU was entrusted the responsibility of defining suitable radiation quantities and units. In the second ICR, ICRU proposed Röntgen as a unit of the quantity exposure (then known as exposure dose, though it had not been unambiguously defined at that point). In the second ICR, most of the physicists accepted one-hundredths of an erythema dose per month as the tolerance dose.

Another important development at the second ICR was the establishment of the International X-ray and Radium Protection Committee (IXRPC), which later became the International Commission on Radiological Protection or ICRP. Around the same time (in 1929), the US Advisory Committee on X-ray and Radium Protection was formed to deal with the radiation protection issues in the US. In the year 1946, it was renamed as the National Committee on Radiation Protection and finally in 1964 it became the National Council on Radiation Protection and Measurements (NCRP). There are many similarities in the recommendations of ICRP and NCRP from the earlier days as one of the pioneers in radiation protection, Dr. L.S. Taylor, was a founding member of these important organizations and served in all three bodies for a very long time. He died in 2004, at the age of 102.

The objective of NCRP was to develop the concept of radiation protection and concepts of radiation quantities, units, and measurements, and issue recommendations at the national level. The national and state level regulatory bodies later adopted these recommendations as presented, or with modifications, for framing radiation safety regulations. In the initial stages, it was supported by the National Bureau of Standards (NBS), now known as the National Institute of Standards and Technology (NIST) and its publications were issued as NBS handbooks. The first publication of NCRP, titled *X-ray Protection* dealing with safeguards against X-ray hazards, appeared as NBS Handbook 15. The NCRP publications continued to appear as NBS handbooks until the 1960s, before NCRP started issuing their own reports.

In the third ICR meeting held in Paris in 1931, ICRP took up the issue of tolerance dose, as the first radiation protection measure for occupationally exposed staff. Since Röntgen had been accepted as a unit of dose in the 1928 meeting, researchers worked on correlating the erythema dose to a dose in röntgens. It was estimated that the skin reddening dose amounted to 600 R, according to the new unit. In the SI units introduced much later, 1 R approximately equaled 10 mSv (as will be explained later in this book). So, the erythema dose turns out to be 6000 mSv, which is about 2 mSv (or 0.2 R) per day, the first ever radiation protection recommendation for the radiation workers based on exposure units. NCRP suggested the same value in its recommendations in 1931. In the fourth ICR meeting held in Zurich in 1934, ICRP recommended this tolerance dose of 0.2 R/day, based on the observation of no ill effects on the studied population.

ICRP went on to state that the same tolerance dose might be used as a guide for radium protection. NCRP, however, revised it to 0.1 R/day in 1936, as the safe whole body tolerance dose per day for hard X-rays. The erythema dose for penetrating gamma rays like radium gammas was estimated at 340 R, against 600 R for kV X-rays, leading to a daily tolerance dose of around 0.1 R.

3.7 Concept of Maximum Permissible Dose

The basis of tolerance dose, equating the absence of observable damage to safety, was still being questioned by many, even for X-rays. So, the ICRP recommendation of tolerance dose was not extended to radium protection. In fact, ICRP explicitly mentioned in its 1934 report that "no similar tolerance dose is at present available in the case of radium gamma rays." The gradual manifestation of radium hazards showed the risk could be much more pronounced compared to X-rays.

Radium (^{226}Ra, decay product of ^{238}U) gamma rays are much more penetrating than the X-rays used in diagnosis and therapy, and are more likely to affect the inner organs. Because it was expensive the medical use was very limited and the exposed population was much smaller. Therefore, skin injury by X-rays appeared to be the more immediate threat. The usage of radium started increasing in the 1920s following a drop in the price of radium. Radium dial painting was the first major cause of radium exposure. This increased the population occupationally exposed to radium. Radium ingestion caused bone disintegration, bone marrow damage leading to leukemia, anemia, diseases of the blood, and cancer of other organs. All the women working in dial painting factories died young due to radium effects. However, radium effects were much slower to appear but much more severe compared to X-ray effects. In due course, radium was replaced by much cheaper mesothorium (^{228}Ra, decay product of ^{232}Th) in paint to produce novelty products like luminous door knobs, switches, etc.

R.D. Evans studied the victims of radium ingestion to establish the maximum permissible body burden of radium. By counting the gammas emitted by the victim, he could estimate the amount of radium in the body. He also studied the radium burden of victims who had lived long without any ill effects. From these studies and allowing a margin of safety, he concluded that 0.1 µCi was the maximum permissible body burden (MPBB) for radium. (0.1 µCi ≈ decay rate of 0.1 µg of Ra). This standard was published in NBS Handbook 27, in 1941. While studying the radium dial painters, he had estimated the average amount of radium ingested by the dial painter as 150 µg/week! Even for the dial painter who had ingested the least amount of 7.5 µg/week, three months of steady work would have caused a body-lethal burden, according to Evans.

Tolerance dose and MPBB implied that there is a threshold for harmful effects of radiation and that the cells would be able to repair any damage below this tolerance dose or MPBB. Since such thresholds are common in biological systems while dealing with toxic agents, few believed it could be any different with X-rays or gamma rays. However, growing evidence, particularly the studies of Herbert J. Muller on the mutations induced in fruit flies by X-rays, showed there was no such threshold for this effect. There was a direct proportion between the quantity of radiation and the number of mutations, without any threshold limit. This seemed to suggest that perhaps the so-called tolerance might mean detriment at a more subtle level and not detectable, such as with the skin injury. In any case, Muller's results and its implications for germ cells led to a reconsideration of the tolerance dose, to set a lower value for it. It also created an opinion in the scientific committee that perhaps there is no real threshold for some of the radiation effects, though there could be a threshold for noncancer somatic effects such as skin injuries or eye cataracts. This led to changing the terminology from tolerance dose to maximum permissible dose (which may not be tolerable but the harm is acceptable).

3.8 Risks Other than Skin Injuries

In the early days of radiation use, radiation risk only implied skin injuries. Looking at the 1930s and 1940s the safe dose limit was meant to avoid erythema or skin injury. No other risk was being considered. The X-ray kV's were also much lower and not penetrative enough to deliver significant doses to the interior organs. With the passage of time, mortality risk gradually became conspicuous. The mortality rates for leukemia and other solid cancers began to rise, showing the large latency between exposure and the manifestation of the effect. This led to clear distinctions between cancer and noncancer effects like skin injuries for the first time. With higher kilovoltage X-ray machines and radium coming into more widespread use in therapy, and following the internal overexposures of the radium dial painters, biological effects and the protection of inner organs assumed greater significance. Also, the extremities were still at risk due to the handling of radioactive sources in various applications. This led to recommendations limiting doses to both the so-called critical organs (to limit the cancer risk) and to the extremities, particularly the hands and feet (to avoid skin injuries). Thus, two different dose limits started to emerge.

The genetic effects weighed heavily on the minds of the scientific community in the 1950s. The genetic concern also led to the recommendation of dose limits to the public. ICRP arbitrarily assigned a factor of 1/10 to set dose limits for the public from the occupational dose limits. This genetic concern was also one of the reasons for the gradual reduction in the dose limit.

3.9 Introduction of Weekly Dose Limits

There were not many radiation protection activities undertaken by the advisory bodies during World War II. Following the war, in 1949, the US, Great Britain, and Canada, based

TABLE 3.1

Weekly Dose Limits (ICRP 1954)

Dose Limits	1954 Recommendations
Critical organs*	0.3 rem/wk **(150 mSv/y)**
Thyroid	0.6 rem/wk **(300 mSv/y)**
Extremities	1.5 rem/wk **(750 mSv/y)**
General population	$(1/10)^{th}$ of radiation worker limit

* Critical organs were the blood forming organs, lens and gonads. Dose limits in brackets are by the author and did not appear in ICRP.

on conclusions reached in a joint meeting, issued a report recommending a dose limit of 3 mSv/wk (about 150 mSv/y) for bone marrow and 6 mSv/wk (300 mSv/y) for the extremities. For the first time we had two limits: one to limit the cancer and hereditary effects, and another to avoid skin reactions. In 1954 ICRP issued the above recommendations (Table 3.1) for the maximum permissible weekly doses (published by BJR as Supplement 6). It may be noticed that the practice of recommending annual dose limits came much later.

Permissible weekly dose was defined as "a dose of ionizing radiation accumulated in one week of such magnitude that, in the light of present knowledge, occurring at this weekly rate for an indefinite period of time, is not expected to cause appreciable bodily injury to a person at any time during his lifetime." The term bodily injury was not defined, but the inclusion of blood forming organs as priority organs for dose limitation suggests that cancer risk (leukemia) was the main consideration. Limiting the population dose to one tenth of the radiation worker limit was stated as being mainly out of genetic concerns toward the population.

3.10 Concept of RBE Dose

The quantity absorbed dose (measured in units of rad) was defined in 1953 and was used for the first time in the recommendations, in place of röntgens. The radiobiological studies had indicated that different radiation types (e.g., photons, alphas, neutrons) exhibited different degrees of tissue damage for the same absorbed dose, because they deposited different amounts of energy per unit path length traversed in the medium. When the density of ionization (the cause of energy deposition) is higher, the recovery from tissue damage decreases. So, to determine a biologically equivalent dose for all radiation types, for certain tissue damage, a new quantity called the relative biological effectiveness (RBE) dose was defined.

The 1954 recommendations referred to this dose by the name dose equivalent (DE)—dose equivalent = dose × RBE. Since RBE is dimensionless, to distinguish between dose and DE, a new unit rem was used to refer to DE. This was the first attempt to come out with a radiation protection quantity that is not a physical dose quantity, but a biological dose quantity relating to the index of biological harm to the tissue. The report also gave a table of RBE versus linear energy transfer (LET) along with typical values for different particle types such as photons and neutrons.

The recommendations also stated that "every effort be made to reduce exposures to all types of ionizing radiations to the lowest possible (practicable) level." This shows the realization that there might not be a threshold for certain types of radiation effects, which could be the reason for such a recommendation.

3.11 Introduction of Quarterly and Annual Dose Limits

In 1958, ICRP published its first report through Pergamon Press. The weekly limits were replaced by quarterly limits. A cumulative dose limit was introduced for the first time using the formula

$$D_N = (N\text{-}18) \times 5 \text{ rems or } (N\text{-}18) \times 50 \text{ mSv}$$

N = age; 18 = starting radiation work as an adult.

From the above, we can see that at the age of 25, the radiation worker should not have accumulated a dose greater than 35 rems. If a radiation worker works from the age of 18 to 65, a common age of retirement, they would have received a cumulative maximum dose of 235 rems (2.35 Sv).

If it is assumed that radiation workers receive the dose at a uniform rate, this equates to an annual dose of 5 rems (50 mSv), and a weekly dose of 0.1 rem (1 mSv). This shows a reduction of the annual dose limit by a factor of 3 compared to the previous recommendations. The quarterly dose limit, however, remained about the same as the previous recommendation (i.e., $0.3 \times 13 \approx 4$ rems/13 wk). However, there was no explicit annual dose limit recommendation, only an inferred one from the above explanation. The cumulative dose limit was also recommended by NCRP in the same year. Also, there was no change in the dose limit to the extremities (1.5 rem $\times$ 13 $\approx$ 4 rems/quarter or but 80 rems/y − 800 mSv/y −) compared to the 1954 recommendations. Table 3.2 summarizes the 1958 recommendations.

During the period 1950 to 1960, increasing knowledge of radiation dosimetry and radiobiology led to several ICRP publications dealing with both external and internal exposures. (The story of internal exposures is not considered here since they are not of concern in radiation oncology). The analysis of atom bomb survivors' data in the 1960s showed that the

TABLE 3.2

Annual Dose Limits [6]

Dose Limits	1957 Recommendations
Critical organs	3 rems in 13 weeks
Quarterly limit	Subject to the annual limit.
Annual MPD*	Expressed as cumulative dose not to be exceeded: $5 \times (N–18)$ rems, at age N
Thyroid	0.6 rem/wk (**300 mSvR/y**)
Extremities	20 rems rem/13 wk (**800 mSv/y**)
General population	$(1/10)^{\text{th}}$ of radiation worker limit

* MPD = Maximum Permissible Dose.

genetic risk was overestimated in the earlier analysis, and that cancer incidence was slowly manifesting in the survivor population, following a latent period.

The period 1960 to 1970 saw the evolution of a scientific basis of radiation protection, and clarification of its objectives. Many recommendations for limiting the internal exposures were published during this period. Quantitative estimates of cancer incidence started becoming available in the follow-up of Hiroshima and Nagasaki victims. This enabled the estimate of risk factors based on estimated doses received by the Japanese survivor population. No hereditary effect could be observed in the survivor population.

3.12 Radiation Risk–Based Recommendations

The objectives of radiation protection were clearly stated in ICRP reports as prevention of acute radiation effects, and limitation of risks of late effects to acceptable levels. Though radiation protection limits must be based on acceptable radiation levels, when radiology and radiotherapy started, neither radiation doses received nor the risks were known. It took nearly 30 to 40 years to gradually understand the radiation risks and also develop methodologies for measurement of dose. Though dosimetry became well developed it required a long follow-up period to observe the late effects of radiation such as cancer, and to estimate the risks quantitatively. So, until the early 1970s, no statement or factors regarding risk were mentioned in any of the radiation protection recommendations. The first ICRP report to base the recommendations on risk rather than dose was ICRP 26, issued in 1977. This report was updated in 1991 (ICRP 60) and 2007 (ICRP 103). These are important recommendations and incorporated a new thinking in radiation protection. Before discussing these recommendations in some detail, we will consider the radiation effects and risk estimation.

3.13 Radiation (Stochastic) Effects and Stochastic Risk

Historically, two effects were observed in the radiologists' population. Skin injuries (especially the hands) leading to ulceration and amputations, and early death of the radiologists, and others overexposed to radiation, from cancer. The 1928 recommendations listed injuries to superficial tissues, derangements of internal organs and changes in the blood as "the known effects of radiation." The 1950 recommendations added malignant tumors and other deleterious effects (e.g., cataracts, impaired fertility, and reduction of lifespan) and genetic effects to the list of radiation effects. ICRP 1 mentioned somatic effects that occur in exposed individuals, and genetic effects in the descendants, as those to consider. The somatic effect term refers to both the acute and long-term effects in the exposed individual. ICRP 9 speaks of prevention of "acute effects" (referring to threshold noncancerous effects like skin burns) and limiting the risk of "late effects" (cancer and hereditary effects). ICRP 26 (1977) mentions the prevention of both; this report was the first to classify these radiation effects as stochastic and nonstochastic, which require some explanation.

Certain effects are probabilistic and can be predicted only on a statistical basis for an exposed population, and cannot be predicted at individual level (e.g., risk of crossing the road, chance of winning a lottery). Radiation exposure-caused cancer belongs to this category. For instance, a person receiving a very low dose may be affected by cancer while another receiving much higher dose may not get the disease at all, since it occurs randomly. With a higher dose there is a higher probability of getting the disease but no certainty. But for an exposed population, a larger dose would lead to greater incidence of the disease (since it is probability times the number exposed). The incidence of the effect in a population can be reduced only by reducing the number exposed, and by reducing the dose.

All the studies clearly show that at high doses, radiation can cause cancer in most parts of the body and also leukemia with a latency period of few years to few decades. The mechanism of cancer formation is also fairly well understood as mainly due to irreparable damage caused to the double strands of DNA. The survival of these cells escaping the cancer detection and rectification mechanisms can lead to malignancy. However, all the radiation exposure data available are at high doses (at least above a few hundred mSv) and the risks have been established for these data from meticulous follow-up studies on the exposed population. However, it is not possible to register the radiation-caused cancer at very low dose levels (i.e., the order of dose received by the majority of the radiation workers, or the dose received from natural background), because the additional increase in cancer incidence (due to these low levels of radiation exposure) is much smaller compared to the natural incidence of cancer or the fluctuations in the natural incidence. (There is no means of distinguishing radiation-induced cancers from naturally occurring cancers). So, the other alternative, to estimate risk at very low doses, is to extrapolate the risk versus dose response curve to occupational radiation exposure levels. The studies carried out in the 1950s on cancer induction or hereditary effects suggested no threshold for these effects to occur.

3.14 Introduction of LNT Hypothesis

In 1956, the US National Academy of Sciences (NAS) Committee on Biological Effects of Atomic Radiation (BEAR I) Genetics Panel issued the important recommendation that the carcinogenic and hereditary risks at low doses must be assessed using *a linear, no-threshold hypothesis*. No-threshold means any dose received, however small, will result in a biological effect. In other words, there is no threshold for the stochastic radiation effects, as may exist for toxic substances. The BEAR committee believed, based on all evidence available then, that no other assumption regarding the relationship between dose and response was justified. Soon, other national and international advisory committees also adopted that assumption.

A Linear No-threshold (LNT) model curve is shown in Figure 3.4 along with a linear curve that exhibits a threshold effect.

The problem faced in radiation protection is that the dose response (risk) data is available only for high doses (at least 0.1 Sv), estimated from atom bomb survivors and other overexposure cases in radiation treatment, or accidents like Chernobyl. The dose–response curve needs to be extrapolated to low doses levels (around 1 mSv/year) encountered in radiation protection activities, to estimate radiation risk at these levels. Since the data at high doses have large uncertainties, both extrapolations shown in Figure 3.4 can be considered valid. Which one is adopted for risk estimation has a lot of bearing on radiation protection. The linear threshold response curve implies that there is no risk below a threshold dose.

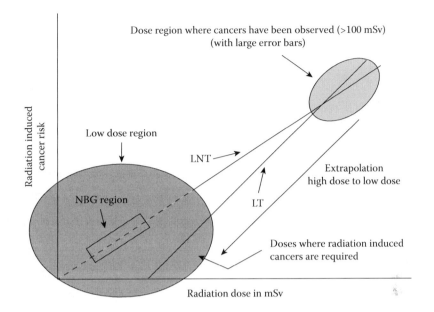

FIGURE 3.4
Dose–effect relationship (based on LNT assumption).

The LNT response curve implies a small risk, however small the dose may be. In the absence of direct verifiable evidence, it is only prudent to adopt the LNT model for extrapolating the dose response curve to protection dose levels to estimate the risk. So, even after nearly 60 years until today, radiation protection recommendations of all the advisory and regulatory bodies are based on the LNT hypothesis. Since theoretically a single mutation may lead to cancer and the natural incidence of cancer is rather high (around 25%–30%), many believe there may not be a threshold for this effect. For reasons mentioned earlier, it is impossible to establish, with statistical significance, if there is any threshold for cancer induction at very low doses, say near natural back ground. So, this no-threshold assumption is always going to be a point of controversy.

On the other hand, there is enough scientific evidence showing that certain types of cancers (leukemia and breast cancer) and hereditary effects follow a linear quadratic dose response. So, accepting the LNT hypothesis for all cancers is being questioned by many individuals and committees. ICRP still believes that the LNT hypothesis is the most conservative assumption and must be adopted in radiation protection until evidence against this assumption is unambiguous. The risk coefficients have been established by ICRP and adopted by the regulatory bodies using this model, and no change has been recommended so far.

One factor that needs to be considered is the pattern of dose received. While the atom bomb survivors and accident victims received high radiation doses as an acute dose, we are concerned with small doses (as received in medical imaging) or small doses received over a long period of time (e.g., natural background radiation exposure or occupational exposures). The low doses observed over longer periods of time allow sufficient time for the damaged cells to repair the damage. So, the radiogenic cancer factors, obtained by projecting the high dose data to radiation protection dose levels, require a correction for dose and dose rate differences. This factor is usually referred to as the dose and dose rate effectiveness factor (DDREF). ICRP has recommended a DDREF factor of 2 to modify the extrapolated risk.

3.15 Hereditary Effects

Like cancer, mutations in germ cells can also cause hereditary diseases, which are considered stochastic in nature. The experiments of Muller on fruit flies and similar effects found in the irradiation of barley and maize conclusively established radiation-induced hereditary effects. Radiation-induced genetic mutations have been observed in animals at doses >20 mSv. From these results it is inferred that hereditary diseases could be caused in humans as well, by exposure to ionizing radiation. However, meticulous follow-up studies carried out on the children of atom bomb survivors did not reveal any hereditary risks in humans. The BEIR VII report, published in 1998, states that at low or chronic doses of low-LET irradiation, the genetic risks are very small compared to the baseline frequencies of genetic diseases in the population.

The stochastic effect refers to cancer development in exposed individuals, owing to mutation of somatic cells, or heritable disease in their offspring owing to mutation of reproductive cells. The manifestation of cancer occurs following a long latent period that may stretch to decades.

The aim of radiation protection is to recommend dose limits to reduce the stochastic risk (and corresponding dose limit) to an acceptable level. The current dose limits will be discussed in the next chapter. Looking back at the history of dose limits, the limits have consistently come down over the years (see Figure 3.5) though there has not been any drastic change in the last couple of decades.

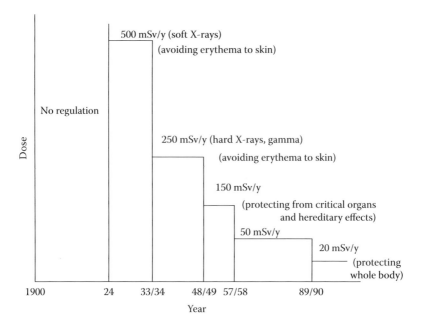

FIGURE 3.5
Progressive reduction in dose limits over a century.

3.16 Nonstochastic Effects

Only cancer and hereditary effects belong to the stochastic category. Stochastic effects are believed to arise from injury to a single cell or a small number of cells that result in the survival of the damaged (mutated) somatic or reproductive cells. The aim of radiation protection is to reduce the stochastic risk to as low as reasonably achievable levels.

ICRP 26 defined other somatic effects as nonstochastic. This usually occurs when a large number of cells in the organ or tissue are killed in a relatively short time and there is no time for eventual repair or replenishment of cells, leading to manifestation of injury. Radiation accidents or prolonged and repeated fluoroscopy of patients in interventional cardiac procedures result in nonstochastic effects. These effects exhibit a threshold and are manifested only after a threshold dose value is reached. Since some repair also takes place the manifestation of the injury depends on the exposure and the course of the injury, and the latency period for manifestation (days to weeks). The latent period for the manifestation of the tissue reactions, which is the presently accepted term for the nonstochastic or deterministic effects, depends on the cell proliferating rate. Fast proliferating cell tissues (intestine, skin, esophagus, etc.) manifest the damage earlier than slow proliferating cell tissues (lung, spinal cord, bladder, etc.).

A typical dose response curve for a nonstochastic effect is shown in Figure 3.6.

It may be noticed in Figure 3.6 that the threshold dose is not exactly the same for all individuals. It occurs over a dose range depending on the biological differences between the individuals. The threshold is defined as the estimated dose that is required to cause a specific, observable effect in 1% of the exposed individuals. It is a practical value for radiation

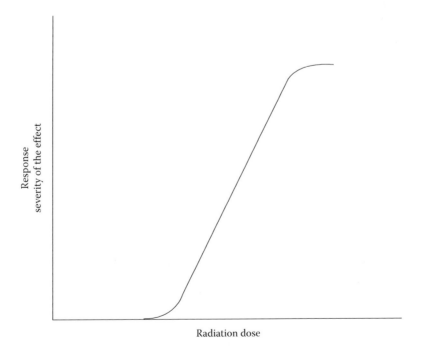

FIGURE 3.6
Threshold response characteristic for deterministic effects.

TABLE 3.3

Threshold Doses for Some Important Tissues

Threshold Effect	Time for Manifestation	Acute Exposure Gy	Chronic Exposure (Gy)
Skin erythema	1–4 weeks	<3–6	30
Skin burns	2–3 weeks	<5–10	35
Eye cataract	>20 years	≈0.5	≈0.5

protection purposes. The manifestation of the tissue reactions following exposure (i.e., the latent period) can be days to weeks, or months to years, depending on the tissue type, its radiosensitivity, DNA repair capacity, proliferation rate, etc. Beyond the threshold, the severity increases with the dose due to increased cell killing. Cataracts, impaired fertility, and hair loss are a few examples of nonstochastic threshold effects. The thresholds for acute or single high doses are generally less compared to chronic exposures. So, thresholds for acute exposures are of interest in diagnostic radiology while occupational exposures are chronic in nature—doses are small increments spread over a long period of time. Eye cataracts occurring in some interventional cardiologists, after many years of fluoroscopy practice, is an example of threshold for chronic exposures. Table 3.3 gives the threshold for deterministic effects, for tissues of importance in occupational radiation protection (ICRP 118, 2012).

The risk to the eyes is of particular importance and will be considered briefly here. Radiation-induced eye cataract risk was known since the discovery of X-rays, and so eye protection has been stressed since then in radiation protection recommendations. Earlier experiments had suggested a threshold of 5 Sv for chronic exposures and 0.5–2 Sv for acute exposures to the eyes. Based on this, the annual dose limit to avoid any eye opacity was set at 150 mSv/y in ICRP 26 (1997). However, a number of studies carried out since then and the appearance of eye cataracts with increasing frequency among interventional cardiologists, and the Chernobyl cleanup workers, suggested that the threshold for chronic exposures could be much lower. The effect showed up with longer latent periods in the case of smaller doses received by the eyes and hence were not accounted for in the earlier studies. In ICRP 103 (2007) the threshold for eye opacities was reduced to 0.5 Sv. The annual dose limit was also brought down to 20 mSv/y for a five-year average and 50 mSv/y for any single year maximum. If future follow-up shows that eye injury can occur at even lower thresholds, in fact down to zero, it may well turn into a stochastic effect instead of a nonstochastic one. This risk, however, is not of much relevance in external beam therapy, but in interventional procedures that involve prolonged high dose fluoroscopy procedures, physicians are at great risk unless special precautions are taken to protect the eyes.

The term nonstochastic is perhaps not the right word for this effect since all radiation effects at the cellular level are stochastic (random) and it is the killing of large numbers of cells, without adequate time for recovery, that results in a clinically observable effect, giving it a nonstochastic character at macroscopic level. ICRP 60 replaced this term with "deterministic effects," causally determined by the preceding events.

Later, this term was also dropped in favor of "tissue reactions" because the effects are not completely determined by the preceding events, as stated earlier, but can be modified even after the exposure (ICRP 103, 2007). Also, not all threshold reactions are as a result of cell killing. Some effects may arise due to tissue responses to radiation; the tissue responses during the latent period can mitigate the tissue damage. Terms such as cancer effects and noncancer effects can also be used to distinguish between the two types. The terminology,

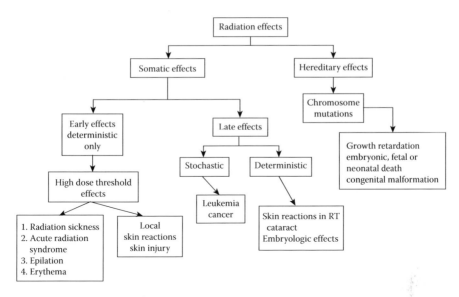

FIGURE 3.7
Summary of radiation effects.

over the years, has changed from nonstochastic effects (1984) to deterministic effects (1991) to tissue reactions (2007), although the previous terms are also still very much in use.

The atom bomb survivors of Hiroshima Nagasaki were exposed to high radiation doses to the whole body over a short period of time. The clinical effects exhibited by these people soon after the exposure are referred to as acute radiation syndrome (ARS). ARS refers to the symptoms and effects following high radiation doses received by the whole body over a short period of time. The details of this can be found in any textbook on radiobiology. The order of doses involved can be understood from the lethal dose concept. $LD_{50/30}$ refers to the lethal dose that would cause 50% deaths in 30 days in an exposed population. $LD_{50/30}$ is in the range of 4–5 Gy whole body dose. However, this is only of academic interest in radiation oncology.

Figure 3.7 summarizes the radiation effects (both stochastic and nonstochastic) that may be expected following radiation exposures.

3.17 Risk Estimates

ICRP 26 gave risk estimates for stochastic effects for the first time although it did not mention any figures for the nonstochastic effects. The whole body mortality risk for radiation-induced cancer was stated to be 5×10^{-2} / Sv or 5×10^{-5} /mSv, an average for both sexes of all ages. The hereditary risk was stated as 0.4×10^{-2}/Sv, as expressed in the first two generations. The report also listed cancer risks for some individual organs of the body. The risks were expressed relative to the whole body (WB) risk. If R_T is the risk for the organ or tissue type T, the total or the whole body risk = $\sum_T R_T$

The relative risk for the tissue T is then given by

$$r_t = R_T / \sum_T R_T = W_T$$

ICRP 26 refers to these relative risk factors as tissue weighting factors, evaluated from the individual organ risks estimated by the commission. The stochastic WB risk includes stochastic risk to all organs including the gonads, which gives the risk of serious hereditary effects in the next two generations.

Example 3.1

The fatal lung cancer risk for a dose of 10 mSv to the lung is 20 deaths in a population of 10^6. The whole body cancer risk for a uniform dose of 10 mSv to all the organs is 165 per million. Calculate the tissue weighting factor for the lung.

$$W_T = R_T/\sum_T R_T = 20/165 = 0.12$$

Example 3.2

W_T for breast tissue = 0.25. The whole body risk factor is 165×10^{-6} /10 mSv. How many cancer deaths can be expected for a breast tissue dose of 5 mSv?

$$W_T = R_T/\sum_T R_T = 0.25$$

$$\sum_T R_T = 165 \times 10^{-6}/10 \text{mSv}$$

$$R_T = 0.25 \times 165 \times 10^{-6}/10 \text{ mSv} = 41 \times 10^{-6}/10 \text{ mSv}$$

So 10 mSv dose to breast tissue would cause 41 deaths in a million, or about 20 deaths can be expected for a 5 mSv dose.

The stochastic risk due to uniform exposure of the whole body is given in Table 3.4, taken from ICRP 60 and ICRP 103.

It can be seen that 5%/Sv is the approximate value of the WB risk for cancer and hereditary effects for the whole population.

Example 3.3

What is the risk of cancer mortality for a radiation worker receiving an annual effective dose of 2 mSv? Compare the risk with the conservative figure of natural incidence of cancer (25% for adults).

$$\text{Cancer mortality risk coefficient} \sim 5\%/\text{Sv or } 5 \times 10^{-2}/1000 \text{ mSv}$$

For 2 mSv, risk of cancer death = $2 \times 5 \times 10^{-2} \times 10^{-3}$ = one death in 10,000. Natural incidence of cancer in the adult population of 10,000

$$= 25 \times 10^{-2} \times 10,000 = 2,500 \text{ deaths.}$$

TABLE 3.4

Stochastic Risk Coefficients at Low Dose and Dose Rate Levels

Exposed Population	Cancer $\times 10^{-2}$/Sv		Hereditary Effects $\times 10^{-2}$/Sv		Total Detriment $\times 10^{-2}$/Sv	
	ICRP 60 1991	ICRP 103 2007	ICRP 60 1991	ICRP 103 2007	ICRP 60 1991	ICRP 103 2007
Whole	6.0	5.5	1.3	0.2	7.3	5.7
Adult	4.8	4.1	0.8	0.1	5.6	4.2

3.18 Radiation Effects (or Risks) to the Embryo and Fetus

Based on comprehensive studies of atomic bomb survivors, studies of patients who had been treated with radiotherapy when they were children, and animal studies, ICRP has come to the following conclusions:

- The pre-conception irradiation of either parent's gonads does not result in increased risk of cancer or malformations in children.

As for effects on the embryo/fetus:

- At doses under 100 mGy, in the preimplantation period of embryonic developments, the probability of lethal effects will be very low.
- There is a true dose threshold of around 100 mGy for the induction of malformations (radiation risks, particularly induction of malformations, are most significant during organogenesis and in the early fetal period, somewhat less in the second semester and least in the third semester).
- For the manifestation of mental retardation, during the most sensitive prenatal period (8–15 weeks post-conception), there is a dose threshold of at least 300 mGy (based on atom bomb survivors data).
- For IQ reduction, following any *in utero* exposure, a dose of <100 mGy, is of no practical significance.
- Radiosensitivity is maximum during the period of major organogenesis.

Regarding cancer risk, there is no threshold for cancer incidence (LNT assumption). The carcinogenic risk is assumed to be constant during the whole period of pregnancy; animal data suggest a strong tendency for carcinogenic effects in the later stages of fetal development. Lifetime cancer risk from prenatal radiation exposure is not very well known. Lifetime cancer risk from childhood radiation exposure is believed to give a good approximation of the prenatal risk.

3.19 Medical Exposure of Pregnant Patients/Pregnant Radiation Workers

While considering the radiation risk to the fetus of a patient due to medical exposure, the magnitude of risk due to natural causes, evidence of negligible risk below 100 mSv, and the risk to the patient for the nonconduct of the procedure, all these factors must be kept in mind. Prenatal doses from properly performed X-ray diagnostic procedures or nuclear medicine diagnostic examinations (which fall well within 100 mSv) present no measurably increased risk of prenatal death, malformation, or mental impairment. The conclusions are also valid for occupational exposure of pregnant women, since the occupational doses are much less compared to medical exposures.

General advice for pregnant patients:

- The risk factors are threshold effects like stunted growth, deformities, retardation, or stochastic effects like cancer, leukemia, and hereditary diseases.

- If the exposure is not to the pelvic area, the scatter dose received by the fetus is minimal and there is no cause for concern.

- In the case of direct exposure of the pelvis, <100 mSv fetus dose is generally safe and there is no reason for the termination of pregnancy based on radiation exposure alone.

- For doses between 100 and 500 mSv, an informed decision must be made based on the circumstances.

- For fetal doses >500 mSv, however, there can be significant fetal injury depending on the dose and the stage of pregnancy. A fetus is particularly sensitive to radiation during its early development, between weeks 2 and 18 of pregnancy.

References

1. R.H. Clarke and J. Valentin, The history of ICRP and the evolution of its policies, *ICRP 109*, 75–110, 2008.
2. J.S. Walker, *A History of Radiation Protection in the Twentieth Century*, University of California Press, Berkeley, CA, 2010.
3. Los Alamos Science Number 23 1995, *A Brief History of Radiation*. http://permalink.lanl.gov/object/tr?what=info:lanl-repo/lareport/LA-UR-95-4005-04 (accessed on April 2016).
4. ICRP (International Recommendations on Radiological Protection), Revised by the International Commission on Radiological Protection and the 6th International Congress of Radiology, *British Journal of Radiology*, 24, 46–53, 1951.
5. ICRP (International Recommendations on Radiological Protection), *Recommendations of the International Commission on Radiological Protection*, British Journal of Radiology, supplement No. 6, London, UK, 1955.
6. ICRP Publication 1, *Recommendations of the International Commission on Radiological Protection*, Pergamon Press, London, UK, ICRP, 1960.
7. *Report of Committee II on Permissible Dose for Internal Radiation*, 1959, ICRP Publication 2, Pergamon Press, London, UK.
8. ICRP, Recommendations of the International Commission on Radiological Protection, ICRP Publication 26, *Annals of the ICRP*, 1(3), 1–53, 1977.
9. ICRP, 1990 Recommendations of the International Commission on Radiological Protection, ICRP Publication 60, *Annals of the ICRP*, 21, (1–3), 1–201, 1991.
10. ICRP, The 2007 Recommendations of the International Commission of Radiological Protection, ICRP Publication 103, *Annals of the ICRP*, 37(2–4), 1–32, 2007.

Review Questions

1. Who discovered X-rays and radioactivity?
2. Who invented fluoroscopy, and in which year?
3. What radiation effects did the early radiologists exhibit?
4. Why is Dr. William Herbert Rollins known as the father of radiation protection?
5. What is erythema dose? Express it in mSv. What is the tolerance dose in terms of erythema dose?

6. Why was the term tolerance dose later replaced by maximum permissible dose?

7. What was the risk faced by radium dial painters?

8. How was the maximum permissible body burden established for Ra?

9. What two important radiation effects, other than skin reactions, became of concern while studying the radiation effects?

10. What are somatic and genetic effects?

11. What are stochastic and nonstochastic effects?

12. What is the significance of LNT hypothesis in radiation protection?

13. What are early and late effects?

14. What effects can be expected as a result of radiation-induced chromosome mutations?

15. Assuming a cancer mortality risk of 5% per Sv, how many of a population of 10,000 radiation workers, receiving a whole body dose of 2 mSv, are expected to die of radiation-caused cancer? How does this risk compare with risk of people working in other so-called safe industries?

4

Radiation Protection Quantities, Units, and Standards

4.1 Introduction

The dosimetric quantities exposure and absorbed dose have been in use since the 1920s and 1950s, respectively. They quantify the charge released (exposure) and the amount of energy locally dissipated (absorbed dose) in a physical medium due to the radiation interactions occurring in the medium. Exposure continued to be used in the radiation protection field to monitor the radiation levels in workplaces and the radiation dose received by the staff in workplaces. In the 1970s, special dosimetric quantities were defined (ambient dose equivalent and personal dose equivalent) for area monitoring and personal monitoring, respectively, and the methodology for realizing these quantities was developed by the International Commission of Radiological Protection (ICRP) and the International Commission on Radiation Units and Measurements (ICRU). This chapter gives a brief introduction to these developments.

4.2 Physical Dosimetric Quantities

The use of physical dosimetric quantities preceded radiation protection quantities, and they are still the fundamental quantities and standards on which all radiation protection quantities and measurements are based. So, the basic dosimetric quantities and their standards will be discussed below in some detail before discussing the protection quantities.

In Chapter 2, the physical dosimetric quantities (kerma and absorbed dose) were defined as arising naturally from the interaction of radiation with matter. This section will discuss how these quantities are measured. In order to measure any quantity, it is necessary to define a unit for the quantity. Standards are then established, as per the definition of the unit, at the national level in a primary standards laboratory, and the unit is then disseminated through one or more regional calibration laboratories to the users. For instance, NIST (USA), NPL (UK), PTB (Germany), and NRC (Canada) are well-known primary standards dosimetry laboratories [1–4] in the world for establishing the primary standards for the dosimetric quantities—exposure, air kerma, absorbed dose, air kerma strength (or reference air kerma rate), etc,. Since there are many countries without adequate expertise to set up primary standards dosimetry laboratories (PSDLs), the International Atomic Energy Agency (IAEA), Vienna, Austria, established a network of secondary standards dosimetry laboratories (SSDL) and most of the member countries are members of this SSDL network. IAEA disseminates the

Dosimetric quantities	SI units (name)	Primary standard
Exposure	C/kg(–)	FAIC/GCIC
Air kerma	J/kg(Gy)	-do-
Absorbed dose to graphite	J/kg(Gy)	Graphite calorimeter
Absorbed dose to water	J/kg(Gy)	Water calorimeter

FAIC : Free air ionization chamber
GCIC : Graphite cavity ion chamber

FIGURE 4.1
Dosimetric quantities, their units, and the primary standards.

standards from the PSDL to the members of the SSDL network and these SSDLs in turn disseminate the units to the users in their countries either directly or through regional calibration laboratories.

Figure 4.1 lists the dosimetric quantities, their units, and the primary standards that represent the units for the quantities.

4.3 Primary Dosimetric Standards

4.3.1 Absorbed Dose Standards

The quantity of interest in radiation oncology is tumor dose but since it cannot be directly measured and is not a suitable quantity for dissemination, "absorbed dose to water" is the most suitable quantity for dissemination. The tumor dose can then be determined by applying patient inhomogeneity corrections and relative dose data to the absorbed dose, measured in a water phantom at a reference depth. This is the standard procedure adopted in radiation oncology, and requires the establishment of a water calorimeter for measuring absorbed dose to water. Some PSDLs prefer to establish a graphite calorimeter and transfer the dose to the water medium through standard correction factors. Due to the complexity of the procedures involved, the water calorimeter has been established in the PSDLs only in the last couple of decades. Many national dosimetry laboratories do not have a calorimetric standard of absorbed dose, but still maintain only an exposure or air kerma standard and disseminate these quantities to the users. Hence in Figure 4.1, both the standards have been mentioned for ^{60}Co beams.

Absorbed dose measurements are not of any interest in radiation protection, so this quantity will not be discussed any further in this chapter.

4.3.2 Exposure/Air Kerma Standards

Free-air ionization chambers and graphite cavity ionization chambers are being maintained as primary standards of exposure or air kerma in the PSDLs, in the respective photon energy ranges. These quantities are widely used in clinically dosimetry, but their use in radiation protection has considerably declined in the last decade in favor of the new protection quantities—dose equivalents (DEs). However, these protection dosimetry quantities are not directly measurable and have to be realized by relating them to exposure or air kerma by Monte Carlo computations. So, exposure and air kerma are also of direct interest in radiation protection for realizing the protection quantities. The quantities exposure and air kerma will be discussed here in brief before describing the protection quantities and later relating them to exposure or air kerma.

Chapter 2 explained how collision kerma and absorbed dose appear as a consequence of the interaction of radiation with matter. However, from the measurement point of view, it is easier to measure ionization produced by the particles rather than the particle energy. Measurement of ionization dates back to the 1900s and so historically the first dosimetric unit defined was in terms of ionization produced in air. Collision air kerma at a point in air represents the collision part of kinetic energy of the charged particles, produced by photons at the point in question. These charged particles also lose energy by ionizing air when they come to the end of their ranges. The unit for the quantity exposure was defined in the later part of the 1920s.

4.3.2.1 *Primary Standard of Exposure or Air Kerma at kV Energy Range*

Free-air ionization chamber (FAIC) is the primary standard of exposure or air kerma for kV X-ray beams. Figure 4.2 illustrates the design of a FAIC. It is a simple parallel plate design with a guard but some FAICs have a cylindrical design.

More details on its construction and working details can be found in other text books (e.g., Attix's book [4]). The basic principles are very briefly explained below.

The measurement volume δV is defined by the length of the collecting plate and the field lines (i.e., by the shaded volume in Figure 4.2). The ionization occurring in the guard region along the axial direction is not collected by the central electrode, since we have charged particle equilibrium (CPE) condition in this direction and the outflow of electrons and their energies are compensated by the inflow of electrons and their energies. However, along the radial direction, the radiation field is confined only to the defined volume δV and so the ionization along the radial direction must be completely collected by the collecting electrodes. FAIC elegantly demonstrates air kerma measurement under both conditions, with CPE along the axial direction, and without CPE along the radial direction.

Exposure at the center of the FAIC (denoted by a small circle), X(c) is given by

$$X(c) = (\delta Q / \delta m) = (\delta Q / \rho \delta V)$$

$$\delta V = A_c \times L$$

However, it is difficult to measure the collecting volume cross section at the center of the FAIC. It is easier to measure A_P at the FAIC entry point P with much higher accuracy. By referring the exposure measurement point to P,

$$X(P) = (\delta Q / \rho \ A_P \ L)$$

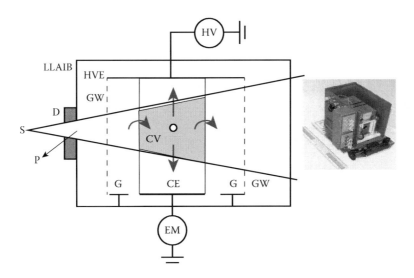

HV: High voltage   CV: Collecting volume
HVE: HV electrode   CE: Collecting electrode
G: Guard   GW: Guard wires
S: Source   P: Point of measurement
LLAIB: Lead-lined A1 box   D: Diaphragm

FIGURE 4.2
Primary standard of exposure—air kerma in kV range.

To realize air kerma in a PSDL, the measurement is taken the same way but uses the following equation to derive air kerma:

$$K_a(P) = X \ (W/e)/(1–g)$$

For instance, if the FAIC measures in a PSDL, an exposure rate of 30 R/min at point FAIC entry point P, along the beam central axis, it can be said that as an air kerma standard, the FAIC measured at the point P gives an air kerma rate of 30 × 0.879 cGy/min.

Example 4.1

How will you show that the exposure measurement in a FAIC refers to the exposure at the entry point and not at its center, where the ionization is measured?
Referring to the center of the collecting volume,

$$X(c) = (\delta Q/\delta m) = (\delta Q/\rho \delta V) = \delta Q/\rho \ A_c \ L$$

Since exposure (or the radiation intensity) varies as per the inverse square law,

$$X \ \alpha \ 1/d^2$$

$$X(P) \times (d_P)^2 = X(c) \times (d_c)^2$$

$$X(P) = X(c) \times (d_c)^2/(d_P)^2$$

where d_P and d_c refer to the distance of the source from the point P and from the center of the sensitive volume, along the beam central axis, respectively. (See Figure 4.3).

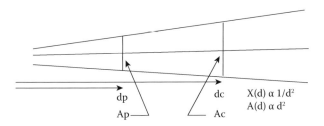

FIGURE 4.3
Beam geometric divergence makes $X \propto 1/d^2$ & $A \propto d^2$.

On the other hand, the area varies as d^2

$$A_P/A_c = (d_p)^2/(d_c)^2$$

Using this relation in the above equation for X(P),

$$X(P) = X(c) \times (A_c/A_P)$$

$$= (\delta Q/\rho \; A_c \; L) \times (A_c/A_P)$$

$$= (\delta Q/\rho \; A_P \; L)$$

which was the equation used to measure the exposure using the FAIC.

4.3.2.2 *Primary Standard of Exposure or Air Kerma at MV (^{60}Co) Energy Range*

It may be noticed that the sensitive volume of FAIC must be surrounded by air on the axial direction for CPE condition and the electrons along the radial direction must spend all their energies in ionization before reaching the collecting electrode (since the radiation field is confined to δV and no CPE compensation can occur in the radial direction). This requires that the dimensions of the FAIC be roughly twice the range of the electrons. So, beyond a few hundred kV, a FAIC will become too large to manage with increasing corrections and their uncertainties, deviating a great deal from "point measurements." For higher photon energies, the dimensions of the FAIC can be minimized (by the ratio of their densities) if air can be replaced by an air equivalent material, but exact air equivalence, for all photon energies, is difficult to obtain. Instead, for higher energy photons (e.g., ^{137}Cs or ^{60}Co γ rays), a graphite ion chamber, rather close to air in terms of effective atomic number in the photoelectric region, is used for exposure realization and a non-air equivalency correction is applied to this chamber to derive exposure or air kerma.

A graphite ion chamber works on Bragg–Gray (B–G) cavity principle, which can be used to relate the chamber ionization to graphite kerma. The B–G cavity principle states that for a small gas cavity (i.e., electron ranges long compared to cavity dimensions) surrounded by a wall material, when the *wall spectrum is not perturbed by the cavity* (i.e., all the wall electrons spectrum is able to cross the cavity with the cavity just sampling the spectrum and there is no electron generation in the cavity by direct photon absorption), the wall dose, in the vicinity of the cavity, and the cavity dose are related by the B–G equation, given below.

$$D_{wl} = D_{air} \; (_mS_{wl.cav}) \; \text{where} \; _mS_{wl,cav} = {_mS_{wl}}/{_mS_{cav}}$$

S refers to spectrum weighted average values of the stopping power of wall and cavity materials (see reference [4] for more details). D_{air} is the dose in the air cavity given by $\delta\varepsilon/\delta m = \delta Q/\delta m \times (W/e)$ or number of ions per unit mass $\times$ average energy required to produce an ion pair in air, W.

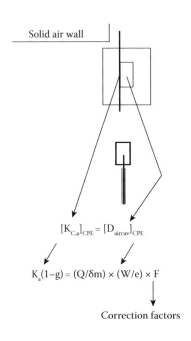

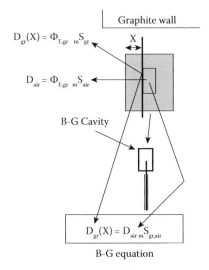

FIGURE 4.4
Bragg–Gray cavity principle.

For a graphite cavity ionization chamber (GCIC), the B–G equation can be written as

$$D_{gr}(x) = D_{air}\, (_mS_{gr.air}) = (\delta Q/\delta m)\, (W/e)\, (_mS_{gr.air}) = (\delta Q/\rho\delta V)\, (W/e)\, (_mS_{gr.air})$$

Figure 4.4 illustrates the concept of B–G cavity principle.
When x = CPE thickness (e.g., for ^{60}Co, 500 mg/cm^2 thickness is about the maximum range of cobalt electrons in graphite, giving the CPE thickness).

$$D_{gr}(x) = K_{c,gr}(x)$$

The above equation requires a small correction β to correct for the finite attenuation of the photons in the wall thickness x.

$$D_{gr}(x) = K_{c,gr}(x)\, \beta$$

This correction is important difficult to evaluate and is one of the reasons for not extending this concept to higher energies to realize primary standards. The correction is of the order of 0.5% at Co-60 energy and can be neglected for radiation protection purposes.
Under equilibrium thickness,

$$K_{c,a}(x) = K_{c,gr}(x)\, (_m\mu_{en})_{a,gr} \text{ or } K_a = \{K_{c,gr}(x)/(1-g)\}\, (_m\mu_{en})_{a,gr}$$

Combining the above three equations,

$$K_a = (\delta Q/\rho\delta V)\, (W/e)\, (_mS_{gr.air})\, (_m\mu_{en})_{a,gr}/(1-g)\, \pi k_i$$

$$K_a = (\delta Q/\rho\delta V)\, (W/e)\, k_{nae}/(1-g)\, \pi k_i$$

where $k_{nae} = (_mS_{gr.air})\, (_m\mu_{en})_{a,gr}$ is known as the non-air equivalency correction factor and can be thought of as a correction factor to the graphite chamber, saying to what extent it is not air equivalent. If it is exactly air equivalent, $k_{nae} = 1$ and it becomes a FAIC, with a solid air wall; πk_i is a product of small corrections that need be applied for realizing a primary standard.

The GCIC is also an exposure standard and measures exposure using the following equation:

$$X(P) = K_a(P) \, (W/e)/(1-g)$$

A graphite PP chamber approximates most closely to a B–G cavity while a cylindrical graphite cavity chamber has a small cylindrical electrode inside the cavity for ion collection, which slightly perturbs the B–G behavior. The primary standards of air kerma/exposure realized in a PSDL, based on the above principles, are shown below from the National Metrology Institute of Japan (https://www.nmij.jp/english/org/lab/5/) and the Australian Radiation Protection and Nuclear Safety Agency (http://www.arpansa.gov.au/services/calibration/cavity.cfm) (Figure 4.5).

4.3.3 Isotopic Neutron Source Standards

There are various sources for neutrons [5]. However, only isotopic neutron sources (Figure 4.6) are generally used in calibration laboratories for the calibration of neutron monitors. They are small in size, relatively less expensive, maintenance free, and have highly accurate and reproducible output (yield or emission rate) to facilitate neutron monitor calibrations.

Japanese primary standard

Australian primary standard

FIGURE 4.5
GCIC, primary standards of air kerma at Co-60 energy.

Am-Be sources

^{252}Cf sources

FIGURE 4.6
Alpha and spontaneous fission neutron sources.

There are three main types of isotopic neutron sources:

1. Transuranium element that emits neutrons by spontaneous fission
2. α emitters that produce neutrons by (α, n) reactions in a low Z element
3. γ emitters that produce neutrons by (γ, n) reactions in a low Z element

These sources are generally referred to as spontaneous fission neutron source, alpha neutron source, and gamma or photo neutron source, respectively. Californium (^{252}Cf) is the most common transuranium radioactive element that emits neutrons by spontaneous fission reactions. It is produced in high flux nuclear reactors by neutron absorption. It mostly decays by α emission (97%) and only about 3% of the decays result in neutron emission. Its neutron spectrum resembles the reactor fission spectrum with a mean energy of around 2.1 MeV. A moderated ^{252}Cf source is also used for the calibration of neutron survey meters. Here, the source is placed inside a 30 cm diameter stainless steel sphere filled with heavy water. The sphere is covered with a thin Cd shell. The whole system becomes a moderated ^{252}Cf source. The degraded neutron spectrum of the neutrons coming from the sphere more resemble the field conditions in the reactor environment, and hence are a more suitable source for neutron survey meter calibrations. The reason for using D_2O (and not the easily available water) is to avoid neutron loss by neutron capture that occurs in H_2O.

To produce neutrons by (α, n) or (γ, n) reactions, an α or γ emitting radioisotope is intermixed with the low Z target material (having low Coulomb barrier for the alphas). The α emitters historically used in neutron sources are ^{226}Ra, ^{210}Po, ^{239}Pu, and ^{241}Am, and the γ emitter most commonly used is ^{124}Sb. Ra sources are not being used these days due to the radium hazard and high gamma background. The low Z nuclides used for neutron sources are historically B, Be, or Li, which have low Coulomb barriers and an α with moderate energy will be able to enter and excite these nuclides. Be is the most common nuclide used due to its higher neutron yield compared to other low Z target nuclides. α must have sufficient energy to overcome the Coulomb barrier or the reaction threshold (in the case of endothermic nuclear reactions), whichever is higher, to cause the (α, n) reactions in the target nuclides. To have close contact with Be, powdered Be metal is mixed with an oxide of alpha emitter and then compressed into a cylindrical source. To prevent the risk of any radioactivity escaping into the ambient, the sources are doubly encapsulated in stainless steel. In the case of a Pu–Be source, Pu can be alloyed with Be to produce a solid (α, n) Pu–Be source. Here, the risk of escape of any radioactive particles is small. Yet, as a precautionary measure, Pu–Be or ^{241}Cf sources are also encapsulated in stainless steel. One concern with the alpha neutron sources is the buildup of helium gas inside the capsule. An empty space is provided inside the inner capsule to allow the buildup of He gas within it. In the case of (γ, n) sources, intimate contact between the γ emitter and Be is not necessary since γ's are very penetrating. Here, the neutron-emitting target surrounds the gamma emitting core. Both the gamma-emitting core and the neutron-emitting envelope are encapsulated in Al. In the case of an Sb–Be source, the core is natural antimony that is activated in a reactor to produce ^{124}Sb by neutron absorption. The ^{124}Sb core is surrounded by a Be shell to form the Sb–Be source.

In (α, n) and (γ, n) sources, neutrons are produced in the following reactions with ^{9}Be:

$$(\alpha, n) \text{ sources: } {}^9\text{Be} + {}^4\text{He} (\alpha) \rightarrow {}^{12}\text{C} + {}^1\text{n} + \gamma$$

$$(\gamma, n) \text{ sources: } {}^9\text{Be} + \gamma (1.7 \text{ MeV}) \rightarrow 2 \, {}^4\text{He} + {}^1\text{n} (\text{few MeV})$$

The threshold for the (γ, n) reaction in ^{9}Be is 1.67 MeV. There are only very few gamma emitters (^{124}Sb, ^{90}Y, and ^{226}Ra) that have a long half-life with emitted gammas above this threshold. Of these, the less expensive sources are Sb–Be and Y–Be which are commonly used. One disadvantage of the photoneutron sources is the high gamma background compared to the alpha neutron sources. Photoneutron sources can emit near monoenergetic neutrons when the emitter emits monoenergetic gammas. This is in contrast to the alpha neutron sources which always emit a spectrum of neutron energies. One more characteristic that distinguishes these two types of sources is the low energy (in keV) of the photoneutrons compared to the alpha neutron sources. Alpha neutron sources are the ones that are most commonly used compared to other neutron sources.

The emission rate (also known as the yield) of a typical ^{252}Cf source is 10^7 to 10^9 n/sec and for (α, n) neutron sources, it is about 10^6 to 10^8 n/sec. The yield and half-life of the α or γ, n neutron sources are governed by the activity and half-life of the α or γ emitters used in the neutron source. So, the neutron yield is not a constant with respect to time and a decay correction must be applied to obtain the yield on a specific date. In the case of Ra–Be and Am–Be neutron sources, the half-lives of the two radioisotopes are 1600 and 432 years, respectively, and so the decay is negligible, and they have the same output, for all practical purposes. On the other hand, ^{252}Cf has a half-life of 2.56 years and so its output slowly changes with time. Since the yield increases with the source activity, it is usually expressed in terms of specific yield—yield or emission rate per unit activity expressed in Ci or Bq. Since 1 Ci = 37 GBq, specific strengths of curie sources are specified per GBq and mCi sources are specified per MBq. Whichever convention is adopted, it is easy to convert from one unit to another. In the case of a ^{252}Cf source, the specific yield is expressed per unit mass (i.e., per g or per mg) instead of per unit activity.

4.3.4 Concept of DER Constant (Γ_n)

The DER constant is defined as the DER at 1 m from a point source of unit activity in free space (in the absence of air). Using this definition, the DER at any distance d from the source can be determined from the following equation:

$$DER = dH/dt = A\ \Gamma_n/d^2$$

For radiation protection purposes, the above equation can also in low scatter geometry (e.g., at the center of a large room more than 1.5 m height from the floor). The DER constant is also sometimes referred to as specific DER constant since it refers to the DER per unit activity. The equation can be used for shielding purposes, say for deciding on the thickness of a storage container to house a neutron source or for room shielding.

The source activity and the specific DER constants are only approximate values. The neutron yield does not depend just on the activity of the source. The construction of the alpha neutron source and the Be-alpha emitter ratio also play an important part. For radiation calibration purposes, the source strength (emission rate) of the neutron sources must be accurately determined using a primary standardization method, as explained briefly in the next section. Knowing the accurate value of the source strength S, the neutron fluence rate at a distance d from the source in a low scatter geometry, can be determined using the inverse square law variation of fluence with distance.

$$\Phi(d) = S\ \Delta t/(4\ \pi d^2); \quad d\Phi(d)/dt = S/(4\ \pi d^2)$$

Δt is the duration of exposure integral measurement. The above equation characterizes the neutron field in terms of fluence and fluence rate.

TABLE 4.1

Alpha and Fission Neutron Source Characteristics

Source	Half Life	Mean Energy (MeV)	Specific Neutron Yield, S (#/MBq.s)	Specific Neutron DER Constant Γ_n (Sv m³/MBq·h)
^{252}Cf	968 days	2.4	$(2.3 \times 10^{12})^a$	22
^{252}Cf+D₂O	968 days	0.54	2×10^{12}	5.2
^{241}Am–B(α, n)	432.7 years	2.8	16	1.8×10^{-10}
^{241}Am–Be(α, n)	432.7 years	4.4	66	7×10^{-10}

Note: DER, dose equivalent rate.
^{a} For ^{252}Cf, the yield and dose rate constant are per g and not per Bq.; ^{252}Cf + D₂O refers to a moderated source (source in a moderating sphere, shielded with 1 mm Cd).

TABLE 4.2

Photoneutron Source Characteristics

Source	Half Life	Mean Energy (keV)	Specific Neutron Yield, S (#/s. MBq)	Specific Neutron DER Constant Γ_n (Sv m²/MBq·h)
^{124}Sb	60 days	26	1.35×10^8	6×10^{-11}

Note: DER, dose equivalent rate.

The important characteristics of the neutron sources are given in Tables 4.1 and 4.2.

Example 4.2

Calculate the DER at 1 m distance from a 20.6 Ci Am–Be source ($\Gamma_{n,\ Am-Be} = 7 \times 10^{-10}$ Sv m²/MBqh)

$$1 \text{ Ci} = 37 \text{ GBq or } 37 \times 10^3 \text{ MBq}$$

$$dH/dt \text{ at } 1 \text{ m} = A\ \Gamma_n/d^2 = A\ \Gamma_n \text{ Sv/h MBq}$$

For a 20.6 Ci source,

$$dH/dt \text{ at } 1 \text{ m} = (20.6 \times 37 \times 10^3) \times 7 \times 10^{-10} \text{ Sv/h}$$

$$= 0.53 \text{ mSv/h}$$

Example 4.3

^{252}Cf source output is usually expressed in terms of mass instead of activity. Given for ^{252}Cf ($T_{1/2} = 2.645$ Y; Avogadro no. $= 6.022 \times 10^{23}$; mass number A = 252).

1. What is the activity of a ^{252}Cf source of mass 1 mg?
2. If 3% of the disintegrations result in spontaneous fission and about 3.7 neutrons are emitted per fission, calculate the neutron emission rate of 1 mg of ^{252}Cf.
3. Also calculate the DER at 1 m from the source ($\Gamma_n = 22$ Sv m²/g h)

4. Californium activity: dN/dt (Bq) $= N \lambda$ s^{-1}; $N = 0.693/T_{1/2}$ s^{-1}

Atoms in 1 mg of mass: $N = (N_A/A) \times 10^{-3}$

$$= (6.022/252) \times 10^{-3} = 2.39 \times 10^{18}$$

(A is the mass number. A gram will contain Avogadro number of atoms, N_A).

$$T_{1/2} = 265 \text{ Y} = 265 \times 365 \times 24 \times 3600 \text{ s} = 8.341 \times 10^7 \text{ s}$$

$$\lambda = 0.693/T_{1/2} = 8.308 \times 10^{-9} \text{ s}^{-1}$$

$$dN/dt = N \lambda = 2.39 \times 10^{18} \times 8.308 \times 10^{-9} = 1.986 \times 10^{10} \text{ Bq}$$

In curies: $1.986 \times 10^{10}/3.7 \times 10^{10} \approx 0.5$ Ci

So, the activity of a mg of ^{252}Cf source is about 0.5 Ci.

5. The emission rate = (disintegrations/s) × (fission fraction) × (neutrons/fission)

$$= 1.986 \times 10^{10} \times 0.03 \times 3.8$$

$$= 2.3 \times 10^9 \text{ n/s per mg.}$$

6. DER at 1 m from 1g ^{252}Cf source $= 22 \times 10^{-3}$ Sv/h $= 22$ mSv/h ($= A \Gamma_n$ 22 Sv m^2/g h).

Example 4.4

The specific DER constant of a 1.5 Ci Sb–Be source is 0.05 μSv m^2/GBq h. Calculate the DER, dH/dt, at 1 m from the source.

$$dH/dt \text{ at } 1 \text{ m} = (A \Gamma_n/d^2)_{\text{Sb–Be}} = (A \Gamma_n)_{\text{Sb–Be}} \text{ for } d = 1\text{m}.$$

Since Γ_n is given in terms of GBq, we must convert the activity given in Ci into GBq.

$$1 \text{ Ci} = 37 \text{ GBq}; 1.5 \text{ Ci} = 37 \times 1.5 \text{ GBq}$$

$$dH/dt \text{ at } 1 \text{ m} = [0.05 \text{ μSv}/(\text{h GBq})] \times (37 \times 1.5 \text{ GBq}) = 2.8 \text{ μSv/h}$$

Example 4.5

What is the approximate yield of a 2 μg point source of ^{252}Cf ? $\Gamma_n = 2.3 \times 10^{12}$ n/s.g. The yield of a 2 μg point source of ^{252}Cf $= 2 \times 2.3 \times 10^6 = 4.6 \times 10^6$ n/s.

4.3.5 Primary Standard of Neutron Emission Rate

Manganese sulfate bath is the primary standard for measuring the neutron emission rate for radionuclide neutron sources [6]. The standard consists of a spherical stainless steel container (about a meter in diameter) filled with an aqueous solution of $Mn(SO)_4$. The neuron source is placed at the center of the container. The neutrons are thermalized in the medium and are captured by ^{55}M—^{55}Mn(n, γ)^{56}Mn—and to very small extent by other nuclides (H, S, and O). A small fraction of the neutrons leaks out of the system and a small fraction is attenuated in the source itself. Neutrons are also lost in capture reactions in nuclides other than ^{55}Mn. The solution is continuously stirred and pumped in a closed

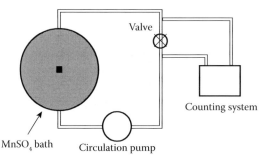

FIGURE 4.7
Manganese bath technique.

circulation past a detection system (NaI scintillation system) which measures the activity of ^{56}Mn (see Figure 4.7). The activity is measured when it reaches saturation (after an irradiation time of 15–20 hours).

$$\text{Emission rate } Q = \text{saturation activity of } ^{56}\text{Mn} \times \pi \, k_i$$

where $\pi \, k_i$ refers to the product of minor correction factors that must be applied to the measurement.

A variety of neutron sources, in the form of a capsule, are calibrated this way and supplied to calibration laboratories for the calibration of neutron monitors.

4.4 ICRP Radiation Protection Quantities

While the dosimetric quantities are useful in clinical dosimetry, they were found wanting in the radiation protection field, where the biological effects depend not only on dose but also on the pattern of dose distribution in the tissue.

ICRP therefore defined a radiation protection quantity, DE, by weighting the tissue dose by a risk factor that relates the dose to the degree of harm to the body tissues.

4.4.1 Concept of Dose Equivalent

The dose equivalent, H, was defined (ICRU, 1970) as:

$$H = D \times Q \times N$$

where D and Q are the dose and quality factor defined *at a point* in tissue. "N" is the product of all other modifying factors due to other influencing factors like dose rate, fractionation, biological end point of interest, which can change the biological response of a biological system. Generally, for exposures involved in radiation protection, N is taken to be unity. H (or DE) is then averaged over the tissue or organ. So, for radiation protection purposes, the Q factor takes into account the biological effectiveness of the radiation (to cause radiation damage). H is also the first measure of detriment introduced in ICRP 26 (1977) [7].

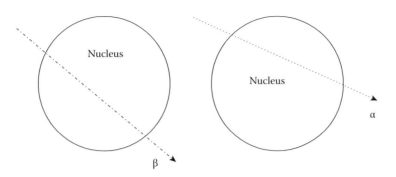

FIGURE 4.8
Dose distribution pattern in a cell nucleus.

For a given physical dose, the differences in biological effectiveness of different radiations arise as a result of the different densities of ionizations or energy depositions (known as the linear energy transfer, LET) produced by these radiations. For example, the ionization produced by alpha radiation is much denser (high LET, LET_H) compared to ionization produced by betas or photons (low LET, LET_L), in the cells traversed by radiation (Figure 4.8). Denser ionization results in double strand breaks of chromosomes, leading to a lower probability of repair and higher probability of tissue damage. So for the same energy absorption, alpha (or proton or neutron) can cause several times more damage to tissues compared to betas (or X-rays or gamma rays). The Q factor accounts for this behavior. Since Q has no dimensions, D and H have the same units—J/kg. So, to distinguish between D and H, the two quantities were given different names—Gray (Gy), and Sievert (Sv). To correlate to biological harm, H must be measured, and not D. Though H is, in principle, immeasurable, it can be determined from measured dose and Q factors, or measured fluence and dose-to-fluence conversion factors.

For a long time, radiation protection monitors were calibrated in terms of exposure or air kerma and many countries still continue to do so. When demand arose for DE calibration of monitors, which is the relevant quantity for monitoring staff doses, the calibration fields characterized in terms of exposure or air kerma rate (AKR) were simply converted to DE using the recommended Q factors (e.g., photons and betas 1, neutrons 10), and the monitors were calibrated using these DE values. More precise definition of DE quantities, namely ambient dose equivalent and personal dose equivalent came into the field much later, as will be described in another chapter. To correlate organ doses to the organ risk, ICRP defined the equivalent dose concept and tissue weighting factors, to characterize organ doses and the whole body dose. These will be discussed in the following sections.

4.4.1.1 *History of Radiobiological Effectiveness*

To account for the variation in biological effectiveness of different radiation types in causing tissue damage for the same organ absorbed dose, a quantity called the RBE (radio biological effectiveness) dose was introduced in the 1950s.

$$\text{RBE dose (mSv)} = \text{organ dose (mGy)} \times \text{RBE}$$

This quantity was later named the dose equivalent.

RBE was defined with respect to a low LET (LLET) reference radiation, which is 250 kVp X-rays, or ^{60}Co radiation. (LET actually refers to the charged particles since uncharged particles like X-rays or neutrons do not lose energy continuously as do charged particles. Here, LET of neutrons or X-rays actually refers to the LET of the spectrum of charged particles produced by these radiations while interacting in the medium).

$$RBE_{HLET} = D_{LLET}/D_{HLET}$$

for a chosen biological end point (e.g., cell killing).

ICRP 2 (1959) defined [8] the values of RBE for different particle types as in Table 4.3.

For equal damage to the organ the RBE doses must be compared, and not the organ doses.

Since RBE is a more precise quantity in radiobiology and its value depends on the end point chosen in the experiment, the term was replaced by Q factor in the later recommendations. The Q value depends on the LET (linear energy transfer or ionization density) of the charged particles in water, as shown in Table 4.4, which gives the LET dependence of Q values (ICRP 2).

Because the distribution of LET is not always known, approximate Q values were recommended for typical radiations in ICRP 2, as shown in Table 4.5.

NOTE: The special units for absorbed dose and dose equivalent are rad and rem, respectively, rem representing "Röntgen equivalent man." "rad" is not an acronym for anything (as pointed out by L.S. Taylor) though it is generally believed it stands for "radiation absorbed dose."

TABLE 4.3

ICRP 2 RBE Values

Radiation	RBE
γ, β^-, β^+, e^-	1
Low energy β^-, β^+, e^- ($E_{max} \leq 0.03$ MeV)	1.7
Alphas	10
Recoil nuclides	20

Note: RBE, radio biological effectiveness.

TABLE 4.4

LET Dependence of Q (ICRP 26)

L_α in Water (keV/μm)	Q
≤ 3.5	1
7	2
23	5
53	410
≥ 175	20

Note: LET, linear energy transfer.

TABLE 4.5

Q Values for Different Radiations

Radiation	Q
X, γ, electrons	1
Neutrons, protons, and singly charged particles of rest mass greater than one amu of unknown energy	10
α particles and multiply charged particles (and particles of unknown charge) of unknown energy	20

I rad is defined as an energy absorption of $100 \text{ ergs/g} = 10^5 \text{ ergs/kg} = 10^{-2} \text{ J/kg}$. The unit of absorbed dose in SI units is Gray (Gy) and 1 Gy = 1 J/kg.

From the above two expressions it can be seen that

$$1 \text{ rad} = 10^{-2} \text{ Gy or 1 cGy.}$$

The unit of dose equivalent, H, in SI units is the Sievert (Sv).

$$1 \text{ rem} = 10^{-2} \text{ Sv or 10 mSv}$$

The use of special or unconventional units is strongly discouraged and the student should learn to think only in terms of the SI units from the beginning.

In its next recommendation (ICRP 60, 1991) ICRP made a few changes [9]. The absorbed dose was not defined at a point in tissue, as in 26, but redefined as an average dose to the tissue or organ, and the tissue detriment was related to the average organ absorbed dose. For penetrating radiations, the dose distribution is sufficiently uniform in the organs and tissues, and for radiation protection purposes, the organ dose as defined in ICRP 60 is a good indicator of organ or tissue detriment. This quantity was redefined as equivalent dose to the tissue or organ, H_T (instead of the earlier quantity dose equivalent). However, the point quantity DE was still very much in use as a suitable quantity for calibrating radiation protection instruments, until new quantities were introduced in the field.

4.4.2 Mean Absorbed Dose to Organ

The organ dose, D_T, is defined as the mean absorbed dose in a specified tissue or organ T of the human body and is given by

$$D_T = \varepsilon_T / m_T$$

where m_T = organ mass and ε_T = total energy imparted to the organ.

NOTE: While absorbed dose is a point quantity in clinical dosimetry defined for infinitesimal mass, in radiation protection, organ doses are macroscopic quantities with doses averaged over the organ volume, which are better suited for evaluating the dose–effect relationships, or organ risks.

4.4.2.1 Radiation Weighting Factor (W_R)

The quality factor (or older RBE) was redefined as the radiation weighting factor, W_R, R referring to the radiation type. The radiation weighting factor W_R accounts for the effectiveness of the radiation to cause biological harm in the tissue of interest, for radiation of type R, normalized to the detriment of 200 keV reference photon radiations. So, the radiation type weighted organ or tissue dose becomes the equivalent dose, H_T.

The recommended values of W_R, for different types and energy of radiation, are given in Table 4.6 [10]:

The W_R values are based on experimental data of the RBE, and are expressed relative to photon irradiation. So, W_R is a dimensionless number (≥ 1) and depends on the LET of the radiation (see Figure 4.9). It is important to remember that for photons, electrons, and positrons of all energies, the radiation weighting factor has the value 1; for alpha radiation the

TABLE 4.6

Radiation Weighting Factors

Type of Radiation	Value of w_R
X-rays	1
γ rays	1
β particles	1
Neutrons	2.5 to 20 (continuous curve depending on radiation energy)
Charged protons and pions	2
α particles	20

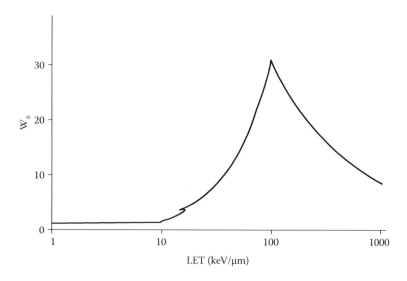

FIGURE 4.9
LET dependence of radiation weighting factor.

value is 20; for neutron radiation, the value is energy dependent and varies between 5 and 20. (In the past this factor was known as the quality factor, Q).

For radiation types and energies not listed in Table 4.6, the following curve can be used to calculate a weighting factor from the LET of the radiation.

Typical LET values for common radiation qualities are given in Table 4.7.

4.4.2.2   Neutron W_R Values

W_R of neutrons are based on the RBE of neutrons following external exposure. The weighting factor is strongly energy dependent because of the variation in LET of the charged particles produced—the recoil protons from hydrogen, other charged particles from other nuclei, and nuclear spallation reaction products produced in the medium depending on the neutron energies. To determine the neutron equivalent dose, the neutron spectrum and the energy dependence of W_R must be known.

The recommended values of W_R for different neutron energies, are given in Table 4.8 (ICRP 60 and ICRP 103).

The functional dependence of W_R on neutron energy as per ICRP 60 and ICRP 103 [9,10] is shown in Figure 4.10.

The formula for calculating W_R is given in ICRP 103. The spectral data of neutrons is required for computing the neutron W_R values for a spectrum.

TABLE 4.7

Typical LET Values for Different Radiations

Radiation	LET (keV/μm)
250 kVp X-rays	2
CO-60 beam	0.3
14 MeV neutrons	12
10 MeV protons	4.7
150 MeV protons	0.5
2.5 MeV α (heavy CP)	166

Note: LET, linear energy transfer.

TABLE 4.8

Neutron Weighting Factors (ICRP 103, 2007)

Neutron Energy Range	W_R
<10 keV	5
10 keV to 100 keV	10
>100 keV to 2 MeV	20
>2 MeV to 20 MeV	10
>20 MeV	5

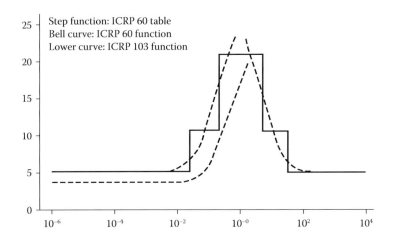

FIGURE 4.10
Functional dependence of W_R on neutron energy.

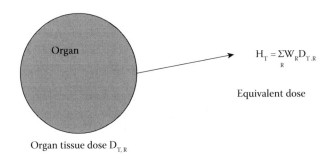

FIGURE 4.11
Concept of organ dose and equivalent dose.

4.4.3 Equivalent Dose

The equivalent dose to an organ, when it is exposed to more than one type of radiation, is computed as follows (Figure 4.11):

When only one type of radiation is involved,

$$H_T = W_R \, D_{T,R}$$

Example 4.6

The dose to an organ is 1 Gy from X-ray exposure and 1 mGy from neutron exposure. Calculate the resultant equivalent dose to the organ. ($W_R = 1$ and 10 for X-rays and neutrons, respectively).

$$H_{T,x} = 1 \text{ mGy} \times 1 = 1 \text{ mSv} : H_{T,n} = 1 \text{ Gy} \times 10 = 10 \text{ mSv};$$

$$H_{T,(x+n)} = 11 \text{ mSv}$$

It may be noted that for the same organ dose, a neutron is 10 times biologically more damaging to the organ, compared to X-rays. So, the neutron equivalent dose is 10 mSv and not 1 mSv.

Example 4.7

The dose to an organ is 1 rad from X-ray exposure and 1 rad from neutron exposure. Calculate the resultant equivalent dose to the organ in Sv. (W_R = 1 and 10 for X-rays and neutrons, respectively).

$$H_{T,x} = 1 \text{ rad} \times 1 = 1 \text{ rem: } H_{T,n} = 1 \text{ rad} \times 10 = 10 \text{ rem;}$$

$$H_{T,(x+n)} = 11 \text{ rem} = 0.11 \text{ Sv}$$

$$(1 \text{ rad} = 100 \text{ ergs/g} = 10^5 \text{ ergs/kg} = 0.01 \text{ J/kg} = 0.01 \text{ Gy or 1 cGy}$$

$$\text{So, 1 rem} = 0.01 \text{ Sv or 1 Sv} = 100 \text{ rem)}$$

4.4.3.1 Tissue Dependence of Equivalent Dose

The radiation risk to a tissue is not a unique function of the equivalent dose to the tissue. This is because of the differing radiosensitivity of various tissues. To take this into account, ICRP (1977) introduced a tissue weighting factor, W_T, which expresses the tissue risk (for radiation-induced cancer, R_T) as a fraction of the whole body risk, R_{total}.

$$W_T = R_T/R_{total}$$

$$\sum_T W_T = \sum_T (R_T/R_{total}) = (1/R_{total}) \sum_T R_T = R_{total}/R_{total} = 1$$

So, by definition, $\sum W_T = 1$.

The W_T values, as given by ICRP 26, ICRP 60 and ICRP 103, are given in Table 4.9 for comparison.

TABLE 4.9

Changes in W_T Values from 1977 to 2007

Organ	ICRP 26	ICRP 60	ICRP 103
Gonads	0.25	0.2	0.08
Bone marrow	0.12	0.12	0.12
Breast	0.12	0.12	0.12
Thyroid	0.15	0.05	0.12
Bone surfaces	0.03	0.05	0.04
Reminder	0.03	0.01	0.01
Colon	0.03	0.05	0.12
Stomach		0.12	0.12
Bladder		0.12	0.04
Liver		0.05	0.04
Esophagus		0.05	0.04
Skin		0.05	0.01
Salivary glands		0.05	0.01
Brain		0.01	0.01

4.4.3.2 Reminder Tissues

The remainder tissues were considered, for which the cancer risk of radiation were not explicitly evaluated. ICRP-26 assigned the $W_T = 0.06$ for each of five unspecified tissues (i.e., 0.30/5). ICRP-60 assigned $W_T = 0.01$ for the reminder (listed as adrenals, brain, upper large intestine, ("lower large intestine" is the colon), small intestine, kidney, muscle, pancreas, spleen, thymus, and uterus). ICRP-103 excluded the brain and upper large intestine from the reminder and listed them as individual organs with separate risk factors. For the reminder ICRP-103 listed 14 tissues (adrenals, extra-thoracic tissue, gallbladder, heart, kidneys, lymphatic nodes, muscle, oral mucosa, pancreas, prostate, small intestine, spleen, thymus, and uterus/cervix) in the reminder and assigned a $W_T = 0.12$.

The following points can be noted from Table 4.9 regarding the most recent recommendations:

- Recent ICRP 103 recommendations include W_T for a larger number of organs, compared to the earlier ICRP reports.
- W_R values have been updated.
- The latest recommendation of stochastic risk, considered now as realistic by the commission, is based on cancer incidence (rather than cancer fatality) weighted for lethality and life impairment.
- The lowering of the tissue weighting factor for gonads suggests that the risk of hereditary effects is much lower than previously thought.

4.4.4 Effective Dose

The effective dose was earlier referred to as effective dose equivalent in ICRP 26 (since, for the organs, it used the word dose equivalent instead of equivalent dose).

Figure 4.12 illustrates the concept of effective dose, as the summation of the risk weighted organ doses.

The effective dose received by the radiation worker is given by

$$E = \sum_T W_T H_T = \sum_T W_T \sum_R W_R D_{T,R}$$

The summation over R refers to the situation where more than one radiation type may be involved in occupational exposure. For instance, in a radiation oncology department with high-energy X-ray linacs, the total external exposure would include the neutron exposure as well (along with X-ray exposure). W_R and W_T are mutually independent—tissue weighting factors are independent of radiation type, and vice versa. When only one type of radiation is involves, as is the normal case,

$$H_T = W_R D_{T,R}$$

This effective dose concept helps in determining the average whole body dose in the case of nonuniform dose distribution in the body. When the whole body is uniformly irradiated (say in the case of staff in a control room of a linac treatment delivery room), H_T is a constant and

$$E = H_T \sum_T W_T = H_T$$

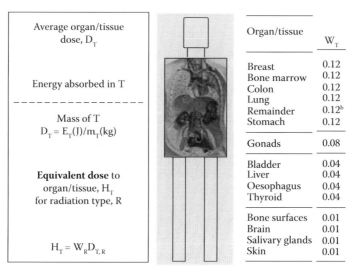

Organ/tissue	W_T
Breast	0.12
Bone marrow	0.12
Colon	0.12
Lung	0.12
Remainder	0.12[b]
Stomach	0.12
Gonads	0.08
Bladder	0.04
Liver	0.04
Oesophagus	0.04
Thyroid	0.04
Bone surfaces	0.01
Brain	0.01
Salivary glands	0.01
Skin	0.01

Average organ/tissue dose, D_T

Energy absorbed in T
- - - - - - - - - - - - - - - -
Mass of T

$$D_T = E_T(J)/m_T(kg)$$

Equivalent dose to organ/tissue, H_T for radiation type, R

$$H_T = W_R D_{T,R}$$

Effective dose $E = \Sigma_T W_T H_T$

FIGURE 4.12
Concept of effective dose.

These quantities of D_T, H_T, and E are known as the radiation protection quantities.

The weighting factors are based on judgment, many simplifications, epidemiological studies of cancer induction in the organ, radiobiology, and experimental genetic effects, and are averaged over both sexes and all ages. Because of these variables, though E is a measure of stochastic risk it cannot be used to estimate the risk of any individual worker. Also, because of the approximations, W_T, and E are applicable only for low doses (as in radiation protection), to evaluate the risk of radiation workers or patients.

The effective dose equation implies that at low doses (encountered in radiation protection), the total radiation detriment to the exposed person (risk of cancer, and genetic harm) is given by the sum of radiation detriments to the single organs. The annual dose limits for the occupational and public exposures actually refer to the effective dose.

Effective dose is used to set limits, constraints, or reference levels in radiation protection for radiation workers, patients, or the members of the public, and for optimization of radiation protection using the as low as reasonably achievable (ALARA) concept. It is also used for planning purposes. The advantages of this protection quantity are:

1. It can be used for estimating effective dose (or equivalent whole body risk for partial body exposures). For example, if "only lung exposure" and whole body exposure gives the same effective dose, the stochastic risk is the same in both cases. The only difference is, in the case of lung exposure, the risk is only lung cancer risk while in the case of whole body exposure, cancer can be manifested in any organ of the body.

2. It can be used for determining the total dose from both internal and external exposures.

3. It can be used for determining the total dose when more than one type of radiation exposure is involved (e.g., neutron and photon exposures).

The assessment of effective dose for radiation workers is essential to show compliance with regulatory dose limits, so it is mandatory in radiation oncology to monitor staff doses.

4.5 Internal Exposure and Committed Dose

External exposure occurs only as long as the radiation worker is exposed to the external sources of radiation. In the case of internal exposure, which occurs as a result of radioactivity incorporated into the body, the body continues to receive radiation dose from the decaying radioactive substances, even when the sources of internal exposure cease to exist, as for instance, when the radiation worker leaves the workplace.

Figure 4.13 illustrates the concept of external and internal exposure of a radiation worker in a workplace.

In the case of internal exposure, the radioactive dust or particles in the ambient get deposited in the body. It is not eliminated immediately from the system, but with a biological half-life, depending on the physical and metabolic characteristics of the radioactive substance. So, once a radiation worker takes in some radioactivity (through inhalation, ingestion, or any skin wounds) the worker is committed to certain dose. In the case of radiation workers, the main entry route is by inhalation into the respiratory system, a fraction of which may reach the throat and get absorbed in the gastrointestinal tract. For occupational (internal) exposure, the committed effective dose to the worker is calculated as the cumulative dose to the radiation worker, for a 50 year integrating period following the intake, irrespective of the age of the worker.

If a radiation worker receives both external and internal exposures (e.g., a person working partly in radiotherapy and partly in nuclear medicine) they would receive external exposure due to photons and neutrons, and internal exposure from the nuclear medicine department. The worker whole body dose limit applies to the total dose E received from both types of exposures (external and internal exposures) and both must be taken into account to determine the annual dose limit compliance. This applies for partial body exposures as well.

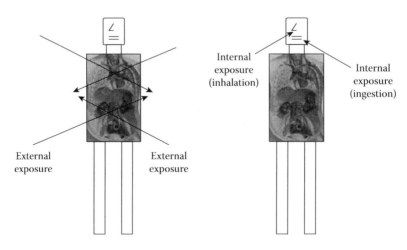

FIGURE 4.13
Concept of external and internal exposure in a workplace.

Internal exposure situation will not be further discussed in this book since only external radiation exposures are involved in a radiation oncology department. Radioactive sources used in radiation oncology are doubly encapsulated, and the likelihood of the release of radioactivity into the workplace is very small. The integrity of the source, however, must be checked periodically to reduce the possibility of source rupture and release of radioactivity.

Example 4.8

The annual effective dose limit for a radiation worker is 20 mSv. If the worker receives dose only to the lung, how much dose can that person receive?

The question implies: How much can the radiation worker receive so the lung dose will put him at the same risk as the whole body risk corresponding to 20 mSv. This requires organ risks expressed relative to whole body risk, W_T. W_T for lung = 0.12 (expressed as a fraction of whole body risk).

So a lung dose of 20/0.12 = 166 mSv will give a risk of lung cancer equal to the whole body risk (or risk of cancer to any part of the body) corresponding to the effective dose of 20 mSv (whole body exposure).

Example 4.9

Each year inhalation of radon in the air results in an "equivalent dose" to lung tissue of 10 mSv which involves some risk. What is the effective dose corresponding to this risk?

$$E = \sum W_T H_T = 10 \times 0.12 = 1.2 \text{ mSv}$$

D_T and hence H_T to other organs is zero, since they are not exposed.

So a lung dose of 10 mSv and a whole body dose or an effective dose of 1.2 mSv entails the same risk to the individual. The probability of death from lung cancer as a result of a 10 mSv lung dose is the same as the probability of cancer to any organ of the body (or genetic detriment in the next two generations) from a whole body dose of 1.2 mSv.

4.6 Reference Phantom for Effective Dose

It is not possible to take measurements of organ doses inside an exposed individual. So, for the purpose of estimating the organ doses and effective doses from radiation exposures at occupational exposure levels, ICRP has developed reference male and reference female (computational) phantoms [11] that simulate an average sized male and female, respectively. They were constructed based on the computerized tomography (CT) data of real patients, close to the reference phantom size and then scaling the sizes to the ICRP reference male size. (By the segmentation of the CT slices, the body organs and tissues, and the human anatomy can be rebuilt in 3D from 3D voxel data of CT). The reference male weighs 73 kg and is 176 cm tall; reference female weighs 60 kg and is 163 cm tall, for the western population. Figure 4.14 shows the typical reference person phantom of ICRP. The skeletal and organ details of the phantom are given in ICRP 116, and are used for the computation of the effective dose for typical irradiations geometries.

The various geometries of irradiation are anterio-posterior (A/P), posterio-anterior (P/A), left lateral (LLAT), right lateral (RLAT), rotational (ROT), and isotropic (ISO), as shown Figure 4.14. The primary dose limiting quantities are calculated for the reference phantoms for monoenergetic parallel beam of ionizing radiation incident on the body for the various geometries. The first three geometries (A/P, P/A, and LAT) best correspond to situations

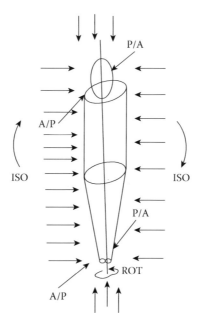

FIGURE 4.14
Mathematical phantom defined for effective dose evaluation.

where a person has their body aligned all the time in some of these geometries. If the person walks around the radiation field in a random manner (as in a control room of a linac treatment room) this situation may better correspond to rotation geometry.

4.6.1 Effective Dose Computation for Reference Person

ICRP computes the effective dose to a reference person, by averaging the equivalent doses to the organs of the male and female phantoms, as shown in Figure 4.15.

The effective dose is determined by averaging the equivalent doses of the male and the female phantoms and hence is applicable to any radiation worker, regardless of sex.

All the computations of effective dose corresponding to the workplace radiation field conditions apply to the adult reference person defined by ICRP. As stated earlier, these cannot be applied to any particular radiation worker since they come in different weights, shapes, and sizes, and do not necessarily have the same organ radiosentivities (W_T) as assumed in the concept of effective dose. Similarly, the stochastic risk cannot be ascribed to a particular radiation worker or the public. It will be used as a conservative estimate of the dose received by radiation workers to ensure regulatory compliance to the legal dose limits, or for relative comparisons like comparing stochastic risk from different diagnostic examinations, or for optimizing doses in occupational exposures.

4.6.2 Dose Conversion Coefficient

The Dose Conversion Coefficients (DCCs) relate the protection quantities (D_T, E) to the incident radiation exposures for the ICRP reference person. The incident exposures can be measured in terms of exposure, air kerma (for photons), absorbed dose (for electrons or protons), and neutron fluence (for neutrons).

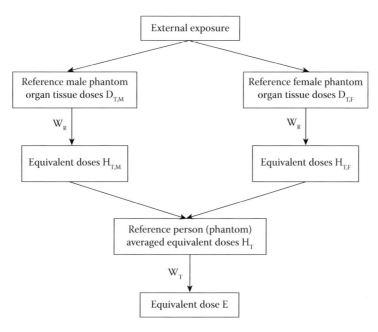

FIGURE 4.15
Effective dose to ICRP reference person.

The dose coefficients are defined as:

$$DCC_X = E/X;\ DCC_{Ka} = E/K_a;\ DCC_{\Phi n} = E/\Phi_n$$

Similar coefficients can also be calculated for organ doses, since organ dose computation is a prerequisite for the effective dose computation.

The dose coefficients are tabulated in ICRP 110, for each of the exposure geometries listed earlier. Using these tables, the effective dose can be computed from the workplace radiation exposures.

Example 4.10

The incident air kerma for the ICRP reference phantom is 1 mGy (for ^{60}Co radiation exposure). Calculate the effective dose for AP and RLAT irradiations of the phantom. The following DCC_{Ka} data are given in (mSv/mGy):

Beam quality	AP	PA	LLAT	RLAT	ROT	ISO
^{137}Cs	1.02	0.839	0.685	0.635	0.813	0.697

$$DCC_{Ka}\ (AP) = E/K_a = 1.02$$

$$E_{AP} = 1.02 \times K_a = 1.02 \times 1 = 1.02\ mSv$$

$$DCC_{Ka}\ (RLAT) = E/K_a = 0.635$$

$$E_{RLAT} = 0.635 \times K_a = 0.635 \times 1 = 0.635\ mSv$$

The above example strikingly demonstrates that the effective dose depends on the direction of incidence of the radiation. For a typical workplace exposure situation, the effective dose will generally lie between the minimum and the maximum values of E. So, by using the maximum E value, a conservative estimate of the effective dose can be computed by monitoring the workplace, say in terms of exposure or air kerma.

4.7 Monitoring of Occupational Radiation Exposures

In order to control occupational exposures, radiation levels in workplaces should be monitored using area monitors, and the dose received by the radiation workers assessed through individual monitoring. Controlling radiation levels in the workplace will automatically control the doses to the radiation workers in that area. Any unexpected changes in the radiation levels will immediately indicate the existence of an unplanned situation, which can be immediately investigated (e.g., source not going back to the parking position in a ^{60}Co teletherapy treatment room or in a high-dose rate (HDR) treatment room). On the other hand, personal monitors are read only once a quarter by the service provider and so any immediate changes will go unnoticed. Radiation levels around radiology equipment or radiation oncology equipment must be within regulatory limits, and must be ensured by measuring the radiation levels around the equipment. So, operational quantities are required for

1. Area monitoring
2. Individual monitoring

The two types of monitoring for external exposures are illustrated in Figure 4.16, which depicts a typical workplace where the radiation in the ambient does not conform to the typical irradiation geometries that we defined in the earlier section.

NOTE: Sometimes only the zone monitors that are permanently installed on the walls of work areas are referred to as area monitors. Here, area monitor refers to all the monitors that are used for workplace monitoring (including the permanently installed monitors).

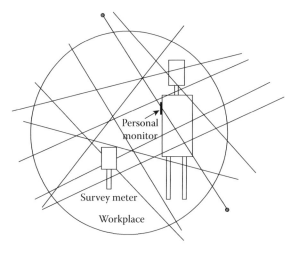

FIGURE 4.16
Concept of workplace and individual monitoring.

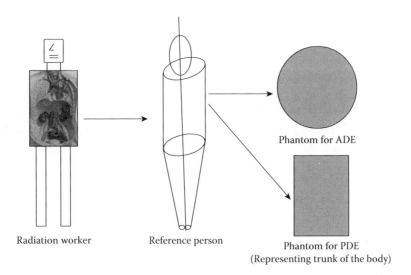

Phantom for ADE

Phantom for PDE
(Representing trunk of the body)

Radiation worker   Reference person

FIGURE 4.17
Phantoms for defining operational quantities in radiation protection.

4.7.1 Defining Phantoms for Operational Quantities in Area Monitoring

Figure 4.17 illustrates the ICRP definition of a reference person to represent the radiation worker and the much simplified reference (calibration) phantoms defined by ICRU to represent ICRP's reference man [12,13].

The primary objective of both monitoring methods (area monitoring and individual monitoring) is to correlate them to the dose received by the radiation worker. Therefore, both the operational quantities are defined in a phantom that represents a typical radiation worker.

The ICRU phantom is a sphere of 30 cm diameter with a density of 1 g/cm^3, and a mass composition of 76.2% oxygen, 11.1% carbon, 10.1% hydrogen, and 2.6% nitrogen.

4.7.2 Characterizing Radiation in Terms of Penetration

Low energy X-rays (say <15 keV), alphas, and betas with less than a few MeV energy have negligible penetrating power. They will deposit much of their energy in the skin and are referred to as weakly penetrating. High-energy X-rays, gammas, electrons, and neutrons with energies greater than a few MeV are more penetrating and are referred to as strongly penetrating. These radiations typically deposit the maximum energy at around a 10 cm depth.

4.7.3 Operational Quantities for Area Monitoring

Typical depths in the ICRU sphere are 0.07 mm (the dead layer of the skin) and 10 mm, recommended for measuring the DE of weakly penetrating and strongly penetrating radiations, respectively, for the purposes of area monitoring.

In a radiation oncology department, the effective energy of the photons in the control areas is around 1 MeV or less, and considering there is a small component of higher energy leakage of photons, 10 mm is a typical depth of maximum dose in the body for workplace radiation exposures. Effective dose, E, is the appropriate quantity for dose limitation in this case.

The operational quantities are defined to give a conservative estimate of the protection quantities. These quantities for the strongly penetrating and weakly penetrating radiations, for area monitoring, are represented by $H^*(10)$ and $H'(0.07, \Omega)$, and are referred to as ambient dose equivalent and directional dose equivalent.

4.7.3.1 Definition of Ambient Dose Equivalent

The ambient dose equivalent at a point in a radiation field is defined as the dose equivalent that would be produced by the corresponding *aligned and expanded radiation field* in the ICRU sphere at the depth d = 10 mm for $H^*(10)$ at the radius opposite to the direction of the field [12–14]. The measurement point is shown by a small white dot in Figure 4.18, which illustrates the aligned and expanded field for computing the ADE (measured in sieverts (Sv)).

The *aligned and expanded radiation field* part of the definition of ADE is quite confusing and requires some clarification. The expanded field is understandable since the radiation field must cover the receptor (ICRU sphere) to include the phantom scatter in the DE measurement, but the *aligned field* requires some explanation. Figure 4.19 shows a typical field in a workplace configured as an aligned field.

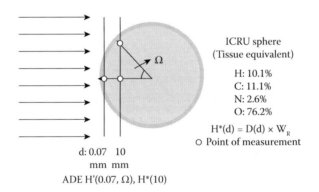

FIGURE 4.18
Definition of ADE $H^*(10)$ and $H'(0.07,\Omega)$.

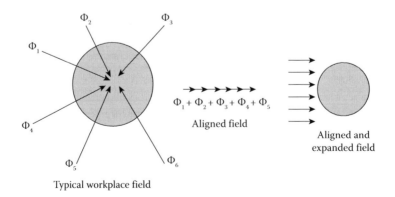

FIGURE 4.19
Concept of aligned and expanded field.

As a result of the alignment of the realistic field in a single direction, the contributions of radiation from all directions add up. Assuming this can be done, the value of H*(10) becomes independent of the directional distribution of the radiation in the actual field. In practice, the alignment is not of a real field but compute H*(10) for an aligned field, which can be realized in a calibration laboratory and calibrate the area monitor in the same field. The area monitor, however, is designed to have an isotropic response so that the aligned field calibration holds good for measuring in a real workplace field. So, isotropic response is an important requirement for area monitors to measure ADE in workplace monitoring. Such a monitor will give the same response in the workplace situation (which is an expanded field but not an aligned field) or in a monodirectional beam (which is an expanded and aligned field), if all the rays shown in the workplace situation are turned around and aligned along a single direction. This is the important point which must be understood in the definition of ADE for area monitoring. Because of the design of the monitors (with different shapes, handle for holding, etc.) it is difficult to design a perfectly isotropic instrument but the less the isotropic response of the instrument, the more is the uncertainty of the monitored dose.

While area monitors measured the ambient fields in terms of exposure or air kerma for a long time, the newer monitors these days measure the ambient dose equivalent.

4.7.3.2 Directional Dose Equivalent for Skin and Eye Lens

The DDE is used for measuring skin dose or dose to eye lens [12–14] due to weakly penetrating radiation (αs, βs, and low energy X-rays) for deterministic risk to organs lying close to the body surface (e.g., skin, eye lens) since they cannot penetrate and deliver much dose at, say, 10 mm depth. The skin dose is very much directional dependent. The DDE monitor has a very thin entry window and would similarly exhibit directional dependence. While monitoring the radiation levels, the DDE monitor must be rotated and the maximum reading noted.

Since this quantity is directional dependent, it is defined as a function of the angle of incidence of the radiation with respect to the specified orientation of the sphere (as shown in Figure 4.20). The directional dose equivalent, H'(d,Ω), at a point of interest in the actual

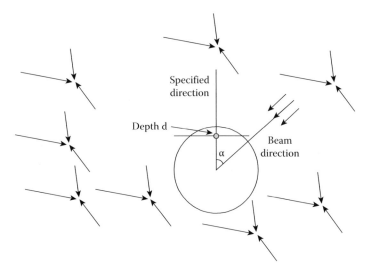

FIGURE 4.20
Concept of directional dose equivalent H'(d,Ω).

radiation field, is *defined* as the dose equivalent that would be produced by the corresponding expanded radiation field, in the ICRU sphere at a depth d, on a radius in a specified direction or orientation Ω (with respect to the radiation direction of incidence). This is shown in Figure 4.20 where α is the orientation angle in a plane. It should be noted in the definition of DDE that the field is expanded to cover the phantom but *not aligned*. So the value of the DDE can strongly depend on the orientation of the sphere in the expanded radiation field, typical of low energy radiations.

The directional dose equivalent is defined at depth d = 0.07 mm, the thickness of the dead layer of the skin, for limiting the equivalent dose to the skin through area monitoring of weakly penetrating radiations. It is denoted by $H'(0.07, \Omega)$ or H_{skin}. In many situations (e.g., while handling brachytherapy sources like ^{125}I seeds, injecting radiopharmaceuticals for imaging purposes, or for alpha monitoring), extremities are at risk where this quantity becomes relevant for limiting the risk of skin reactions, rather than H*(10).

In some situations (e.g., while conducting fluoroscopic procedures on a patient), the physician's eyes are at risk. The depth of eye lens is about 3 mm. The directional dose equivalent $H'(3, \Omega)$, H_{lens}, is the operational quantity for monitoring the eye lens dose.

However, area monitoring is rarely carried out for limiting eye or skin exposures.

4.7.4 Operational Quantities for Personal Monitoring (or Individual Monitoring)

Personal dose equivalent (PDE, $H_p(d)$) is the operational quantity for individual monitoring. Initially, ICRU had defined the same phantom as for ADE for defining PDE, but for practical reasons (e.g., calibrating a batch of personal monitoring (PM) badges on a plane phantom surface is more convenient, shape of the trunk of the body is cylindrical rather than spherical), ICRU, in a later recommendation [13] defined a tissue-equivalent slab phantom of size 30 cm × 30 cm × 15 cm having the same density and composition as the ICRU sphere. So, $H_{p,slab}(d)$ is a surrogate for the quantity $H_p(d)$ in the body of the person.

Both the quantities ADE and PDE are supposed to represent the DE received by the radiation worker, but the sphere quantity is defined for workplace monitoring and so areas monitored are calibrated free in air (i.e., without a backing phantom) and are also used the same way in air for ADE monitoring. On the other hand, PDE is defined for individual monitoring where the monitor is worn on the body, to assess the radiation dose received by the radiation worker in a radiation field. To simulate this situation, the PDE monitors are calibrated on a PDE phantom (as it is not possible to do the calibration on a real person) but used in the field (by wearing the monitor on the body of a person) for individual monitoring.

As in the case of ADE, the operational quantities for individual monitoring, PDE, should also allow the effective dose and the equivalent dose to be assessed as a conservative estimate, under nearly all irradiation conditions. In individual monitoring, the whole body dose (or effective dose), dose to eye lens, and the skin dose (e.g., to legs, hands, or fingers) are the doses to be monitored depending on the workplace exposure situation. $H_p(10)$, $H_p(3)$ and $H_p(0.07)$ generally provide a conservative estimate of these quantities. $H_p(10)$ is a conservative estimate of E, even in cases of lateral or isotropic radiation incidence on the body. In the case of unidirectional radiation, the radiation worker must face the source direction to get a conservative estimate of the E. If the exposure is from the back, the PM badge will not properly assess E since the dosimeter is worn on the front side.

4.7.4.1 Definition and Concept of Personal Dose Equivalent

Figure 4.21 illustrates the concept of Personal Dose Equivalent (PDE). The personal dose equivalent H_p (d), is the dose equivalent in soft tissue, at an appropriate depth, d, below a specified point on the body. $H_p(d)$ is measured with a monitor which is worn at the surface of the body and covered with an appropriate thickness of tissue-equivalent material corresponding to the defined depth d. It is measured in sieverts (Sv). The PDE $H_P(d)$ is defined, for both strongly and weakly penetrating radiations, as d = 10 mm for strongly penetrating radiations for effective dose monitoring, d = 0.07 mm for weakly penetrating radiations for monitoring dose to the skin, hands, and feet, and d = 3 mm for monitoring the dose to the eye lens.

The slab phantom defined in the context of PDE serves as a surrogate for the radiation worker, as shown in Figure 4.21. Initially, the ICRU sphere was defined as the surrogate for the radiation worker, but considering the practical difficulties in calibrating PM badges on a spherical phantom, the ICRU tissue slab phantom was later defined.

So, for the purposes of type testing or calibration of personal dosimeters,

$$[H_p(10)]_{ICRUslab} = [H_p(10)]_{person}$$

$$[H_p(3)]_{ICRUslab} = [H_p(3)]_{person}$$

$$[H_p(0.07)]_{ICRUslab} = [H_p(0.07)]_{person}$$

The directional dependence of PDE for weakly penetrating radiations is defined in Figure 4.22. (Other phantoms have been defined which will be considered in Chapter 6.)

Most of the individual monitors are calibrated for $H_P(10)$ or $H_P(0.07)$, and it is generally assumed that limiting the skin dose will limit the eye lens dose as well. However, with the recent ICRP recommendation (ICRP 2012) lowering the annual dose limits to eyes from 150 mSv to 20 mSv/y (average), assessment of eye lens dose, $H_p(3)$, assumes special significance [15].

Table 4.10 summarizes the operation quantities for radiation protection monitoring.

Figure 4.23 illustrates how area monitoring and personal monitoring is carried out in a workplace.

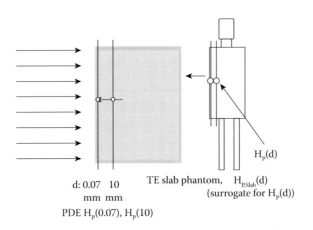

d: 0.07 10
mm mm

PDE $H_p(0.07)$, $H_p(10)$

TE slab phantom, $H_{p,Slab}(d)$
(surrogate for $H_p(d)$)

$H_p(d)$

FIGURE 4.21
The concept of personal dose equivalent.

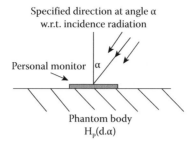

FIGURE 4.22
Concept of $H_p(d,\alpha)$.

TABLE 4.10

Summary of the Operational Quantities

		ICRU Operational Quantity for	
External Exposure	**Limiting Quantity**[a]	**Area Monitoring**	**Individual Monitoring**
Strongly penetrating	Effective dose	H*(10) Ambient DE	$H_p(10)$ Personal DE
Weakly penetrating	Skin dose[b]	H′(0.07,Ω) Directional DE	$H_p(0.07)$[c] Personal DE
	Eye lens dose[b]	H′(3,Ω) Directional DE	$H_p(3)$ Personal DE

[a] Limiting quantities are the radiation protection quantities defined by ICRP.
[b] Actually refer to dose equivalent (DE).
[c] Hp(0.07) is also used for monitoring the doses to the extremities from all ionizing radiation.

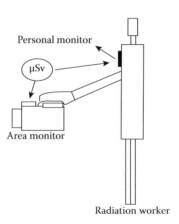

FIGURE 4.23
Area monitoring and personal monitoring in a workplace.

4.8 Relations between ADE and Physical Dosimetric Quantities

Knowing the photon fluence of a mono energetic photon beam, incident on the hypothetical ICRU sphere (see Figure 4.18), the incident air kerma, incident exposure, and the ambient and directional DE quantities can be computed.

$$K_a = \Phi_E \, dE \, E \, [_m\mu_{en}(E)]_{air}$$

$$X = K_a \, (1-g)/(W/e)$$

$$H^*(10) = \int \Phi_E \, (10) \, dE \, E \, [_m\mu_{en}(E)]_{ICRUtissue}$$

Conversion coefficients: $h_\Phi = H^*(10)/\Phi$; $h_x = H^*(10)/X$; $h_{Ka} = H^*(10)/K_a$

The conversion coefficients are fluence to DE, exposure to DE, and air kerma to DE conversion factors. So, for a typical incident photon fluence of monoenergetic photons, the conversion coefficients h_Φ, h_x, h_{Ka} can be computed, using Monte Carlo codes. These data are available in ICRP 74 and ICRU 51 for photons, electrons, and neutrons [16,17]. For weak penetrating radiations, the quantities are directional dependent and hence are tabulated as a function of angle.

For well-defined radiation fields, the exposure, or air kerma, or neutron fluence at a point on the central axis of the beam can be determined using an appropriate reference instrument calibrated in terms of these quantities.

$$X = N_x \, M \, K(T,P); \; K_a(P) = N_K \, M \, K(T,P); \; \Phi = N_\Phi \, M$$

Now using the conversion factors, the fields characterized in terms of exposure, kerma, or neutron can now be characterized in terms of operational quantities as shown by the following equation:

$$H^*(10) = h_x \, X = h_{Ka} \times K_a = h_\Phi \, \Phi$$

So in principle, if X or K_a can be measured using a reference standard, the ambient dose equivalent and calibrated exposure, or air kerma rate measuring field instruments can be computed, in terms of $H^*(10)$. More details on this can be found in the section dealing with the calibration of protection monitors.

NOTE: In fact there is no ICRU sphere or primary standard of ADE existing in any standards lab to measure the ADE. ICRU phantom is not a physical phantom but a computational phantom (like the ICRP reference male phantom) and the computed ADE can be conveniently used for ADE calibration of area monitors. The monitor display can also be directly marked in terms of ADE instead of exposure or air kerma.

Some comments on the protection and operational quantities:

The computation of conversion factors is based on kerma approximation (i.e., kerma ~ absorbed dose). This is a reasonable approximation for energies up to 2 MeV for photons, and about 35 MeV for neutrons. With the increasing need for protection at energies beyond this limit (e.g., particle therapy and exposure of aircraft crew), the conversion coefficients need to be recomputed without the kerma approximation. Beyond 10 MeV, for photons, the computed ADE is an underestimate of the effective dose.

The definition of operational quantities would have made more sense if they are measurable since the ICRU sphere is a simpler phantom compared to the ICRP male or female phantom. However, such a tissue composition phantom cannot be fabricated and a quantity like $Q \times D$ cannot be measured experimentally. So, when both (effective dose and dose equivalent quantities) are not measurable but only computable it is not clear why it was felt necessary to replace one nonmeasurable quantity by another, to obtain a conservative estimate of E. The ICRP conversion factors—E/X or E/K_a factors—could have been directly used to relate the dosimetric quantities to E instead of through H^*.

ICRP is presently thinking of defining operational quantities [18] based on protection quantities (effective dose-to-fluence conversion factors). These conversion coefficients for the protection quantities are already available (ICRP 2010). Adopting these conversion coefficients may make radiation monitoring and dose assessment simpler and more straightforward. More details on this issue can be found in the ICRP 116 report [19].

4.9 Relations between Personal Dose Equivalent and Physical Dosimetric Quantities

Knowing the photon fluence of a monoenergetic photon beam incident on the ICRU slab phantom, the incident air kerma, incident exposure, and the personal dose equivalent quantities Hp(d) can be computed for any depth d in mm; d=0 corresponds to air kerma or exposure on the phantom surface. (The air kerma on the surface is incident air kerma, K_a(P) × backscatter factor at the phantom surface.)

$$K_a (0) = \{\Phi_E \; dE \; E \; [_m\mu_{tr}(E)]_{air}\}_{d=0}$$

$$X (0) = K_a (0) \; (1–g)/(W/e)$$

$$H_p(10) = \int \{\Phi_E \; dE \; E \; [_m\mu_{en}(E)]_{ICRUslab}\}_{d=0}$$

Conversion coefficients:

$$h_{\Phi,slab} = H_p(d)/\Phi(P); \; h_{x,slab} = H_p(d)/X (P); \; h_{Ka,slab} = H_p(d)/K_a(P)$$

The conversion coefficients are fluence to DE, exposure to DE, and air kerma to DE conversion factors. So, for typical incident photon fluence of monoenergetic photons, the conversion coefficients h_Φ, h_x, h_{Ka} can be computed, using Monte Carlo codes. These data are available in ICRP 74 and ICRU 54 for photons, electrons, and neutrons

For well-defined radiation fields, the exposure, air kerma, or neutron fluence at a point on the central axis of the beam can be determined using an appropriate reference instrument calibrated in terms of these quantities, as explained in the previous section.

$$X (P) = N_x \; M \; K(T,P); \; K_a(P) = N_K \; M \; K(T,P); \; \Phi (P) = N_\Phi \; M$$

Now using the conversion factors, the fields characterized in terms of exposure or kerma or neutron fluence can now be characterized in terms of operational quantities as shown by the following equation:

$$H_p(d) = h_x \; X(P); \; H_p(d) = h_{Ka} \times K_a (P)$$

So, in principle, if X or K_a can be measured using a reference standard, the personal dose equivalent can be computed and exposure or air kerma rate measuring field instruments calibrated in terms of H_p(d). More details on this can be found in the section dealing with the calibration of protection monitors.

NOTE: In fact there is no ICRU slab phantom or primary standard of PDE existing in any standards laboratory to measure the PDE. ICRU slab phantom, like the ICRU sphere, is not a physical phantom but a computational phantom, and the computed PDE can be conveniently used for PDE calibration of personal monitors. The monitor display can also be directly marked in terms of PDE instead of exposure or air kerma.

Figure 4.24 illustrates the relationship between the various physical and operational dosimetric quantities, discussed in the previous sections.

The protection quantities and the operational quantities are related by the conversion coefficients E/H^* and E/H_p. Figure 4.25 shows that $E/H^*(10) <1$ or $H^*(10) >E$, and so the measured ADE for a radiation worker is a conservative estimate of E [18], and compliance with respect to the operational quantity $H^*(10)$ ensures compliance with the regulatory quantity and the effective dose, up to 10 MeV photon energy. The figure is an approximation for the actual computed curves in the literature (e.g., ICRP 116 [19]). In the energy range considered, $H_p(10)$ is also a conservative estimate of E.

In the case of neutrons, the operational quantities are conservative overestimates of the effective dose up to 3 MeV, but underestimate E between 3 and 12 MeV, and the underestimate increases further with an increase in neutron energy. However, in conventional radiation oncology (involving linac therapy), the neutron spectrum in the control room is heavily degraded and the mean energies are much less than 1 MeV. The operational quantities give a conservative estimate of the effective dose.

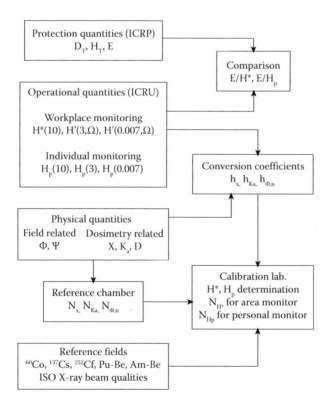

FIGURE 4.24
Relationship between protection, operational, and dosimetric quantities.

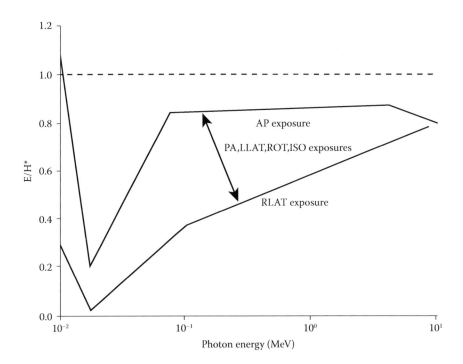

FIGURE 4.25
Energy dependence of ADE to effective dose conversion factor for photons.

We do not have to consider therapeutic electron beams for radiation protection monitoring, as they do not exit the shielding walls.

The operational quantities and the dosimetric quantities are related by the conversion coefficients h_x, h_K, and $h_{\Phi,n}$. These conversion coefficients help to realize the operational quantities from the dosimetric quantities exposure, air kerma, or neutron fluence. The dosimetric quantities have primary standards at the primary standards dosimetry laboratories and can be disseminated to the regional calibration laboratories through well-designed reference standards. The regional calibration laboratories, by measuring the exposure, air kerma, or neutron fluence in the calibration laboratory, with the reference standard for a reference calibration field, can characterize their calibration fields in terms of the dosimetric quantities or operational quantities, and offer calibration to the protection monitors in terms of either quantities.

4.9.1 Energy Dependence of the Air Kerma to H*(d) Conversion Factors for Photons

Figure 4.26 [20] shows the energy dependence of the conversion factor, $H^*(10)/K_a$:

While monitoring the workplace in terms of exposure or air kerma, the monitors were designed to have a flat energy response so that the monitor calibration, done at select beam qualities, holds for the spectrum existing in the typical workplace. Now, the protection monitors must be designed to have a flat energy response for $H^*(10)$.

4.9.2 Energy Dependence of the Fluence to H*(d) Conversion Factors for Neutrons

The energy dependence of the conversion factor for neutrons is shown in Figure 4.27 [21,22].

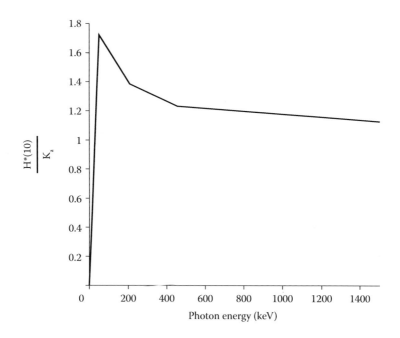

FIGURE 4.26
Energy dependence of ADE to air kerma conversion factor for photons.

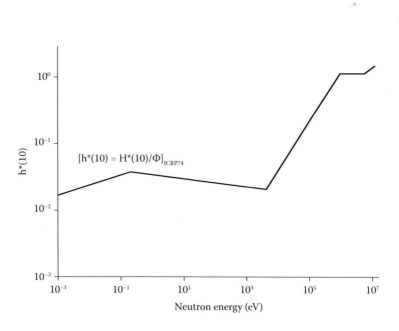

FIGURE 4.27
Energy dependence of fluence to ADE conversion factors for neutrons.

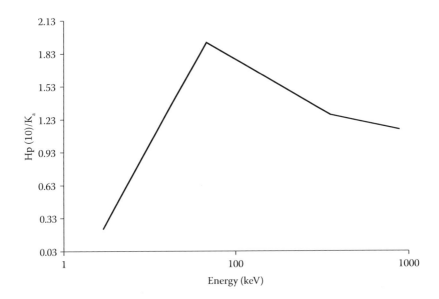

FIGURE 4.28
Air kerma to PDE conversion coefficients for photons. The conversion factor curves for $H_p(3)$ and $H_p(0.07)$ also show a similar trend.

4.9.3 Energy Dependence of the Air Kerma to $H_p(d)$ Conversion Factors

Figure 4.28 shows the typical energy dependence trend of the PDE conversion factor [23].

In earlier times, similar to workplace monitoring, the PMDs were also designed to have a flat energy response curve for exposure or air kerma measurements. Now, to measure the new operational quantities, the personal monitoring devices should be designed so their PDE response is not significantly energy dependent, by taking into account the energy dependence trend of the conversion factor.

References

1. M. McEwen, *Primary Standards of Air Kerma for ^{60}Co and X-rays and Absorbed dose in photon and electron beams*, Paper presented at American Association of Physicists in Medicine (AAPM) Summer School, Colorado, 2009.
2. M. O'Brien, Comparison of the NIST and PTB Air-Kerma Standards for Low-Energy X-Rays, *Journal of Research of the National Institute of Standards and Technology*, 114(6), 321–331, 2009.
3. L. Büermann and D.T. Burns, Air Kerma standards, *Metrologia*, 46(2), 24–38, 2009.
4. F.H. Attix, *Introduction to Radiological Physics and Radiation Dosimetry*, Wiley, Hoboken, NJ, 1986.
5. Neutron Sources. A presentation posted at the NRC site. http://www.nrc.gov/docs/ML1122/ML11229A704.pdf (accessed on May 2016).
6. Activities of the neutron standardization at the Korea research institute of standardization and science. 13th Presentation at CCRI Meeting, BIPM, France, 2005. http://www.bipm.org/cc/CCRI(III)/Allowed/16/CCRI(III)05-13.pdf (accessed on May 2016).
7. ICRP, The 1977 Recommendations of the International Commission on Radiological Protection, ICRP Publication 26, *Annals of the ICRP*, 1(3), 10, 1977.

8. ICRP 2, *Permissible Dose for Internal Radiation*, ICRP Publication 2, Pergamon Press, London, UK, 1959.
9. ICRP 60, The 1990 Recommendations of the International Commission on Radiological Protection, ICRP Publication 60, *Annals of the ICRP*, 21(1–3), 25–32, 1991.
10. ICRP, The 2007 Recommendations of the International Commission on Radiological Protection, ICRP Publication 103, *Annals of the ICRP*, 37(2–4), 1–166, 2007.
11. ICRP, Adult Reference Computational Phantoms, ICRP Publication 110, *Annals of the ICRP*, 39(2), 5–36, 2009.
12. ICRU, *Determination of Dose Equivalents Resulting from External Radiation Sources*, Part 1, Report 39, Oxford University Press, London, UK, 1985.
13. ICRU, *Determination of Dose Equivalents Resulting from External Radiation Sources*, Part 2, Report 43, Oxford University Press, London, UK, 1988.
14. ICRP draft 2005, *Basis for Dosimetric Quantities used in Radiological Protection*, 2005. http://www.icrp.org/docs/physics_icrp_found_doc_for_web_consult.pdf (accessed on May 2016).
15. ICRP, Statement on tissue reactions/early and late effects of radiation in normal tissues and organs—Threshold doses for tissue reactions in a radiation protection context, ICRP Publication 118, *Annals of the ICRP*, 41(1/2), 116–138, 2012.
16. ICRP, Conversion coefficients for use in radiological protection against external radiation, ICRP Publication 74, *Annals of the ICRP*, 26(3–4), 1–205, 1996.
17. ICRU, *Quantities and Units in Radiation Protection Dosimetry*, Report 51, Oxford University Press, London, UK, 1995.
18. N.E. Hertel and D. Bartlett, *Revision of ICRU Operational Quantities*, ICRP Symposium on Radiological Protection Dosimetry, Tokyo, 18 February 2016.
19. ICRP, Conversion coefficients for radiological protection quantities for external radiation exposures, ICRP Publication 116, *Annals of the ICRP*, 40(2–5), 1–258, 2010b.
20. V. Ramzaev et al., On the relationship between ambient dose equivalent and absorbed dose in air in the case of large-scale contamination of the environment by radioactive cesium, *Radiation Hygiene*, 8(3), 2015.
21. ICRU, *Conversion Coefficients for use in Radiological Protection against External Radiation*, Report 57, Oxford University Press, London, UK, 1998.
22. ICRP, *Conversion Coefficients for use in Radiological Protection Against External Radiation*, ICRP Publication 74, Pergamon Press, London, UK, 1997.
23. G.F. Gualdrini and B. Morelli, *Air kerma to Personal Dose Equivalent conversion factors for the ICRU and ISO recommended slab phantoms for photons from 20 keV to 1 MeV*, ENEA report, Centra Ricerche Ezio Clementel, Bologna, 1996.

Review Questions

1. What are the SI units of the quantities exposure, air kerma, and absorbed dose?

2. Name the primary standards of exposure and air kerma for kV energy X-rays and for ^{60}Co beams.

3. Explain in brief what types of radioactive neutron sources are generally used in neutron standardization.

4. What is the primary method of determining the output of neutron sources in terms of the emission rate?

5. What is the difference between the organ dose and organ equivalent dose?

6. What is the main parameter on which the RBE of any type of radiation depends?

7. Give the typical Q factors of X- and gamma rays, electrons, neutrons, and alpha radiation.

8. What is the SI unit and the special name of the quantity equivalent dose?

9. Which has a higher W_R value, neutrons or photons, and why?

10. The stomach and the esophagus receive the same equivalent dose. Will the two organs exhibit the same cancer risk?

11. What is the significance of tissue-weighting factors?

12. Is there any radiation protection monitor that can measure effective dose? If not, how can the radiation workers be monitored to assess their doses?

13. As per the ICRU recommendations, radiation protection monitoring of workplace is carried out in terms of the quantity _____.

14. What are the operational quantities for area monitoring to limit the dose received by the whole body, the eyes, and the extremities, respectively?

15. How can the values of $H_p(10)$ or $H^*(10)$ be determined in a radiation field, if the air kerma values are known in the field?

5

System of Radiation Protection and Regulations in Radiation Oncology

5.1 Introduction

The International Commission of Radiological Protection (ICRP) is the most important International Organization for issuing recommendations for the protection of the patients, the radiation workers, and the public against unnecessary exposure to ionizing radiation. ICRP works closely with ICRU in defining operational radiation protection quantities and their realization through traceable calibration procedures. ICRP derives its input from the United Nations Scientific Committee on the Effects of Atomic Radiation (UNSCEAR) which periodically reviews the current knowledge on ionizing radiation effects and risks. UNSCEAR is an important source for ICRP recommendations. ICRP also gives practical advice on how to apply their recommendations in the field of medicine. ICRP has official relationships with various other bodies such as the World Health Organization (WHO), the International Atomic Energy Agency (IAEA), and many other international agencies. The ICRP, however, does not have regulatory authority in any country. It is an international recommendatory body, and its recommendations are adopted by many national recommendatory or regulatory bodies for drafting national regulations. For instance, the US laws can be found in the Code of Federal Regulations (CFR) and are enforced by governmental organizations such as the Nuclear Regulatory Commission (NRC), and the Environmental Protection Agency (EPA). Figure 5.1 shows the well known publications of UNSCEAR, ICRP and IAEA.

Many European countries and the US have a long history of radiation protection and have advisory bodies to issue national recommendations on radiation safety. These countries recognize a regulatory body or bodies and issue authority to them, through acts of parliament or government to establish a regulatory framework and enforce the regulations. The regulatory authorities adopt the recommendations of the national advisory bodies. In most countries, there is some kind of regulatory mechanism to ensure radiation safety though not all of them may have national advisory bodies. These countries are assisted by the International Atomic Energy Agency (IAEA) which publishes basic safety standards (BSS) based on ICRP recommendations, and periodically updates them to conform with newer ICRP reports and recommendations. BSS is jointly sponsored by many organizations like the Food and Agricultural Organization, International Atomic Energy Agency, International Labor Organization, Nuclear Energy Agency of the Organization for Economic Co-operation and Development, Pan American Health Organization, and the World Health Organization. BSS apply to all practices, sources of radiation within practices and interventions that may be undertaken in certain circumstances. The regulatory authorities of many of the IAEA member states adopt BSS and bring out their own radiation safety regulations.

Figure 5.2 illustrates how the radiation protection recommendations are disseminated to the users in various countries:

The availability of a well-developed regulatory body alone is not enough to realize radiation safety, in a country. The licensee and the users have equal responsibility to understand the regulations and implement them. IAEA conducts many education and training programs with the member states for this purpose. Current regulations make hospitals accountable for the safety of all staff and patients exposed to ionizing radiation within the hospital for ensuring radiation safety. The regulatory bodies in many countries are authorized to stop any practice that involves unnecessary radiation exposures to

FIGURE 5.1
Publications of UNSCEAR, ICRP, and IAEA
(Basis of radiation protection recommendations & regulations).

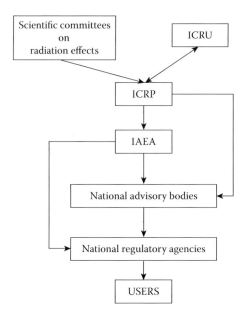

FIGURE 5.2
Dissemination of ICRP/IAEA recommendations.

patients, staff, or the public. Breaches of the regulations are a criminal offence under the Health and Safety Act and can result in criminal and civil liabilities in many countries, so education and training is very important.

USA Today (November 2000) reported an injury award of $1 million to a 57-year-old man who sustained serious skin injury after two coronary artery angioplasties that occurred 5 months apart. The patient filed a medical malpractice lawsuit alleging use of excessive fluoroscopy.

The radiation protection concepts continue to evolve with increasing use of technology and techniques in the uses of radiation in society. In this chapter, we will discuss the system of radiation protection, as developed by ICRP in recent years, and how it addresses the various situations for the safe use of radiation in radiation oncology.

5.2 Natural and Man-Made Sources of Radiation Exposure

There are two major sources of radiation exposure:

1. Natural background radiation (NBGR)
2. Man-made sources of radiation (used in society for beneficial purposes, e.g., in medicine, nuclear industry, agriculture)

While in general, we have no control over NBGR, man-made sources are under strict regulatory control. There is another component of radiation exposure known as "technologically enhanced radiation exposure" from natural sources. This refers to human activities that result in an enhanced dose from the natural sources. One example is the dose received by air crew who receive a much higher dose, due to their activity. This increment in dose can be classified as an occupational dose since the person receives the enhanced dose due to their occupation.

5.2.1 Natural Background Radiation Exposure

The three main sources of NBGR are

1. Cosmic radiation (coming from space)
2. Terrestrial radiation (that emanates from radioactivity in the soil, air, and water resources)
3. Internal (radiation) exposure from the radionuclides in the body

Figure 5.3 illustrates how the public receive radiation exposure from NBG radiation.

5.2.2 Cosmic, Terrestrial, and Internal Radiation Exposures

The dose due to both the cosmic radiation and terrestrial radiation depends on location and altitude. For instance, pilots receive more dose due to cosmic radiation compared to

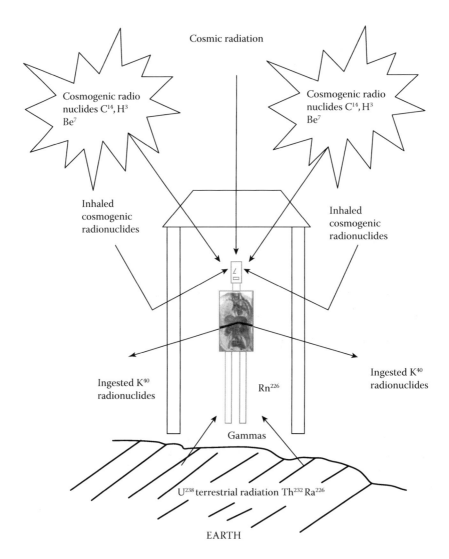

FIGURE 5.3
Natural background radiation exposure.

the population at sea level. The source of cosmic radiation is high energy (10^9 to 10^{20} eV) charged particles (protons, alphas, and heavy nuclei) from intergalactic space (and from solar flares) which interact with earth's atmosphere (mainly oxygen and nitrogen) producing showers of secondary radiation. The radiation reaching the earth is largely composed of muons, electrons, and photons. The photon component is mainly 511 keV photons arising from annihilation of positrons produced in nuclear reactions and gamma ray interactions. Because the earth's magnetic field can deflect the charged particles, the cosmic ray intensity also varies with latitude.

The cosmic ray interactions with the atmosphere molecules leave many nuclides in a radioactive state. They are known as cosmogenic nuclides (mainly H^3 and C^{14}). They enter the body through inhalation and ingestion, through eating and drinking. The contribution of this component to background dose, however, is negligible.

The second component is terrestrial radiation (mainly gamma rays) which arises from the radioactivity in the earth. The sources of radioactivity are the uranium and thorium deposits containing a series of radioelements under secular equilibrium. In certain parts of the world (e.g., Brazil, France, and India), there are large deposits of uranium and thorium giving rise to much higher terrestrial background radiation dose (about 5 to even 50 times the average values).

One component of radioactivity in air is of cosmogenic origin, as stated above. The other (and significant) component of radioactivity in air is of terrestrial origin. Here, the radioactive deposits in the soil contribute to the exposure. Ra^{226} in the uranium deposits transforms to Rn^{222} and Ra^{224}, and the thorium deposits transform to Rn^{220} (known as thoron). Both, being gaseous, seep through the soil and can become airborne. The radon and thoron concentrations in air (say inside homes) very much depend on the activity concentrations in the soil, the building materials used and the amount of ventilation provided. In mines and underground regions, the radon concentration can be very high. Rn^{222} and Rn^{220} emit gamma rays and are, therefore, like gamma ray sources in the ambient medium, giving rise to external exposure to the individual. Terrestrial radioactivity contributes only about 5% of the external exposure. What is of more significance is the airborne daughter products which attach themselves to aerosols and are inhaled by the population, giving rise to internal exposure (lung dose).

Water also contributes to the background dose since all sources of water contain radioactivity. Sea water contains large amounts of K^{40}, ground water picks up radioactivity from the soil, and the rain water from the atmosphere and the soil (uranium, thorium, and their decay products). So, drinking and cooking with water is a source of internal radiation exposure. Since other life forms also take in water (e.g., plants or animals), food is also a source of internal radiation exposure.

It must be remembered that alpha and beta emitters deliver very little dose to the exposed individual as sources of external exposure (because of their very low penetration) but can cause appreciable damage when taken inside by inhalation or ingestion. The main source of external exposure is gamma ray exposure from the radionuclides. In addition, the neutron fluences at ground level (produced by the interactions of cosmic rays with the atmosphere) also give rise to external exposure.

5.2.3 Radiation Exposure from Man-Made Sources

The population also receives a small dose from man-made sources of radiation due to their widespread use in society. In fact about 50% of the total dose received by the population arises from man-made sources. Here, the major contribution comes from imaging procedures in diagnostic radiology and nuclear medicine. A small contribution (about 2%) also comes from other sources like nuclear fallout (e.g., the aftermath of the Chernobyl accidents, atmospheric nuclear tests), consumer products (e.g., cigarettes contain polonium-210), and occupational exposures, radiotherapy practice, etc.

Figures 5.4 and 5.5 show percentage contribution from various sources of human exposures.

The percentage contribution from medical X-ray and nuclear medicine exposures will be higher in richer countries compared to middle- and low-income countries. Also, the radon dose will be higher in colder countries where the house ventilation is poor compared to warmer countries. Terrestrial exposures will be higher in high-background areas because of radioactive deposits. So, each country estimates its own values of NBG radiation for

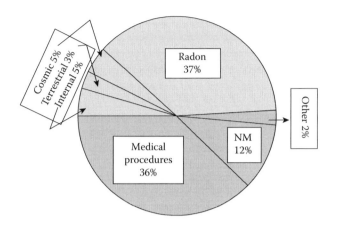

FIGURE 5.4
Radiation exposure due to natural and man-made sources of radiation.

A: Rn, thoron internal exposure; B: Terrestrial gamma exposure
C: Cosmic gamma exposure; D: Internal exposure (K⁴⁰)
E: Medical X-ray exposure; F: Nuclear medicine exposure
G: Other: Consumer products, fallout, occupational exposure

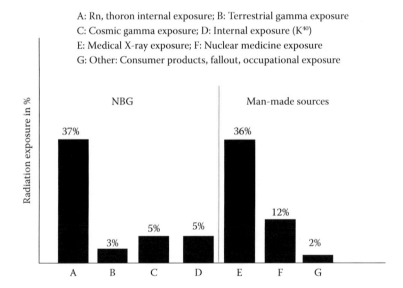

FIGURE 5.5
Percentage contribution of various sources of exposure.
Total dose to the individual (US data for year 2006, NCRP 160).

their population. Radiation accidents (e.g., Chernobyl) can significantly increase the local population exposure.

Figure 5.5 shows the values for US, as given in NCRP 160 [1]. The total dose (national average) due to NBG is around 3.1 mSv in the US, and the dose from man-made sources is about the same. The NBG dose received by the public, as a global average, may generally lie in the range 2 to 3 mSv.

There are regions where the NBG dose can be as high as 10 mSv or even higher, and studies in this region have not shown any statistically significant difference in cancer incidence compared to low NBG regions. Also, people moving from one place to another do not bother about the NBG dose variations. So, the NBG dose or its variations can be taken

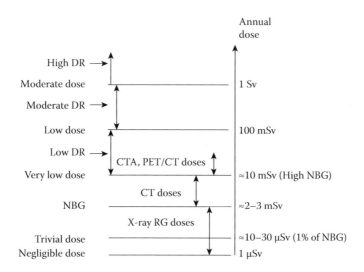

FIGURE 5.6
Concept of low and high dose relative to background dose (DR: dose range).

as a yardstick to judge the doses received by the patients in imaging procedures or occupational exposures, as being low or high. It has been suggested that an effective dose amounting to a few percent age of the NBG dose (say 50 to 100 μSv) can be taken to be a "trivial" dose.

Figure 5.6 compares the dose received during imaging procedures against the NBG dose.

The maximum doses received in diagnostic procedures are comparable with the doses received by the population from the NBG radiation, in high-background areas.

Example 5.1

What is the criterion to classify the effective dose received by an individual as trivial, or moderate, or high?

We are all subjected to an annual dose of about 2 to 3 mSv from the NBG radiation. There are many high-background areas where the NBG doses can reach values 10 to 100 times this value (i.e., about 20 mSv to 200 mSv). The highest NBG of 260 mSv was observed in Ramsar, Iran, and people have lived here for generations without exhibiting any ill effects of radiation. This must be kept in mind while talking of low or high dose in relation to radiation detriment. No health effects are observed for a dose of 100 mSv received over a short period or over a longer period (though there is a small risk of cancer, if we assume a linear no threshold dose response). Based on this, a dose 100 times less than the normal background dose can be considered as a trivial dose and a dose 100 times larger than the NBG dose as low dose. Since acute skin exposure of >1 Sv may exhibit temporary skin reactions, doses higher than 1 Sv can be considered as a high dose. This also helps to look at doses received in diagnostic procedures in a proper perspective.

5.3 ICRP System of Radiation Protection

The whole system of radiation protection, as developed by ICRP and adopted by other radiation protection bodies (applicable to all practices) can be explained from Figure 5.7.

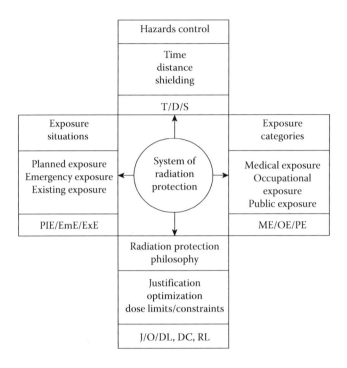

FIGURE 5.7
System of radiation protection.

Each module of the system, as shown in Figure 5.7, will be described briefly here. ICRP discusses radiation protection in the following order [2]:

1. Types of exposure situations
2. Categories of exposure
3. Principles

5.3.1 Classification of Exposure Situations

In the earlier recommendations, ICRP referred to two types of exposure situations—"practice" and "intervention". The activities that increased the radiation exposures to people were referred to as practice and activities that tend to decrease the radiation exposure to the people were referred to as intervention. For example, activities such as the introduction of radiation facilities (nuclear reactors, cancer centers, etc.), which increase radiation exposure to the people involved, were examples of practices. Activities such as trying to reduce the exposure levels from natural background or in an accident situation were examples of interventions.

Figures 5.8 and 5.9 illustrate this concept.

Though these terms are still very much in use, ICRP in the more recent recommendations (ICRP 103) generalized the system by classifying exposure situations to three main categories:

1. Planned exposure situation (PlES)
2. Emergency exposure situation (EmES) and
3. Existing exposure situation (ExES)

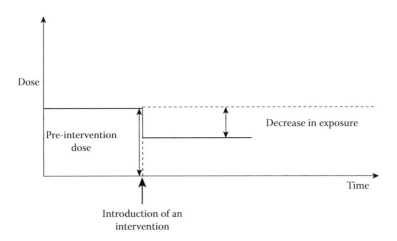

FIGURE 5.8
Concept of intervention (ICRP 60).

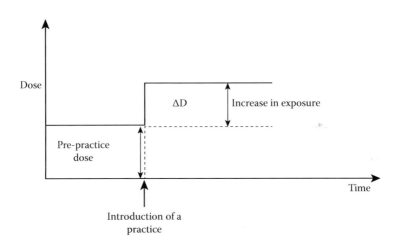

FIGURE 5.9
Concept of practice (ICRP 60).

An exposure situation comprises a source (of radiation), exposure pathways and exposures to individuals. Planned exposure situations are when the source is deliberately introduced and applied. The practices, in the sense of ICRP 60 (e.g., practice of radiology, installation of nuclear reactors), are planned exposure situations and are well under control. The people under all categories of exposure are well protected by explicit rules and regulations, and implementation mechanisms.

When a planned exposure situation gets temporarily out of control, due to an incident or an accident, it leads to an emergency exposure situation. In medical exposure, for instance, many unplanned accidental situations have been addressed in the design and operation of the equipment from past experience, or by anticipating some accidental situations and yet an emergency situation may arise due to operator error or equipment malfunction. Any emergency situation requires immediate action in order to avoid or reduce undesirable consequences. A series of steps can be planned to bring the emergency situation under control (i.e., the dose to the public must be brought below a level that is considered ALARA) taking into account economic and societal factors. A classic example of an emergency situation is the Fukushima nuclear accident that occurred recently in Japan, and it is very interesting to study how the Japanese government addressed this situation [3]. In the case of medical exposures, as in radiation oncology, the scenarios are different and, while most of the incidents are relatively minor in character, a few accidental deaths or severe injuries have occurred in radiation oncology due to overexposures. These will be discussed Chapter 12.

It takes considerable time to return to the near normal situation following an accident.

This is referred to as an existing situation. ExES are when the source already exists when decisions on control are taken. When an EmS becomes an ExS depends on many factors, and how the ExS is defined. Experience shows that the shift is determined by a decision of the national authorities. (Again, a study of the handling of the Fukushima accident will clarify these issues better, although they are not of much relevance in radiation oncology.) Both emergency and existing situations may require intervention to reduce the dose to the exposed population. For example, radiation exposure received from the natural radioactivity or the NBG (e.g., radon doses) are examples of ExES. Conducting nuclear tests can be seen as a radiological emergency and they add to the background radioactivity. This also gives rise to ExES following the declaration that the EmES has ended. There are efforts to address ExES by intervention to bring some control, but the action initiated depends on factors like the doses involved, size of the population exposed, in addition to the social and economic factors. According to ICRP, by adopting this situation-based approach, the ICRP system of protection becomes applicable to any situation of radiation exposure.

Figure 5.10 illustrates the concept of the three exposure situations on a time scale. Following an accident, a PlES (under regulatory control) turns into EmS (temporarily out of control) and when things settle down it turns into ExES.

Figure 5.11 shows the examples of PlExS and EmExS (in radiation oncology) arising from man-made sources of radiation and ExES (arising from NBG radiation). Radiation oncology is an example of a situation fully under regulatory control and any accidental medical exposure refers to the situation arising from temporary loss of control (which is eventually brought under control). NBG radiation is an example of ExES situation, arising from natural sources of radiation and is generally not amenable for regulatory control.

5.3.2 Categories of Radiation Exposure

The ICRP recognizes three categories of exposed individuals:

1. Medical exposure
2. Occupational exposure
3. Public exposure

To explain the exposure situations in a simpler language, medical exposure is exposure of patients for disease diagnosis or treatment. Exposure incurred by those assisting patients (carers) or volunteering for any biomedical research also fall under this category. Occupational exposure is radiation exposure of radiation workers as a result of their

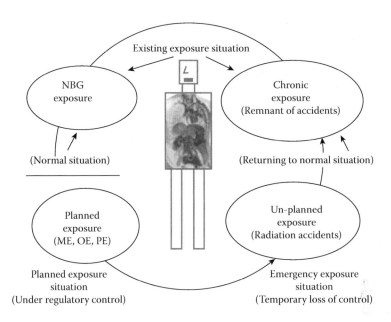

FIGURE 5.10
Exposure situations on a time scale (ICRP 103).

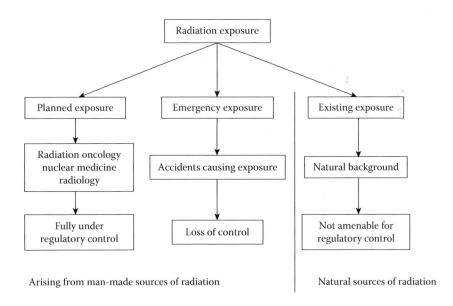

FIGURE 5.11
Illustration of radiation exposure under planned and emergency exposure situations.

occupation. Public exposure refers to the exposure incurred by the public as a result of radiation activities in the society. There is no direct benefit to the radiation worker from the exposure itself but may derive some indirect benefit through gainful employment or societal benefit. (Of course, you may also view this as direct benefit to the radiation worker, though the above way of looking at it looks more reasonable). Similarly, there is no direct benefit to the carers or volunteers from the exposure itself but they may derive an indirect benefit (e.g., a relative holding a child while taking an X-ray and helping in the diagnosis). The patient, however, derives direct benefit from the exposure itself (e.g., disease diagnosis or treatment). There is no direct benefit to the public because of the exposure (e.g., living in the vicinity of a nuclear installation) but there is an indirect benefit (since the society benefits and they are part of the society).

Of the three categories, occupational exposure assumes special significance since the staff or the radiation protection workers work for the societal benefit. Their risks cannot exceed the risk-based dose limits recommended by ICRP or the national regulatory bodies, for the planned exposure situations. This involves staff dose assessment and optimization. ICRP 60 further discusses the concept of "acceptable" and "tolerable" risk with respect to certain reference values defined for various exposure situations. For instance, the background radiation dose can be taken as a reference for discussing the tolerable risk of population exposure.

Sometimes, there is talk of potential exposure. Potential exposure refers to the probability but not certainty of the occurrence of an exposure. For instance, there is a small probability of a radiation worker or member of the public receiving some radiation by entering the treatment room during treatment, if radiation safety rules are not properly observed. The teletherapy source may fall down during the source loading operations giving rise to accidental exposure to the personnel involved. (Such accidents have happened before. See Chapter 12 for more details.)

The primary means for controlling potential exposures is by optimizing the design of installations, equipment, and operating procedures

1. To restrict the probability of events that could lead to unplanned exposures
2. To restrict the magnitudes of the exposures that could result, if such events were to occur

Similarly, the terms "acute exposure" and "chronic exposure" are often mentioned in radiation protection. Acute exposure refers to generally large doses received over a short period of time (say a few minutes to a few days to a few weeks) and chronic exposure refers to generally small doses received over a long period of time (over several years or decades). Acute exposures are only involved in accident situations like nuclear disasters (e.g., the people received acute doses of radiation during the atom bomb explosions that occurred in Japan during World War II). Such exposures can give rise to acute radiation syndrome (ARS) which is generally described in books on radiobiology. High-radiation exposures are highly unlikely in radiation oncology where the "defense in depth" concept (that involves multiple layers of safety) makes the occurrence of such exposures rare.

In radiation protection, the main concern is with a chronic exposure situation (small amounts of dose received regularly over a long period of time). Occupational exposure is a classic example of chronic exposure. This is considered as less detrimental since tissues have time to repair the damage, although presently this is being questioned in certain quarters.

Example 5.2

Who are referred to as comforters, carers, and volunteers in biomedical research? The dose received by them comes under what category of radiation exposure and exposure situation?

The family members, or close friends, or relatives of a patient while helping the latter (say holding a child for a radiograph or nursing a radioactive patient at home) are referred to as carers or comforters. Those receiving a small (recommended) level of exposure as part of a biomedical research program are referred to as volunteers.

These exposures come under planned exposure situation and are classified as medical exposures.

5.3.3 Principles of Radiation Protection

Since radiation stochastic effects (cancer or hereditary effects) at exposure levels below 100 mSv are not unequivocally demonstrable, it is always prudent to assume that any dose of radiation, however low it may be, involves an element of risk proportional to exposure. All the radiation protection bodies make this conservative assumption in their recommendations and regulations. This implies that while radiation can be used for its beneficial effects, all controllable exposures (i.e., exposure under any category of exposure or exposure situation) must be justified and optimized. These aspects are briefly dealt with here.

5.3.3.1 *Justification*

Justification implies that the use of radiation should do more good than harm (benefit vs. risk) to the exposed people, in any exposure category. This is decided by the professional bodies (or national authorities, or the government). In the case of medical exposures, the justification operates at three levels:

1. General level: the use of radiation in medical practices—consensually considered as justified
2. Generic level: at the procedure level—the procedure must be justifiable for positive benefit
3. Individual patient level—the procedure must be justifiable as proper for the patient in question (clinical responsibility)

5.3.3.2 *Optimization*

All radiation exposures must be optimized. In other words, justification of an exposure does not confer the right to give any amount of dose to the patient, the staff, or the public. This simply means unnecessary exposures should be eliminated or reduced to ALARA levels taking into account economic and societal factors. For example, optimization of medical exposures implies the radiation dose (e.g., patient dose) utilized to realize an objective must be as low as reasonably achievable, without compromising the clinical objective. It must be noted that dose reduction of medical exposure, by itself, is not the objective of optimization since radiation practice is necessary for the care of the patient. Achieving the clinical objective is of primary importance. For instance, in the case of medical exposure, if the clinician misses the diagnostic information, however low the patient dose may be, it becomes unjustified exposure. Similarly, in radiotherapy treatment of patients, "tumor dose to the normal tissue dose" must be optimized (Figure 5.12).

Of course, the social and economic factors play an important role and must be taken into account in optimization. If a new technology developed in imaging equipment reduces patient dose compared to the older technologies, it is not practical to insist that all the hospitals must switch over to the newer equipment. The social and economic considerations

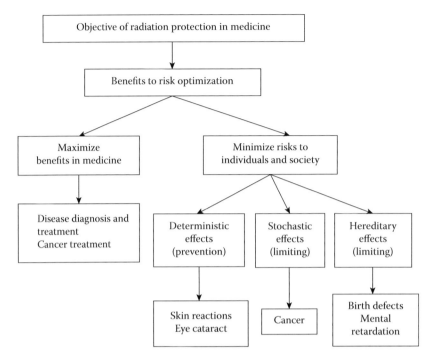

FIGURE 5.12
Concept of optimization in medical exposures.

will dictate the kind of equipment in use. For instance, in the design of radiation installations, efforts must be made to reduce staff doses as much as possible but not to the extent that the costs far outweigh the benefits. In the case of occupational and public exposure, the likelihood of incurring exposures, the number of people exposed, and the magnitude of their individual doses should all be kept as low as reasonably achievable, taking into account economic and societal factors.

ALARA (as low as reasonably achievable) is the key to optimization.

5.3.3.3 Dose Limitation

There is a subtle difference between medical exposures on the one hand and occupational and public exposures on the other. In the case of medical exposures, the benefit and the risk go to the same individual. So, as long as the benefit outweighs the risks, the medical exposure is clinically justified. This is a clinical decision and the physicians working in this area are the competent ones to make this decision. As previously mentioned, in medical practices, the dose to the patient is commensurate with the clinical objective and hence no dose limits are prescribed even when medical exposures are classified under PlExS. The dose limits prescribed for the staff or the public are irrelevant for medical exposures (justification and optimization are of course necessary, as described earlier).

However, medical staff are in the same situation as radiation workers, occupationally exposed in any other PlES. Similarly, public exposures due to medical practices should be managed like public exposure in any other PlES. The occupational dose limit applies to medical staff and the public dose limit applies to public exposures (i.e., exposures to members of the public other than patients) arising from medical practices. These dose

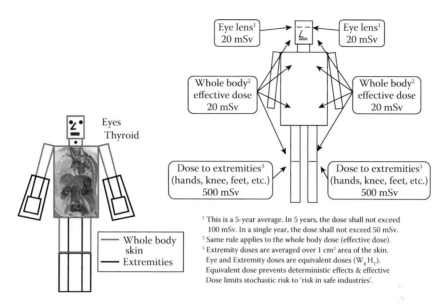

FIGURE 5.13
Annual dose limit to radiation worker (ICRP 103).

limits are illustrated in Figure 5.13. The objectives of the equivalent dose limits are to prevent deterministic effects to sensitive organs such as the extremities and the eyes, and to limit stochastic risk of cancer and hereditary effects.

Several national regulatory bodies have not adopted the ICRP 103 dose limits. Some national regulations still retain the 50 mSv/y dose limit for every year. The regulatory body in India has adopted the 20 mSv/y dose limit but reduced the single year maximum to 30 mSv/y. The users must follow the national regulations since ICRP is only a recommendatory body.

Some of the important characteristics of the dose limits:

- The limits apply to adult radiation workers.
- The 20 mSv dose limit recommended by ICRP in its Publication 60 (1991) was still considered acceptable in its Publication 103 (2007). To allow some flexibility, the 50 mSv limit for any single year was retained but with the provision that the 5-year average would not exceed 20 mSv. However, very few workers ever reach even the 20 mSv limit. In most of the occupations, the annual doses received are one-tenth of this value.
- They apply to only planned exposure situations and not to emergency exposure situations, though they can be used as a guideline for such situations.
- The medical exposure, or NBG radiation exposure, or even emergency handling exposure received by a radiation worker belong to separate "bank accounts" and will not form part of the occupational exposure.
- If the radiation worker receives internal exposure due to any radioactivity taken into the body through ingestion or inhalation, then the limit applies to the total annual dose received from both external and internal exposures. In radiation oncology, any internal exposure (from the teletherapy and brachytherapy sources used for cancer treatment) is very unlikely.

TABLE 5.1

Dose Limits for Radiation Workers and the Public

Dose Limits	Occupational Worker (Staff) (mSv)	Public (mSv)	Apprentices (mSv)
Effective dose	20 (5 y average) ≤ 50 (any single y)	1[a]	6[c]
Equivalent dose to eye lens	20 (5 y average) ≤ 50 (any single y)	15[b]	20
Equivalent dose to hands, feet, and skin	500	50	150

a In special circumstances, a higher value can be allowed in a single year provided the 5-year average does not exceed 1 mSv.

b In the earlier recommendation, staff annual eye dose limit was 150 mSv and one-tenth of that was recommended as the dose limit for the public. The same value is retained here, in the recent ICRP 103 recommendation.

c Apprentice dose was taken as three-tenth of the adult dose, but the adult value itself is retained now, for the apprentice, to protect the eyes.

- For a fetus, the public dose limit of 1 mSv applies for the duration of the pregnancy. So this would limit the dose to a pregnant radiation worker for the period of the pregnancy.

Table 5.1 compares the occupational dose limits with dose limits for the public and 16–18 age group workers (apprentices), as per ICRP 103.

Example 5.3

According to ICRP 103, what is the maximum annual dose permitted in a single year?

The maximum annual dose permitted in a single year is 50 mSv. But this is qualified by the statement that in a 5-year period, the dose should not exceed 100 mSv. This means that on average, the permitted annual dose is 20 mSv but if the dose exceeds this value (say it is 30 mSv) and is justified then an allowance should be made for the excess dose—in the remaining 4 years the cumulative dose should not exceed 100–30 mSv.

Example 5.4

What is the dose limit recommended for a pregnant radiation worker?

The fetus is treated as a member of the public and so the annual public dose limit (1 mSv) applies to the fetus as well, for the period of the pregnancy. At the end of the pregnancy, the normal limit applies. However, the declaration of pregnancy is voluntary. If the radiation worker chooses not to declare her pregnancy, the normal dose limit applies.

5.3.3.4 Dose Constraints/Reference Levels

The principle of optimization introduces some dose constraints in all radiation exposures. The term "dose constraint" has been used with different meanings and connotations in radiation protection. Here we generally refer to dose levels set for performing an action as a dose constraint, which are sometimes given special names. Dose constraints are not legally enforceable dose limits but are some sort of "optimized dose limits" set to initiate follow-up actions to optimize the exposure, when exceeded.

5.3.3.5 Dose Constraints for Occupational and Public Exposure in a Planned Exposure Situation

Optimization has led to significant reduction in individual doses over a period of time and there is a realization that the dose limit is not the border between "harm and no harm" and so ALARA must be practiced at all levels. Dose limits do not play a significant role in the system of radiation protection. For instance, the average dose received by staff in radiation oncology is about 2 mSv/y and any individual receiving even 5 mSv/y is inappropriate from the ALARA point of view, and needs to be investigated. Therefore, 5 mSv/y can be set as an investigation level in radiation oncology departments. This is some form of dose constraint and implies that optimization is still possible. The constraint is an upper bound on the dose in the optimization of protection in occupational exposures. Figure 5.14 illustrates this concept.

The dose limit for the public is related to the individual and should not be exceeded from all the sources and practices permitted in society. Since the public are not monitored, like the radiation workers, source-related dose constraints are often introduced (e.g., an upper bound of about 0.3 mSv dose to public from single practice like putting up a power reactor). Many practices (e.g., a cancer center or an X-ray diagnostic center in the vicinity) contribute very little dose to the public.

In the case of medical exposure (patient dose), there is no dose limit imposed. The justification/optimization requirement reduces the patient doses in diagnostic radiology and normal tissue doses in radiation oncology to ALARA levels. In patient imaging procedures, diagnostic reference levels (DRLs) have been defined which can be thought of as a dose constraint and any dose much lower or much larger than the DRLs need to be clinically justified.

There are two categories of special medical exposures involving comforters and carers, or volunteers in biomedical research. Carers and comforters are individuals (usually the relatives of patients) who may receive some exposure in the support and comfort of the patients (e.g., holding a child while taking a radiograph) or volunteers who would receive some exposure while subjecting themselves to an exposure in biomedical research. There is a third category of exposed people—the relatives in the household of an I-131 patient who has been

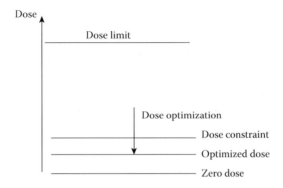

FIGURE 5.14
Concept of dose constraint and optimization in planned exposure situation.

discharged from the hospital with some residual radioactivity in the body. Dose constraints of few mSv is usually recommended for these exposures.

5.3.3.6 Reference Level for Emergency and Existing Exposure Situation

The dose bounds set in EmES and ExES conditions are known as reference levels (RLs). These exposure situations require intervention to bring the situation under control, and the doses to appropriate levels. The process would give rise to occupational and public exposures depending upon the nature and scale of the accident. The intervention needs justification and optimization to reduce doses to radiation workers and the public to ALARA levels. The national authorities or the regulatory bodies set the reference levels. ICRP recommends setting RLs in the band 20–100 mSv for the protection of the radiation workers in the emergency situation, but it depends on the magnitude of the accident. For instance, in the Fukushima nuclear event in 2011, the Japanese authorities set an RL of 100 mSv which was later revised to 250 mSv when it was discovered that the scale of the event was worse than anticipated. While it is difficult to anticipate accidents and set RL values, two dose values that must be kept in mind while setting reference levels—500 mSv below which radiation effects are barely noticeable, and 1 Sv, the threshold for tissue reactions to show up and which should not be exceeded. RLs, like dose restrictions, are subject to the optimization principle. For life-saving measures, no dose level can be assigned. In the case of the public, measures such as evacuation are usually undertaken to prevent any exposure.

ICRP defines three bands of dose ranges, which can be used for setting dose constraints or RLs depending on the situation. Figure 5.15 illustrates the concept.

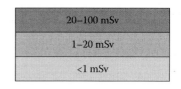

FIGURE 5.15
Bands of dose constraints and reference levels (ICRP 103).

The DC/RLs may be set for the following categories of personnel, from the bands of increasing doses.

- Lower band: <1 mSv

  Individuals need not derive any direct benefit from these exposures. The added dose is less than the dose from NBG. No training or dose assessment is required.

- Mid band: 1–20 mSv

  Individuals derive direct or indirect benefit. Actions include setting occupational exposure dose constraints; setting intervention (RL) to reduce the radon dose in dwellings, and establishing dose constraints for carers or volunteers; setting RL for the existing exposure situation, for the responders/workers following the emergency. Training and dose monitoring are essential.

- Upper band: 20–100 mSv

  Emergency exposure situations—RLs need to be set for interventions to bring the situation under control.

Optimization is an integral part of all these activities as illustrated in Figure 5.16.

Table 5.2 summarizes the discussions.

The EmES and ExES generally apply to disasters such as a nuclear reactor accident (Chernobyl accident) or nuclear bomb disasters, and are not of much relevance to medical exposures where the concern is with planned exposure situations and incidents of over or under exposures. However, these incidents would not be managed using the strategies

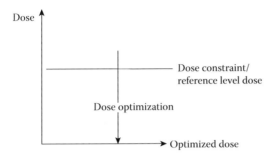

FIGURE 5.16
Optimization in existing and emergency exposure situations.

TABLE 5.2

Exposure Situations and Categories

Exposure Situation	Occupational Exposure	Public Exposure	Medical Exposure
Planned exposure situation	Dose limit dose constraint	Dose limit dose constraint	Diagnostic reference Level[b] dose constraint[c]
Emergency exposure situation	Reference level[a]	Reference level	N/A
Existing exposure situation	N/A	Reference level	N/A

a Long-term measures must be considered as part of planned occupational exposure
b For patients undergoing X-ray diagnostic examinations
c Only for carers, comforters, and volunteers in biomedical research
   N/A: not applicable

and tools provided in the ICRP system for EmES and ExES. The whole discussion on the concepts of radiation protection in the safe use of radiation in society, as developed by ICRP, may be very confusing, especially for the first-time readers.

To refocus on our objectives, Figures 5.17 and 5.18 illustrate the objectives of radiation protection in medical exposures.

5.4 Establishment of Regulatory Body

Today, most countries have developed a framework for regulating the use of man-made sources of ionizing radiation in society. National regulatory agencies produce rules and

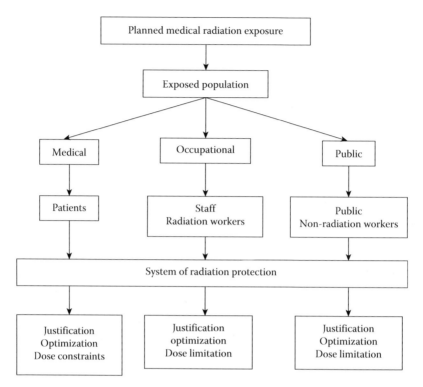

FIGURE 5.17
System of radiation protection in medical radiation exposures.

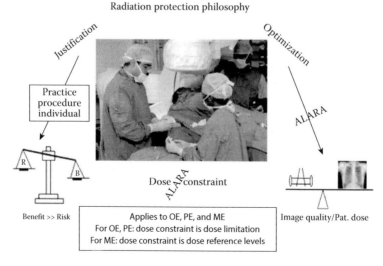

FIGURE 5.18
Concept of justification, optimization, and dose limitation/constraint in radiation oncology.

regulations on the safe and responsible use of radiation in various applications (including in radiation oncology) and enforce them to ensure the safety of all. It is mandatory for the users to follow the regulations issued. Regulations are legally binding and any violation can lead to penal action such as closing down the installation.

The regulatory body must have legal authority to enforce the regulations. This is usually achieved through legislation, for example, a parliamentary act (e.g., as in India and Europe). The legislation would establish a basic framework of the national infrastructure and may also issue additional regulations or rules to control the safe use of radiation in all or in specific applications, for example, in medicine. In some countries, the legislation may sometimes authorize the regulatory body to issue the rules and regulations, instead of issuing them at the government level. The employer is primarily responsible for the radiation protection, and also the safety of the sources, though others too become responsible through the delegation of responsibility (e.g., the radiation safety officer [RSO]).

In the US, the regulatory requirements are given in the Codes of Federal Regulations (CFR). The regulations of relevance here are 10 CFR20 [4] (NRC Basic Standards for Radiation Protection) and 21 CFR 1020 [5] (FDA Performance Standards for Ionizing Radiation Emitting Products). The CFRs are updated annually. In the US, a number of government agencies are involved in implementing the regulations, such as the Nuclear Regulatory Commission (NRC) [6], Food and Drug Administration (FDA) [7], Environment Protection Agency (EPA) [8], and the Department of Transportation (DOT) [9]. NRC controls the use of radioactivity (e.g., Telecobalt unit, HDR unit, brachytherapy sources), and FDA controls the use of other equipment (e.g., linac, X-ray machine). Agreement states (that have entered into agreement with NRC) exercise the regulatory control, in their respective states, over the use of radioactive substances or radioisotope machines, in place of NRC. EPA deals with environment protection and DOT controls the transport of radioactive materials. An FDA-approved body controls mammography. The FDA also sets standards and ensures compliance with the Code of Federal Regulations. However, the US is an exception. In the majority of countries, there is a single body regulating the use of ionizing radiation. Regulatory bodies often publish regulatory guides (e.g., the NRC) which assist in the interpretation of the regulations. Following the procedures, as given by the regulatory agencies or other professional bodies in their published guides, is an easy way of ensuring compliance with regulations.

5.5 Some Basic Regulations and Responsibilities in Radiation Oncology

5.5.1 Consenting Process for Exercising Control

The RB exercises regulatory controls over all the radiation facilities in the country (e.g., a nuclear reactor site or a radiation oncology practice). These radiation facilities require regulatory consent (e.g., license, registration, and authorization) for possession of radioactive sources and operation of radiation-generating equipment. The type of consent granted depends on the type of equipment and the potential risk involved. A license is applicable to the highest hazard radiation sources (such as radiation oncology equipment, CT) and registration to the lowest hazard sources (such as radiography equipment, therapy simulator). The practices and devices having minute quantities of radioactive materials (consumer products like smoke detectors come into this category) are exempted from regulatory approval. The consenting process includes three main stages—pre-licensing,

licensing (during the useful life of the equipment), and decommissioning of the radiation source or facility, at the end of its useful life.

5.5.2 Responsibility for Enforcing Regulations

The licensee or the employer is primarily responsible for enforcing the regulations, but due to the complexity of radiation oncology practice, the responsibilities or the tasks are delegated at various levels with individual or collective responsibility as shown in Figure 5.19.

The principal staff members responsible for radiation protection are (1) the employer, (2) the medical practitioner (radiation oncologist), (3) the medical physicist (radiation safety officer), and (4) the technician (person administering the radiation). The suppliers of the equipment have the responsibility to demonstrate that the equipment meets the required performance standards and the purchase specifications.

5.5.3 RSO Responsibilities

One of the prerequisites for licensing of a radiation oncology facility is the employment of a qualified radiation safety officer (RSO). The RSO should be given the resources and necessary authority to ensure radiation safety rules. The main responsibilities of an RSO are

- Advising the licensee on radiation safety and regulatory compliance matters
- Assisting the licensee in the creation of a radiation safety committee

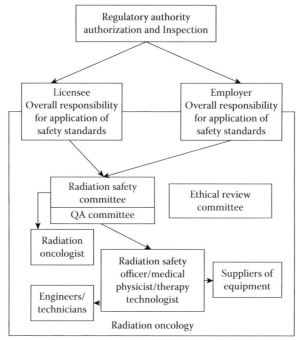

Safety to be ensured for staff, patients, and the public

FIGURE 5.19
Delegation of tasks for enforcing regulations.

- Preparing a radiation safety manual
- Procuring all the radiation protection tools
- Implementing all safety measures to minimize dose to patients, staff, and the public
- Installation, planning, and procuring the approval of the regulatory authority for necessary construction
- Designating areas as controlled/uncontrolled areas
- Posting signs (for control areas)
- Conducting area monitoring and radiotherapeutic equipment survey
- Ensuring area monitor calibration and maintenance
- Installing equipment, arranging acceptance testing, and commissioning data generation
- Conducting equipment and installation surveys
- Monitoring staff using personal monitoring devices
- Maintaining (individual and area monitoring) records
- Investigating overexposures
- Reporting radiation incidents to the radiological safety committee and the regulatory authority, and taking corrective measures
- Ensuring safety of brachytherapy sources
- Emergency response plans

To keep the occupational exposures at ALARA levels, there are some administrative controls exercised in radiation oncology like area designation, investigation of over exposures, area, and personal monitoring which have been listed above.

5.5.4 Area Designation

Many countries define two types of areas—controlled and uncontrolled. The controlled area is also known as a designated area, and the uncontrolled area is a non-designated area. A control area is an area with control of access and is accessible only to radiation workers with work responsibility in the area, and whose doses are subject to an occupational dose limit. The staff authorized to work in controlled areas in radiation oncology are the radiation oncologists, the medical physicists, the dosimetrists, the therapy technicians, and the oncology nurses. Non-classified workers such as maintenance personnel may be authorized to work in the control area under special circumstances with some dose constraints, for example, where the dose does not exceed 3/10 of the occupational dose limit.

In some European countries, the control area is defined as one where the likelihood of radiation exposure will exceed 3/10 of the annual dose limit (i.e., 6 mSv). If the dose received in the area is unlikely to exceed 6 mSv, the area is designated as a supervised area. According to this classification, the control rooms will become supervised areas since the annual dose received by staff in this area amounts to less than few mSv. Generally, personal monitoring is not used in these areas and so the RSO must ensure that those working in these areas will not exceed the specified limits by assessing the doses. Though these areas are not physically segregated, they will be clearly demarcated to recognize the area and the occupational exposures in this area would be kept under review. Non-radiation workers may be allowed in this region through managed access. In radiation oncology, designation of a supervised area does not seem to give any special benefit in radiation protection, and many countries have only two areas designated for radiation protection—controlled and uncontrolled. Of course, the areas must be designated according to the country's regulations. Here we will use only two designated areas—controlled and uncontrolled.

The uncontrolled areas are accessible to the public, so this is where the doses are subject to public dose limits. These areas are occupied by the patients and visitors to the facility, and other employees of the hospital who would be moving around. Of course, all areas beyond the radiation installation are uncontrolled areas. There is no control of access, and no workplace monitoring, or personal monitoring, is required in these areas. Any kind of radiation work is not forbidden in an uncontrolled area but the designated dose limits (for the public) cannot be exceeded in any radiation activity.

Examples of controlled areas:

- All irradiation rooms for external beam radiotherapy
- Remote afterloading brachytherapy treatment rooms
- Operating rooms during brachytherapy procedures using sources
- Brachytherapy patient rooms
- All radioactive source storage and handling areas

Controlled areas

- Are physically segregated
- Have access restrictions
- Require interlocks where appropriate
- Require radiation/radioactive signs
- Require protective equipment and monitoring
- Require staff to follow written procedures

5.5.5 Posting of Areas

For the protection of the public and other staff of the hospital from any inadvertent or unnecessary radiation exposure, appropriate legible notices and warning signs with the radiation symbol must be prominently posted at strategic points in the radiation oncology department. The control room door is the most important place for the posting of the sign. Figure 5.20 shows the signs generally used for this purpose.

5.5.6 Personal Protective Devices

It is mandatory to use protective devices like a lead apron in diagnostic radiology (effective energy less than 100 keV), but such personal protective devices are not much use in radiation oncology, where the photon energies are very high (Figure 5.21).

The lead aprons used in diagnostic radiology have about 0.5 mm of lead equivalent, which transmits <5% at 100 keV but transmits >90% even at 511 keV. In radiation oncology,

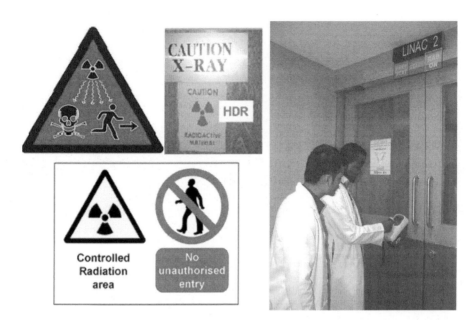

FIGURE 5.20
Posting signs in areas to prevent unauthorized entry.

FIGURE 5.21
Lead apron used for protection in diagnostic radiology.

the safety is incorporated into the design of the installation except in manual brachytherapy where much thicker lead equivalent bed shield offers some protection to the nursing staff. (The nurse can't attend to a brachytherapy patient wearing a lead apron, as they need thick lead shields between them and the patient. Similarly the medical physicist uses an L bench to handle brachytherapy sources.)

The following section illustrates the typical steps involved in starting and running a radiation oncology facility.

5.5.7 Regulatory Requirements for Starting and Running a Radiation Oncology Facility

To obtain regulatory consent for starting a radiation oncology facility [10], the following conditions must be met:

1. Employment of qualified personnel (radiation oncologist, medical physicist, RSO, therapy technologist, etc.)
2. Installation of type-approved equipment or a "no objection" certificate (NOC) for new equipment
3. Layout approval
4. Availability of suitable monitoring and dosimetric instruments
5. Availability of associated equipment and accessory tools (simulator, TPS, etc.)
6. Equipment installation
7. Acceptance testing
8. Commissioning approval
9. License for operation

In the case of a brachytherapy facility, the HDR equipment would require a source, for which the following conditions must be met:

1. Authorization for source procurement
2. Source transport approval
3. Receipt of source advice
4. Source loading

Post licensing:

1. Periodic QA and QA following any repairs or modification
2. Filing of annual radiation safety status to the regulatory agency
3. Advising of any change in working conditions (e.g., changes in occupancy)
4. Reporting of accidents and incidents

Post useful life of the equipment

- Decommissioning/disposal approval

These steps will now be briefly described.

5.5.7.1 Employment of Qualified Personnel

The RB will ensure that suitably qualified and adequate personnel (radiation oncologist, medical physicist, RSO, therapy technician, etc.), as stipulated in the regulations, have been employed by the institution, prior to the granting of approval. The RB in India requires that a minimum of one radiation oncologist, one medical physicist, and two radiotherapy technicians per unit are employed by the institution.

5.5.7.2 Installation of Type-Approved Equipment or NOC for a New Equipment

Before ordering any radiation equipment, it is important to ensure that the unit is type approved by the regulatory body. It is an offence to install non-type approved radiation equipment. The type approval certificate must be physically verified by the user. In the case of new equipment, which requires import from another country, an NOC must be obtained from the regulatory body for the import of the equipment. This requires filing support documents along with the application for NOC, such as specifications of the new equipment, QA report, any clearance from the regulatory agency from the country of origin, any prior installation of equipment. Issuance of NOC does not amount to approval of the equipment. On receiving the NOC, the institution can import and install the equipment, but the supplier must submit the QA reports to the regulatory body for type approval certification. On scrutinizing the report, the regulatory body will type approve the equipment, provided it passes all the criteria regarding performance and radiation safety. Following the type approval, the supplier can import the equipment for other institutions without NOC, until the validity of the approval certificate expires—which is usually three years—following which the type approval must be renewed.

The purchased equipment must conform to applicable standards of the International Electrotechnical Commission (IEC) and the ISO, or to equivalent national standards. All type-approved equipment installed in the country must be acceptance tested to ensure that it continue to meet the standards set.

5.5.7.3 Layout Approval

The site layout drawings and room layout drawings, to the prescribed scale, must be submitted to the regulatory body along with the application form. The RB will grant approval to the institution after evaluating the drawing from the radiation safety point of view, and finding it acceptable. The institution must wait for this approval before starting the construction. An RSO is a mandatory requirement for receiving the approval. The medical physicist can be nominated as RSO if they have passed the RSO exam. (This requirement may vary from country to country. In India, there is only one institution, the Bhabha Atomic Research Center, approved by the RB for offering the RSO exam.)

The machine must be installed as per the approved plan. In case of any modification (e.g., changes in room shielding or changes in the location of the treatment unit), the new plan, based on the changes, needs to be reapproved before starting construction. No modification can be effected without RB consent.

The RB will also ensure that the radiotherapy facility is away from the unconnected facilities and close to the associated facilities (e.g., simulator room, mold room, TPS room, or a minor operating theatre adjacent to the HDR room). It is desirable to have the facility in one corner of the hospital instead of in the middle and shielding becomes easier if the treatment room is in the basement. It must also be convenient to move the equipment to the basement, if this is available.

In the case of radiotherapy simulator or CT simulator installations, the same regulations apply—prior approval for layout and source, or equipment procurement.

The receipt of the source and equipment must be advised to the RB within 2 weeks (or as the country's regulations require) from the date of receipt.

5.5.7.4 *Radiation Protection Monitoring and Clinical Dosimetry*

Two types of instruments must be available with the radiation oncology department (Figure 5.22):

1. Area and personal monitoring instruments
2. Clinical dosimetry instruments

In the case of Telecobalt and Ir-192, or Co-60 HDR equipment, there can be radiation in the room if the source has not completely retracted to the parking position, with the machine turned off. This can be indicated by an area monitor or a zone monitor and so it is necessary to install these in such rooms. This is not necessary for linac rooms, but if the regulation demands it, the institution must comply; such a regulation existed in Canada some years ago. According to some regulatory agencies (e.g., the RB in India), the gamma zone monitor for the Telecobalt unit and remote afterloading brachytherapy unit should be of the autoreset type, whereas, for manual brachytherapy it must have manual reset button. A radiation protection survey meter must also be available for treatment room or radiation equipment monitoring. For pulsed-type radiations (e.g., linac), an ion chamber–based survey meter must be used; a GM-based system is inappropriate and may give wrong values. A GM survey meter can be used for Co-60 or Ir-192 HDR rooms.

Personal monitoring badges must be available for all the radiation workers, during the installation of the unit. A pocket dosimeter or an electronic personal dosimeter (EPD) may be convenient for detecting any abnormal or accidental exposures, or for monitoring daily or weekly exposures. It is essential during source transfer operations and may be insisted upon by an RB. The PM badges must be procured from an accredited or authorized service provider.

To perform clinical dosimetric measurements, clinical dosimeters need to be procured. Plane-parallel chamber types for electron dosimetry, and a Farmer-type chamber for photon dosimetry are mandatory for a radiation oncology department along with a suitable electrometer and temperature/pressure monitors. A re-entrant chamber and electrometer must be available for Ir-192 or Co-60 HDR source calibrations. The dosimeters must have suitable calibration, traceable to an accredited dosimetry calibration laboratory, for measuring the beam output in external beam therapy or the source strength in brachytherapy, according to the national or international dosimetry protocols adopted by the country.

(a)   (b)

FIGURE 5.22
(a) Protection monitoring and (b) clinical dosimetry instruments.

5.5.7.5   Associated Equipment and Accessories

All the necessary associated equipment for the practice of radiotherapy such as the treatment planning system, planning CT or radiotherapy simulator, or mold room unit must be available within the department before the RB can issue a commissioning certificate for patient treatment. The accessories required are the patient immobilization devices, QA tools, movable shields for brachytherapy, etc.

5.5.7.6   Equipment Installation

The installation of a radioisotope unit (e.g., Telecobalt unit or ^{192}Ir HDR unit) involves an additional step of source loading. In the case of the Telcobalt unit, the source transfer operation must be conducted by a qualified and authorized service engineer under the supervision of an authorized medical physicist/RSO. The medical physicist concerned should apply to the RB and obtain prior authorization before initiating the source loading operation. On completion of the operation, the RSO files a source transfer report, in the prescribed format, to the RB.

5.5.7.7   Acceptance Testing

When switching on the radiation for the first time, the first measurement must be a radiation protection survey of the installation, under the maximum scatter conditions, for the secondary walls and maximum output conditions (i.e., without a phantom at the isocenter), and for the primary walls. If the radiation levels in the surrounding areas exceed the regulatory limits, the RB must be advised and all measurement activities have to be suspended until the situation is rectified.

The medical physicist performs acceptance testing of the equipment (after confirming the shielding adequacy of the installation) to ensure that it meets the purchase specifications and also complies with all regulatory requirements. The AT report is kept in the departmental records and must be readily available during regulatory inspections.

5.5.7.8   Commissioning Approval

The licensee applies to the RB to obtain a commissioning certificate before initiating patient treatment. The RB will issue a commissioning certificate after ensuring that the licensee has fulfilled all the required conditions.

5.6.7.9   License for Operation

The licensee is ultimately responsible for all licensed activities including radiation safety, even though the management may delegate the responsibility to the radiation safety committee or to individual staff members. Most of the regulations only say that a licensee must be responsible for "such and such a thing". So, it is important for the licensee to ensure that the personnel with delegated responsibility actually discharge the duties as per the requirement and that all necessary documents are kept ready for regulatory inspections.

In the case of a brachytherapy facility, the HDR equipment requires a source, for which the conditions mentioned in the earlier section must be met.

5.5.7.10   Public Safety

Public safety is generally ensured by the design of the radiation installation and limiting public access to the radiation areas by posting appropriate warning signs. It is mandatory to post the signs in key places to prevent entry of public to radiation areas. A radiation symbol is posted on the treatment room door and also on the door of radioactive source storage rooms.

5.5.7.11 Post Licensing

Post licensing, regular QA testing of the equipment must be performed, as per the regulatory requirements, to ensure that the equipment continues to meet the minimum performance standards required for clinical use. QA testing is also essential following any major repairs or modifications to the equipment.

The licensee must file an annual safety report on the radiation safety status of the department, in the prescribed form.

The licensee must also report any incidents or accidents that occurred in the department, to the RB.

5.5.7.12 Decommissioning

It is important to ensure safe disposal of radioactive sources and segregation and disposal of depleted uranium and contaminated parts. This requires prior permission from the RB for decommissioning, and for transporting sources for safe disposal. The teletherapy head, for example, may contain depleted uranium which has to be treated as radioactive waste and must not reach the public domain.

5.5.8 Basic Regulations in Optimizing Medical and Occupational Exposures

The main objectives of the regulations are to ensure the safety of the staff, the patients and the public, and hence all the regulations concern these main aspects. While some regulations specifically ensure the safety of the staff or the public, others may benefit both or all three categories of exposed individuals. The following section covers the basic regulations in optimizing medical and occupational exposures. It is not comprehensive and it is not exactly in the regulatory language, but much of what follows can be found in any national regulations relating to radiation oncology.

In order to implement the regulations, the licensee needs to provide the necessary resources and form a radiation safety committee, and a QA committee, to delegate responsibility at different levels. These committees will oversee the implementation of the regulations regarding radiation safety and QA, investigate any breach of safety procedures or over exposures, and institute necessary corrective actions. The members of the committees are typically an administrative person from the management (representing the licensee), the chief radiation oncologist, a medical physicist, the RSO, a therapy technician, and the service engineer.

The following sections summarize the regulations that appear in many national regulations, in one form or the other.

5.5.8.1 Staff Education and Training

- The institution shall provide all radiation workers necessary training on protection and safety and health risks, and information to female radiation workers regarding the importance of notifying the employment regarding pregnancy.
- The records of training details shall be maintained in the department for regulatory inspections.

5.5.8.2 *External Beam Therapy Equipment (Teletherapy, Linac Therapy, and Brachytherapy) Performance Standards*

- A multilayer (defense in depth) system of protection shall be available in the equipment commensurate with the risk of potential exposure so that failure at one layer will prevent any radiation incident from occurring.
- If the system fails to terminate the exposure as per the program, a backup timer or a monitoring system shall terminate the exposure.
- The equipment shall be so designed that
    - Failure of a single component of the equipment is promptly detectable so the chances of any unplanned medical exposure can be minimized
    - The probability of human error is minimized
- The treatment delivery equipment (TDE) used in radiation oncology shall conform to the International Electrical Commission (IEC) standards.
- The relevant documents shall be provided in English, and if English is not the local language they shall also be provided in the local language.
- The equipment shall have a safety interlock to prevent clinical use of the machine in conditions other than those set at the control panel.
- The equipment shall have safety interlocks so designed that when they are bypassed, the operation of the machine shall be possible only under service mode using appropriate device codes or keys.
- One or more "beam OFF" switches shall be provided, at convenient locations inside the treatment room, to interrupt exposure from inside the room.
- The TDE shall be provided with necessary safety switches to prevent any unauthorized use. The energizing switch shall be accessible only to authorized personnel.
- The TDE shall never be left unattended with the energizing key in the console.

5.5.8.3 *External Beam Therapy Equipment Commissioning Calibration, QA, and Maintenance*

- The required clinical dosimetry systems and QA tools must be available in the department for this purpose.
- The dosimetry system shall carry a valid calibration certificate (at all times), traceable to the national dosimetry laboratory or a regional calibration laboratory (traceable to the national standards).
- Room survey, acceptance testing, commissioning measurements, and beam calibration shall be carried out for obtaining the approval of the RB or any other authorized body of the country prior to initiating patient treatment. The tests shall be conducted as per the national or international protocols.
- Equipment QA, verification of commissioning data, beam calibrations, and room surveys shall be carried out, by a qualified medical physicist, on a regular basis and after any major repairs, during the useful life of the equipment, as per the regulatory requirements.
- All measurement records with full details of measurements including the names of the physicists responsible for measurements, the systems used, etc., shall be kept

for hospital reference and regulatory inspections. RBs may specify a minimum period for the maintenance of these records.

- A beam output constancy check shall be carried out on a daily basis. If the constancy test output differs from the calibration value by more than 3% (or whatever value is stipulated by the RB), on the comparison date, a recalibration of the beam output shall be performed.

- In the case of the isotopic units (Telecobalt, HDR ^{192}Ir, or gamma knife units), basic safety and functional tests (spot checks) as defined by the regulations shall be carried out on a monthly basis. (In case of any malfunction, the control console shall be locked in the OFF position unit and shall not be operated for clinical use, until the defects are put right).

- The licensee shall keep a record of all QA test procedures and QA test results along with the details of the treatment unit, instruments used, name of the physicist or the technician performing the tests, etc., for a minimum period of 3 years or as recommended by the regulatory body.

- Full beam calibration (defined as the beam output measurement, under reference conditions, with a dosimetric system with a valid traceable calibration factor) shall also be carried out following any equipment repair that is likely to affect the beam quality or beam output, or on an annual basis, or as per the regulatory requirements.

- The calibration shall be undertaken independently by two qualified medical physicists preferably using different dosimetry systems, and the results compared to eliminate any human error.

- In brachytherapy, the source strength shall be measured in terms of air kerma strength or reference air kerma rate, and the measured value shall be used in the TPS for tumor dose calculations. (The output should be mathematically corrected for physical decay at intervals consistent with 1 % physical decay).

- All brachytherapy source calibration data (including the measurement dates, dosimeter details, details of the person carrying out measurements, etc.) shall be retained by the department for regulatory inspection for the period stated by the RB.

5.5.8.4   *Additional Requirements for the Brachytherapy Sources and Applicators*

- Brachytherapy sources shall be leak tested once every 6 months, by a wet wipe test. Any removable contamination amounting to >185 Bq (5 nCi) would imply loss of source integrity and shall be withdrawn from use.

- The sources shall not be used beyond the manufacturer's recommended working life.

- In the case of treatment delivery equipment housing a radioactive source (teletherapy or HDR units), the accessible surface near the source (e.g., the collimator interior of a Telecobalt unit or in the afterloading drive assembly of the HDR unit) shall be wipe tested, during source transfer, for any radioactivity.

- The transfer tubes and couplings shall be visually examined (in HDR brachytherapy) for any damage so the source movement will not be impeded.

- The manual brachytherapy sources shall be used only with metallic or plastic applicators as supplied or recommended by the manufacturer.

- The containers used for source storage and movement shall conform to radioactive transport regulations.
- In the case of LDR brachytherapy (both manual and remote controlled), details of the treatment including patient name, source used, date and time of insertion and removal, nursing requirements, and permitted time for visitors shall be posted.
- Emergency handling instructions (in the case of unplanned removal of the source and applicator) shall be posted.
- Beta sources (eye applicators) shall be supplied with shielding to minimize bremsstrahlung exposure during preparation or use.
- A patient with a temporary implant in their body shall not leave the room unless accompanied by an attendant.
- In the case of brachytherapy procedures, both the patient and the transport container (or source housing) shall be surveyed to ensure the presence of the source in the patient (and not in the container), following an implant.

5.5.8.5 Quality Assurance

- All the necessary equipment QA tools and instruments shall be available in the department for this purpose.
- QA tests shall be carried out on a daily, weekly, monthly, quarterly, or annual basis depending on the stability and the importance of the parameters, as per the QA protocol and the regulatory requirements.
- The radiation leakage from the treatment head or from the source housing (in the case of teletherapy or afterloading brachytherapy units) shall meet the regulatory requirements. Records of procedures and results shall be maintained by the department as per the regulatory requirements of the country.

5.5.8.6 Maintenance

- Only a trained and duly authorized person shall install, maintain, or repair any radiation generating equipment (e.g., ^{60}Co unit, ^{192}Ir HDR unit, gamma knife unit).
- The equipment shall always be under a maintenance contract with the company, or the company's authorized agent, throughout the useful life of the equipment.
- The equipment shall be regularly inspected and serviced by authorized service engineers throughout the licensing period.
- The licensee shall maintain a record of the equipment inspection and servicing details along with dates of servicing, components serviced, the name of the service engineer, etc., for the duration of the license.

5.5.8.7 Workplace Monitoring

- A radiation protection survey meter shall be available in the department with a valid calibration certificate, at all times. This requires calibration of the survey meter, after any major repair of the instrument or at regular intervals of 1–3 years, as per the national regulatory requirements.

- A record of the calibrations and check results, initialed by the technologist performing the check, shall be maintained for a period of 3 years or as recommended by the RB.
- All the radiation equipment and radiation installations shall be surveyed using a survey meter, before commissioning treatment and regularly thereafter on an annual basis, or after major repairs, or after any approved structural modifications, as stipulated by the RB.
- The survey shall be conducted by the medical physicist or RSO or a technician (under supervision).
- The radiation levels around the radiation installations shall meet the regulatory compliance requirements. If the levels are higher than the regulatory requirements, the unit shall not be used until remedial action is taken to ensure compliance.
- Workplace monitoring systems, capable of continuous monitoring of radiation levels inside treatment rooms, shall be installed in all treatment delivery rooms using radioactive sources (e.g., teletherapy, stereotactic radiosurgery, or medium or high dose rate remote afterloading brachytherapy rooms).
- An authorized medical physicist and an authorized user, or any staff under an authorized person and trained in emergency response for the unit, shall be physically present during the initiation of all patient treatments. (An authorized person must also be available at other times during the continuation of treatments).
- Each workplace monitoring system shall be equipped with an independent backup power supply separate from the equipment power supply.

5.5.8.8 *Occupational Exposure and Personal Monitoring*

- Radiation workers in radiation oncology shall be provided with a personal monitoring badge (PMB) issued by an accredited personal monitoring service provider in the country.
- It is mandatory for staff working with radiation to wear the PM badges and avoid unjustified radiation exposures. The monitoring period is as per the national practice, usually every quarter.
- Trainees shall use PM devices and work under the direct supervision of authorized staff.
- A female radiation worker, on declaration of her pregnancy, shall be provided with an additional badge to monitor the dose to her fetus. She shall engage in activities that will deliver a dose, during her pregnancy, not exceeding the limits adopted by the RB (e.g., 5 mSv in the US and 1 mSv in European countries).
- If a radiation worker loses their badge, the institution shall estimate the dose for that quarter from the person's previous dose records and enter it in the dose records.
- The institution shall maintain all the dose records of all the radiation workers including those who may be working in more than one place. Every radiation worker shall enjoy right of access to their own dose records.
- The annual dose received by radiation workers shall not exceed the limits prescribed by the RB and shall remain within the local limits (investigation level) set by the institution. (The investigation level or the dose constraint takes care

of the application of regulatory dose limits for radiation workers qualified by ALARA, leading to optimized dose limits).

- Any staff dose exceeding the investigation level set by the institution shall be investigated by the RSO and if necessary, efforts will be made to optimize the radiation exposures.
- Doses received from emergency interventions or accidents shall be distinguished from normal occupational exposures.
- All the personal monitoring dose data shall be available within the department, and kept for a minimum period recommended by the RB.

5.5.8.9 Safety of the Public

- The licensee shall ensure the safety and the security of the sources (radioactive sources and radiation-generating equipment) at all times, so that no member of the public receives any inadvertent exposure.
- Areas in a radiation oncology practice shall be classified as controlled and uncontrolled areas, or as controlled, supervised, and uncontrolled areas, as per the national regulations.
- The licensee shall ensure control of access to the treatment rooms (controlled areas) at all times so that no member of the public receives any inadvertent exposure.
- Radiation levels in uncontrolled areas (hallway and adjacent patient rooms) shall not exceed 2 mrem (20 µSv) in any 1 hour or 100 mrem (1 mSv) in a year.
- The control panel shall be installed in such a way that the technician is able to see the entrance to the treatment room—an important safety precaution.
- The access to the treatment room door shall be controlled by a safety interlock system that would prevent energizing of the TDE when the room door is open, or trip the machine if the treatment room door was opened during treatment.
- Following an interlock interruption, the beam shall be turned on only after the room door is closed and the beam on–off control is reset at the console.
- In the case of malfunctioning of door interlock, the control console shall be locked in the OFF position, until it is repaired.
- A licensee shall control access to the treatment room by
  - A door at the entrance
  - A red warning light to the interlocked door, interlinked to the control panel in such a way that the light glows when the source is in the ON position

5.5.8.10 Posting Signs in the Treatment Rooms

- The licensee shall fix signs at strategic places in the controlled and supervised areas, and also provide adequate information to the public regarding the control of access to treatment rooms to ensure no unnecessary exposure to the public takes place.
- An emergency storage container shall be available at all times in the brachytherapy treatment room, for storing any dislodged source.

5.5.8.11 Procurement of Brachytherapy Sources

- The licensee shall obtain authorization from the RB for the procurement of a source by submitting all the relevant documents as required by the RB.
- The licensee shall comply with the regulatory requirements in the transport of radioactive materials used in radiation oncology.
- Notification regarding any receipt (or disposal) of a source shall be sent to the RB within the stipulated time period.
- The RB shall issue authorization for a single source for a new HDR brachytherapy unit to enable the facility to carry out acceptance testing and commissioning measurements, and ensure it meets all regulatory requirements.
- On specific request from the licensee, the RB shall issue authorization for a maximum of four ^{192}Ir HDR sources at a time, with the condition that only one source would be ordered and procured at a time.
- The source transfer operation shall be carried out by a qualified and certified service engineer under the supervision of a medical physicist, if the RB so desires.
- During the source loading or source transfer operation, all persons involved in the operation shall wear PM badges.
- It is the responsibility of the RSO to survey the installation, following a source-loading operation, to ensure that radiation levels outside are complying with regulations.

5.5.8.12 Source Storage and Inventory

- The source room shall be used only for the designated purpose by the authorized and trained personnel.
- The source room door shall carry a sign on the door indicating the presence of radioactive material along with the "danger" sign and the contact details of the RSO in case of any emergency.
- Any lost or stolen source shall be immediately notified to the RB.
- Any unused source shall be sent to the disposal authorities complying with all regulations regarding the disposal of sources. (Keeping unused sources in storage increases the risk of a source becoming lost and ending up in the public domain).
- A source shall not be transferred to a third party without proper regulatory clearances. (Any source, not under regulations, is a radiation risk to all).
- A source inventory containing all the relevant details regarding the sources, their identification, their location at any point in time (whether in use or transit or in storage), current activity, etc., shall be maintained.

5.5.8.13 Source Preparation (^{125}I or ^{192}Ir Wires) and Handling

- There shall be shielded source storage for all sources, with marking for easy identification of the sources and their activities.

- The safe shall be located close to the work bench.
- An L bench shall be available on the work bench for the handling of the sources.
- The work bench shall have adequate lighting, a magnifying glass, forceps, a device for threading sources, etc., and a smooth surface for easily noticing and picking up small fragments of ^{192}Ir wires.
- The sources shall be readily identifiable by sight and sources of the same appearance but different activities must be distinguishable by suitable means (e.g., colored threads or beads).
- A sink shall be provided in the room with a filter to prevent any source loss through the drainage.
- Source transport containers shall be provided with long handles for easy movement, securely closable lids to prevent spillage during transport, and the radiation symbol on the outside surface with a "Danger" warning sign.
- A zone monitor on the wall or a survey meter in measuring mode shall be used to visibly monitor the radiation levels during the handling of the sources.
- Storage containers shall be provided for storing decaying sources such as ^{192}Ir.
- Proper handling tools like forceps, cutters, etc., shall be available on the work bench for cutting ^{192}Ir wires with a suitable container for storing waste fragments of ^{192}Ir wires.

5.5.8.14 Brachytherapy Treatment

- The treatment room shall be in a low-occupancy area and ideally house a single patient. In the case of more than one patient room, they shall be adjacent to each other. An area monitor shall be provided at the entrance to detect any patient or source leaving the room.
- In manual brachytherapy rooms, a bedside shield shall be provided, close to the patient's bed, to reduce the radiation exposures to the nursing staff or visitors.
- External beam therapy and HDR brachytherapy shall be performed in the rooms designated for that purpose.
- The treatment room door shall have an interlock which will terminate the treatment if the room door is opened during treatment. The entry room door light going OFF is an indication of treatment termination.

5.5.8.15 Emergency Response

- The licensee shall prepare emergency response plans, in the form of well-written procedures, to be implemented for every conceivable emergency situation.

  Some of the well-known emergency situations are (1) source loss, for example, ^{125}I seeds, (2) source left behind in the patient (^{192}Ir wire implant patient), (3) source not returning to the parking position in the remote controlled units, (4) death of a brachytherapy patient during treatment, (5) selection of wrong mode or wrong energy for treatment, (6) wrong dose or wrong treatment site during treatment.

- Emergency procedures, in the case of teletherapy and remote afterloading brachytherapy, shall be displayed prominently near the control console.

- The T-rod for handling teletherapy emergency situations shall be kept in the control room at all times, in an easily accessible spot.

- Necessary emergency handling training for recognizing abnormal situations and initiating necessary action to restore the normal situation shall be imparted by the licensee to the appropriate radiation workers.

- Necessary tools (cutter, tongs, workplace monitoring, and personal monitoring devices, etc.) shall be readily available in the room.

- In case of any emergency in HDR brachytherapy, the physician concerned and a medical physicist, or staff trained for the HDR emergency situation, shall be present to remove the catheter and promptly carry out the emergency procedures. An emergency storage container shall be available in the treatment room to store the source along with the catheter.

- Any emergency during source transfer shall be handled only by the maintenance staff or staff specifically trained and authorized for this purpose.

- The licensee shall advise the RB (within the stipulated time) regarding the emergency, investigate the causes of the accidents, and also file a copy of the investigation report to the RB.

- The institution shall investigate any of the following incidents:

  - Any treatment delivered to a wrong patient, or wrong site, or using wrong beam quality, or source or dose, or dose fractionation different from that planned.

  - Any equipment malfunction

  - Any other type of occurrence not intended that may cause a deviation from the planned treatment

- The licensee shall investigate the incident, compute or estimate the delivered dose and dose distribution in the patient, implement corrective measures to prevent such an occurrence in the future, and submit copies of the investigation report to the safety committee and to the RB, as per the country's regulations.

5.5.8.16 *Safety Relating to the Release of a Brachytherapy Patient*

- Prior to leaving the room, a patient with a temporary implant shall be surveyed to ensure the presence of source in the container (and not in the patient).

- The clothes and bed linen shall also be checked before being taken out.

- Prior to the discharge of a patient with a permanent implant, the dose rate at 1 m from the patient body surface shall not exceed the limits specified by the RB. (These limits, for the three most common permanent implant sources are, in mSv/h, 0.21 (^{198}Au), 0.01 (^{125}I), and 0.03 (^{103}Au), respectively.)

- Written safety instructions shall be provided to the patient concerning contact with family members and other persons.

5.6.8.17 Handling the Death of a Brachytherapy Patient

- Any deceased brachytherapy patient under treatment, containing sources *in situ*, shall be handed over to the family only after all the sources are removed and tallied, and the patient surveyed to confirm that no source is left behind. (Postmortem examinations, if any, shall not be performed until all the sources have been removed from the body and are accounted for).

5.5.8.18 Safety of Brachytherapy Sources

- The radioactive sources used in radiation oncology shall conform to the definition of a sealed source and their design shall comply with the International Standards Organization (ISO)-developed performance standards.
- The source supplier shall supply the source, along with source strength and source leak certificates.
- The physicist shall independently measure the source strength and use the value in the treatment planning system (TPS) for tumor dose calculations.
- The measured source strength shall agree with the certificate value within a tolerance defined by the RB or the brachytherapy protocol (any difference beyond the tolerance needs to be investigated and resolved).
- A calibrated ionization chamber (e.g., a well chamber) with a valid calibration certificate shall be available at the department for the actual source strength measurements.
- Brachytherapy sources shall be leak tested once every 6 months, by wet wipe test. Any removable contamination amounting to >185 Bq (5 nCi) would imply loss of source integrity and shall be withdrawn from use.

5.5.8.19 Security of Sources

- No radioactive source shall be acquired, used, stored, or disposed of without the required regulatory approvals, unless the exposure from such practice or source is excluded from the Standards or the practice or source is exempted from the requirements of the Standards.
- The afterloading brachytherapy treatment room shall have an emergency storage container and a remote handling tool, to store any dislodged source.
- In the case of temporary manual implants, the number of sources leaving the storage facility shall be tallied against the number of sources returning to the facility. A source movement log with a record of the date and time of issue, patient name, room number, the date and time of issue and return, etc., shall be maintained.
- In the case of permanent implants, the number of sources leaving the storage, the number returned to storage, the number permanently implanted in the patient, the patient name, room number, etc., shall be entered in the movement log.
- Immediately following the source placement in, and immediately following source removal from the patient, a patient survey using a survey meter shall be carried out to ensure that no source is left behind in the patient.

- A redundant survey of the room shall also be carried out to ensure that no source is misplaced.
- The survey results shall be maintained by the department for a period of 3 years or as specified by the RB.
- A physical inventory of all sealed sources shall be periodically performed to confirm that all the sources are properly accounted for, and are secure in their locations.
- Source containers, when outside the radiation oncology department, shall carry a warning label which can be recognized as "danger" by the public.
- The source storage facility shall be kept under lock and key all the time, to prevent theft or any unauthorized use.

5.5.8.20 *Transport of Brachytherapy Sources*

- Brachytherapy sources shall not be transported outside the hospital premises without prior approval of the RB.
- The licensee shall not transfer the source to another property that does not have the authorization to receive.
- Any authorized transport of sources shall be carried out only in compliance with the radioactive materials transport regulations.
- On receipt of any radioactive source package, the institution shall monitor radiation levels around the package (to ensure compliance with regulatory requirements) and also check for any removable contamination.

5.5.8.21 *Treatment Planning System*

- The licensee shall perform acceptance testing on the TPS as per the accepted national or international protocol, and the requirements of the RB.
- Treatment planning shall be carried out by a qualified and adequately trained operator. A trainee can do the planning under the supervision of a trained operator, or under a medical physicist.
- Acceptance testing shall not be restricted to the treatment-delivery equipment, but also must be extended to other systems that have implications for treatment accuracy and safety, such as the treatment-planning system.

5.5.8.22 *Treatment Delivery*

- Treatment delivery shall be carried out by a qualified and trained technologist.
- The operator shall check that the machine parameters are as per the treatment plan document, before switching on the machine for treatment delivery.

Here, only the basic regulations in the field of radiation oncology have been summarized. For more details, consult the publications of the national regulatory agencies.

References

1. NCRP Report 160, *Ionizing Radiation Exposure of the Population of the United States,* National Council on Radiation Protection and Measurements, Bethesda, MD, 2009.
2. ICRP Publication 103, Recommendations of the International Commission on Radiological Protection, *Annals of the ICRP*, 103, 81–100, 2007.
3. Atomic Energy Society of Japan, *Effects of Radiation Exposure on Human Body and Radiological Protection Criteria,* AESJ, Tokyo, Japan, 2011. http://www.aesj.or.jp/en/release/Radiological_Protection_Criteria110509.pdf (accessed on August 2016).
4. Part 20 – Standards for protection against radiation, U.S. Nuclear Regulatory Commission. https://www.nrc.gov/reading-rm/doc-collections/cfr/part020/ (accessed on August 2016).
5. Part 1020 – Performance standards for ionizing radiation emitting products, U.S. Food and Drug Administration. https://www.accessdata.fda.gov/scripts/cdrh/cfdocs/cfcfr/CFRSearch.cfm?CFRPart=1020 (accessed on August 2016).
6. Nuclear regulatory commission. https://www.nrc.gov/ (accessed on August 2016).
7. Food and Drug Administration. https://www.fda.gov/ (accessed on August 2016).
8. Environment Protection Agency. https://www.epa.gov/laws-regulations (accessed on August 2016).
9. Department of Transportation. https://www.transportation.gov/ (accessed on August 2016).
10. Requirements and Guidelines to start a radiation therapy facility, Atomic Energy Regulatory Board, India. http://www.aerb.gov.in/AERBPortal/pages/English/t/forms/regforms/radiotherapy/RTGuidelines.pdf (accessed on August 2016).

Review Questions

1. Name three important bodies responsible for issuing recommendations and guidelines for the radiation quantities and units, and safe use of radiation in society.

2. Name the body that periodically reviews the effects of ionizing radiation, and issues updated recommendations.

3. What are the two most common radiation exposures to the public?

4. What is the typical dose received by the public from NBG radiation and what source of radiation forms a significant component?

5. What are the three categories of radiation exposure as per the ICRP system of radiation protection? Which category does not have any dose limit prescribed by ICRP?

6. What are the three categories of exposure situations defined by ICRP? Give an example for each.

7. Give two examples of chronic exposure situations.

8. What are the three types of radiation effects? Give examples.

9. What are the three important concepts that the biological effects of radiation have given rise to in radiation protection?

10. Explain the concepts of justification, optimization, and dose limitation.

11. What are the characteristics of deterministic and stochastic effects?

12. Describe the dose limits for an adult radiation worker as per ICRP 103 recommendations.

13. What is the dose limit for a pregnant radiation worker, and under what circumstances it is applied?

14. A radiation worker receives 3 mSv in a medical exam. How much radiation dose can this person receive in that year in their occupation?

15. Is a radiation worker working in a radiation oncology department justified in receiving a dose of 20 mSv in an year, the legal dose limit?

16. What are the three parameters of importance for radiation hazards control? Explain how they control the radiation dose received by a radiation worker. Which parameter plays the main role in controlling staff exposures in radiation oncology?

17. What are the responsibilities of an RSO in a radiation oncology department?

18. Give examples of controlled and uncontrolled areas. Why is it necessary to designate the areas as controlled and uncontrolled?

19. Why are lead aprons not used by staff in radiation oncology?

20. Can any qualified radiation oncologist start a radiation oncology department? If not, what are the basic requirements to be met?

21. What needs to be ensured before releasing a patient being treated with temporary implants?

22. How is the security of sources used in brachytherapy ensured?

6

Calibration of Radiation Monitoring Instruments

6.1 Introduction

A primary standard realizes a quantity, as per the definition of the unit of the quantity or by measuring something that is numerically equal to the unit. An example is measuring kerma by establishing a charged particle equilibrium condition. The primary standard relates its response to the dosimetric quantity it represents in an absolute sense (i.e., the unit for the quantity can be derived from the response from first principles without any calibration in terms of the quantity). For instance, the primary standard of air kerma, at ^{60}Co radiation beam quality, is a graphite cavity ionization chamber. It relates ionization response to air kerma, at the point of measurement, using the Bragg–Gray (B–G) principles. This is known as absolute dosimetry. Another important characteristic of the primary standard is that it measures the quantity with the highest accuracy. In order to avoid any systematic errors in absolute dosimetry, primary standards dosimetry laboratories (PSDLs) have an elaborate program of intercomparison between the various PSDLs and also with Bureau of International Weights and Measures (BIPM) to detect and eliminate common systematic errors. On the other hand, the users do not have the expertise to perform absolute dosimetry, nor can they detect errors in such measurements. So, dosimetric quantities are always disseminated to the users through a calibration chain from the PSDL to secondary standards dosimetry laboratory (SSDL) to regional calibration laboratories (RCLs). The user instruments (e.g., radiation workplace monitors) are calibrated at any of the above calibration laboratories against their reference standard, then disseminated to the users.

Through such an elaborate dissemination scheme, the user measurements themselves become traceable to the PSDL. There are many countries that do not have the required expertise to develop primary standards for the quantities. These countries do not have PSDLs and their SSDLs are the national standards dosimetry laboratories. The SSDLs of all the member states of International Atomic Energy Agency (IAEA) come under an IAEA SSDL network [1]. IAEA disseminates the unit to the national SSDLs which in turn offer calibration to the regional calibration laboratories, or directly to the local users. The reference standard maintained at IAEA is traceable to the international measurement system that involves intercomparison between the PSDLs and BIPM. Figure 6.1 illustrates the concept of unit dissemination. The scheme also applies to clinical dosimetry but the reference standards, the field instruments, and the calibration beam qualities used are different from those used for protection calibrations. In smaller countries, there may not be a need to establish regional calibration laboratories and one calibration laboratory may be adequate to offer calibration to the users.

When the user procures a monitor for workplace monitoring purposes, it is mandatory for the monitor to have a traceable calibration certificate.

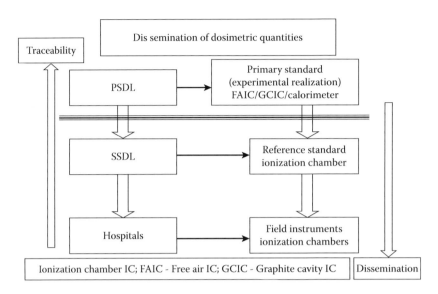

FIGURE 6.1
Dissemination of dosimetric quantities.

Since there is an uncertainty associated with each measurement, the accuracy of calibration gradually decreases from PSDL to RCL but even at the RCL level, the calibration uncertainty is very small compared to the accuracy required for measurements at the protection level and so it doesn't matter whether the instrument procured for protection measurements is calibrated at the PSDL (direct traceability), SSDL, or regional RCL (indirect traceability). Because of the equipment design and totally different workplace field conditions (compared to calibration conditions), additional uncertainties arise in the measurements. These uncertainties are much greater compared to the calibration uncertainties and must be taken into account while considering the accuracy of measurements at the hospital level.

6.2 Principles of Protection Calibrations

Figures 6.2 and 6.3 illustrate typical calibration setups established in a calibration laboratory for calibrating dosimetric instruments in X-ray, ^{137}Cs, or ^{60}Co beam qualities.

Usually the calibration laboratory maintains radioactive sources (^{192}Ir, ^{137}Cs, ^{60}Co) to cover a range of gamma energies for calibration. In the keV region, the calibration laboratory maintains X-ray machines with reference beam qualities covering a range of dose rates, from a few tens of kV to about 300 kV. The beams are heavily filtered to make the spectrum very narrow which approximates to a monoenergetic beam. The beam characteristics for calibration use are defined by ISO [3–5]. A large source to calibration point distance (about 2.5 m to 3 m) and a larger field diameter compared to the reference instrument or the workplace monitor size at the calibration point, approximates the beam to an aligned and expanded field.

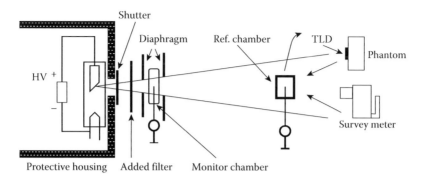

FIGURE 6.2
A typical X-ray beam calibration facility (<250 kV).

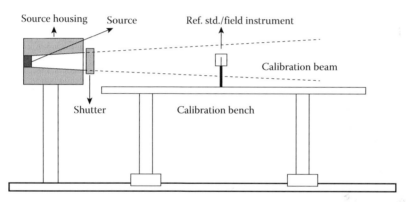

FIGURE 6.3
A typical ^{137}Cs or ^{60}Co calibration facility.

6.2.1 Calibration of Workplace Monitors (Survey Meters)

There are two basic methods followed for the calibration of the protection instruments: (1) source in open air geometry and (2) a collimated beam facility. In the first case, the source is kept in open air geometry, in the center of a large calibration room that reduces scatter reaching the calibration point. The uncollimated beam geometry helps in exposing several personal dosimeters at equal distance from the source, in an arc of a circle, in a single exposure. In the case of a collimated beam, the monitors are generally calibrated by a substitution technique. This involves keeping the reference instrument and monitor at the same point, one after the other, and taking the readings as shown in Figure 6.2). Alternatively, it can be done by simultaneous exposure technique where the reference instrument and the monitor are kept symmetrically on either side of the central point, and the readings are taken in a single exposure.

The source output roughly follows an inverse square law. Thus by measuring the output—in units of absorbed dose equivalent (ADE) or absorbed dose equivalent rate (ADER)—at one distance, the output at other distances can be computed. Alternatively, the output can be actually measured for each distance which would give a higher accuracy. These distances are used for calibrating the survey meter at different ADE rates that the instrument covers. By exposing the monitor for a predetermined time, the monitors can also be calibrated in terms of the integral quantity ADE.

6.2.1.1 Characterization of Photon Reference Fields in Terms of Air Kerma or ADE

This involves two steps:

    Step 1: Characterization of reference fields in terms of air kerma or fluence
               (for neutrons)
    Step 2: Characterization of reference fields in terms of ADE

To characterize the photon fields in terms of air kerma (AK) and air kerma rate (AKR), the AK or the AKR must be actually measured using a reference standard, in the calibration field (as shown in Figures 6.2 and 6.3). The reference standard is positioned at the calibration point (see Figure 6.2) and the reading M of the reference standard noted for the duration of the exposure Δt. The reference standard is an ionization chamber which is open to the atmosphere and so the mass of air in the chamber (and hence the charge collected) is a function of room temperature and pressure. The reference standard calibration, N_K, is usually referred to temperature $T_0 = 20°C$ or $22°C$, and pressure $P_0 = 1013.2$ mbar. The measurement readings M must also be referred to the same (T,P) conditions by noting the room temperature and pressure at the time of measurement and applying the correction $K(T,P)$.

The radiation field is now characterized in terms of AK and AKR, using the following equations:

$$K_a (P) = [N_K \, M \, K(T,P)]_{ref.std} \; Gy$$

$$dK_a (P)/dt = [N_K \, (M/\Delta t) \, K(T,P)]_{ref.std} \; Gy/s \text{ or } Gy/h$$

If the reference standard is calibrated in N_X, then the field will be characterized in terms of exposure, instead of air kerma. It is rather easy to characterize this field, in terms of exposure, using the following equation:

$$K_a(P) = X(P) \, (W/e)/(1-g) \; Gy$$

Considering the other uncertainties in the calibration, the term $(1-g)$ can be neglected since the Bremsstrahlung correction $g \approx 0$ for kV X-rays and $\approx 0.5\%$ at ^{60}Co energy. The average energy to produce an ion pair W is a constant and is independent of energy. $(W/e) = 33.97$ J/C.

A PTW 1000 cc chamber is a typical example of a reference standard used in several calibration laboratories for low-radiation level measurements, and for the calibration of protection-level monitors (see Figure 6.4). The reference standard used for calibrating ADE monitors possesses a flat energy response, high reproducibility, long-term stability, and an N_K calibration factor directly (or indirectly) transferable to a primary standard in a PSDL.

To characterize the photon fields in terms of ADE, the AK to ADE conversion factors for the calibration beam quality in question must be made use of. AK to ADE conversion factors for different reference radiations are given in the literature [6]. Table 6.1 gives the conversion factors for some of the typical calibration beam qualities.

FIGURE 6.4
A PTW 1000 cc chamber reference standard.

TABLE 6.1

Air Kerma to ADE Conversion Factors

Beam Quality	Mean Energy	H*/K$_a$
^{60}Co	1.25 MeV	1.16
^{137}Cs	0.661 MeV	1.21
X-ray: 100 kV/22 mA	83 keV	1.71
X-ray: 80 kV/10 mA	65 keV	1.71

Note: ADE, ambient dose equivalent; (H*/K$_a$), denoted as h*.

The following equation will characterize the photon fields in terms of ADE or ADER, depending on whether the measured quantities are rate quantities or integral quantities.

$$H^*(P) = K_a(P) \times h^*; \; [dH^*(P)/dt] = [dK_a(P)/dt] \times h^*$$

By measuring AKR and ADER at different distances from the source (or X-ray machine focus) using the reference standard, the field is characterized in terms of AKR or ADER across the full range of calibration distance. This makes calibration of protection-level instruments much easier compared to the substitution technique where two sets of measurements, one with the reference standard and one with the field instrument, are required. A third method is to keep the calibration source at the center of a large room with minimum scatter and assume an inverse square law (ISL) of variation of the AKR or the ADER with respect to distance from the source. The field characterization requires only one set of measurements at a standard distance from the source; using the ISL, AKR, or ADER at other distances can be determined. All three methods are in use for protection-level calibrations. The substitution method offers maximum accuracy, multiple-point measurements offer less accuracy, and the ISL method offers the least accuracy. The calibration procedures for the substitution technique are described below.

6.2.1.2 Air Kerma and Air Kerma Rate Calibration of Field Instruments

The field instrument is the monitor used in the field (i.e., at the hospital) by the user. Since the calibration field can be characterized in terms of AK (AKR) or ADE (ADER), as explained in the previous section, the workplace monitors can be calibrated in terms of AK (AKR) or ADE (ADER), in the calibration field.

The ionization-based instruments used for radiation workplace monitoring are usually open to the atmosphere and so the mass of gas inside (and hence the ionization measured) is a function of room temperature and pressure, therefore the chamber response is a function of the temperature and pressure conditions of the room. During calibration, the chamber response (and hence its calibration factor) is usually referred to 20°C and 760 mm or 1013.5 mbar pressure. While taking measurements in a hospital, the room (T,P) values can be quite different even during the course of a day. So, during measurement in a hospital, the room (T,P) values must always be noted and referred back to the calibration conditions (20°C and 1013.5 mbar P), using the following equation:

$$K(T,P) = [(273.2 + T)/273.2 + 20] \times (1013.5/P)$$

In terms of mm Hg pressure, the pressure term would be (760/P). There are some survey meters with pressurized air in the detector to increase its sensitivity to monitor very low radiation levels. Such detectors are sealed and their responses do not vary with the ambient (T,P) corrections. They need no (T,P) correction.

To calibrate a protection-level monitor, usually referred to as field instrument (FI), it is substituted at the calibration point and its average reading (for say six repeated measurements) M is noted, for an interval of time Δt. The room temperature and pressure readings, (T,P), are also noted. The calibration factor of the FI, $[N_K]$, is given by

$$[N_K]_{FI} = K_a(P)/[M\ K(T,P)]_{FI}\ \text{in Gy/C (or μGy/nC)}$$

$$= [N_K\ M\ K(T,P)]_{ref.std}/[M\ K(T,P)]_{FI}$$

$$[N_K]_{FI} = [dK_a(P)dt]/[(M/\Delta t)\ K(T,P)]_{FI}\ \text{in Gy/C (or μGy/nC)}$$

$$= [N_K\ (M/\Delta t)\ K(T,P)]_{ref.std}/[(M/\Delta t)\ K(T,P)]_{FI}\ \text{in Gy/C (or μGy/nC)}$$

There won't be much change in the room temperature and pressure in the couple of hours usually taken to complete the calibration readings, in which case the K(T,P) correction in the numerator and denominator cancel out. If on the other hand, the measurements are spread over several hours, say reference measurements are done in the morning and the field instrument measurements are carried out during the afternoon, there can be significant change in the room temperature and pressure conditions. In this case, the K(T,P) values in the numerator and denominator will be different and so the K(T,P) corrections must be applied to measurements. For protection calibrations, this correction factor can be neglected, appearing as a ratio, without much loss in accuracy. The correction applies to ion chamber detectors that are open to the atmosphere and not for sealed chambers or other types of detectors that may not be affected by (T,P) conditions (e.g., scintillation detectors). The equation also assumes that reference standard and FI readings were obtained for the same duration of exposure Δt, but that need not be so and different intervals of time can be used. The field instrument can also be calibrated directly in the rate mode. In the rate mode, the instrument display directly shows in μGy, in the radiation field. In rate mode calibration, $M/\Delta t$ in the above equation represents the display readings of the reference standard and the field instrument, respectively. Calibration in the integral mode gives better accuracy compared to the rate mode of calibration. However, this makes a difference only in clinical dosimetry where the therapy chambers need to be calibrated for high accuracy. Radiation protection monitors are generally calibrated in the rate mode.

The FIs can also be calibrated in terms of exposure or exposure rate by characterizing the calibration field using the relationship between X and K_a:

$$K_a(P) = [X(P)\ (W/e)]/(1-g)$$

The reciprocal of the calibration factor is known as the sensitivity of the monitoring instrument, S_K. The sensitivity of a detector depends on its sensitive volume. To measure very low radiation levels, as encountered in radiation protection situations, a fairly large volume ionization chamber is required, say a few hundred to a thousand cc, to obtain adequate sensitivity.

$$S_K = (N_K)^{-1}\ \text{C/Gy and}\ S_x = (N_x)^{-1}\ \text{nC/R}$$

6.2.1.3 ADE and ADER Calibration (Photons)

$$H = K_a(P) \times h^*$$

$$[N_H]_{FI} = [K_a(P) \times h^*]/[M'\ K(T,P)]_{FI} \text{ in Gy/C (or } \mu Gy/nC)$$

$$= [N_K\ M\ K(T,P) \times h^*]_{ref.std}/[M\ K(T,P)]_{FI}$$

$$[dN_H/dt]_{FI} = [K_a(P)/\Delta t \times h^*]_{ref.\ std.}/[(M/\Delta t)\ K(T,P)]_{FI} \text{ in Gy/C (or } \mu Gy/nC)$$

$$= [N_K\ (M/\Delta t)\ K(T,P) \times h^*]_{ref.std}/[M/\Delta t\ K(T,P]_{FI}$$

where M is the total measured charge noted on the reader (for a certain duration of exposure), under the room temperature/pressure conditions (T,P). K(T,P) correction refers the reference standard and the FI readings to the standard conditions (T_0,P_0) to which the calibration factor now refers. For the same duration of exposure, the reference standard measures $K_a(P)$ at the calibration point. If the FI is an ion chamber–based system and is hermetically sealed, or a solid state–based system to which no (T,P) correction is necessary, K(T,P) can be assumed to be unity or deleted from the equation. If the FI reader system is directly marked in AK or AKR, then the calibration factor is dimensionless. The ADER calibration can also be carried out directly in the rate mode of the survey meter. In that case, the survey meter display directly reads in $\mu Sv/h$, and $(M/\Delta t)$ in the above equation can be replaced by the display reading. For rate mode calibration, the reading is steady and this reading only needs to be noted, keeping the source ON for an arbitrary time interval.

It may be noticed that K_a and H are related by h and so there is no need to calibrate the field instrument separately in ADE or ADER, if the instruments has been calibrated in terms of AK or AKR. The ADE calibration factor can easily be determined simply by multiplying the AK calibration factor by the tabulated conversion factors h, for the beam quality in question. The calibration laboratories are now offering ADE calibrations and so there is no need for any conversions.

Portable survey instruments are calibrated at least at one point on each measuring range, that is, at approximately half of their full-scale value, or in each decade for an instrument with a logarithmic scale or with digital indication.

The photon area monitors are being increasingly designed as ADE monitors but still some countries prefer to calibrate the monitors in terms of exposure or air kerma. So, many manufacturers offer area monitors marked both in exposure and ADE, or air kerma and ADE (see Figure 6.5).

These monitors therefore need calibration in terms of different quantities. If the area monitor reads only in exposure or air kerma (as is the case with the older instruments) then the survey meter can be calibrated only in terms of these quantities. Such monitors can still be used for workplace monitoring survey and the survey results can be converted to the ADE quantities using the conversion factors given in one of the above tables.

FIGURE 6.5
A commercial survey meter scale meter marked in both mR/h and μSv/h.

According to NRC, US [7] portable radiation survey meters must be calibrated annually to an accuracy of ±20% (*presumably at 1 sigma level*) under calibration conditions. Analog instruments must be calibrated at two points on each range, or for digital instruments at one point per decade over the range of use.

6.2.1.4 Monitor Sensitivity (or Response)

The ADE sensitivity (or response) of a radiation monitor is given by

$$[S_{ADE}]_{FI} = M' \, K(T,P)/[K_a(P) \times h^*]$$

For instance, if the field at the calibration point is 400 μSv/h and the instrument reads 432 μSv/h (corrected for the background reading), the monitor sensitivity is given by 432/400 = 1.08. The instrument calibration factor is the inverse of the response, that is, 0.925. The following example illustrates some of the concepts explained above.

Example 6.1

A ^{137}Cs calibration field has been characterized in terms of air kerma rate. At 1.5 m from the source (see Figure 6.3), along the beam central axis the AKR is 17.706 mGy/h. Characterize the field in terms of ADER. (Given H^*/K_a for ^{137}Cs is 1.21 mSv/mGy).

$$(dH^*/dt) = 17.706 \times 1.21 = 21.424 \text{ mSv/h.}$$

Example 6.2

In a ^{137}Cs calibration field, a reference standard placed along the central axis at 1 m distance from the source, reads 52.892 pC (corrected for the background reading and room temperature and pressure), for an exposure time of 2 minutes. [N_K]$_{Ref.Std.}=$ 25.1 μGy/nC at 22°C and 1013.2 mbar pressure. A survey meter, not open to the atmosphere, kept at the same point, reads 47.64 μSv/h.

a. Calculate the ADE calibration factor of the survey meter.

b. If the survey meter is also designed to read in air kerma and reads a value of 39.37 μGy/h on the air kerma scale, what will be the air kerma calibration of the same survey meter?

c. AKR at the calibration point (AK during δt/δt)

$$(dK/dt) = [N_K \, M \, K(T,P)/\Delta t]_{ref.std}$$
$$= [(25.1 \times 52.892 \times 10^{-3})/2] \text{ μGy/min} = 0.664 \times 60$$
$$= 39.840 \text{ μGy/h}$$

ADE rate at the calibration point (AKR × h*)
$$(dH^*/dt) = 39.840 \times 1.21 = 48.206 \text{ μSv/h}$$

Survey meter calibration factor (ADER/survey meter reading) at calibration point
$$[N_H]_{survey \ meter} = 48.206/47.64 = 1.02$$

The calibration factor is dimensionless since its display is marked in μSv instead of in scale divisions or in coulombs (as can be seen from the units shown on the display of the survey meter). The survey meter reading needs no temperature/pressure correction here since it is a sealed one.

d. To calibrate the survey meter in terms of air kerma, the above conversion is not necessary.

Survey meter calibration factor [(dK_a/dt)/survey meter reading] at calibration point

$$[N_K]_{survey\ meter} = 39.840/39.37 = 1.02\ [(\mu Gy/h)/(\mu Gy/h)]$$

6.2.1.5 Characterization of Neutron Reference Fields in Terms of Neutron Fluence or ADE

Some of the sources that have been used for the calibration of area neutron monitors are ^{252}Cf (f, n), ^{252}Cf–D_2O (fission source–moderated neutrons), ^{241}Am–B (α, n), ^{241}Am–Be (α, n), and ^{238}PuBe (α, n). The bare sources are standardized in terms of yield (emission rate) using the primary standard [8] known as the manganese sulfate bath, as explained in Chapter 4. The properties of the sources are given in Table 6.2.

The neutron fluence rate, at a distance d from the source is given by

$$d\Phi(d)/dt = (dY/dt)/(4\ \pi d^2)$$

where Y is the yield or emission rate (in 4π steradians) of the neutron source (#/s).

The fluence, at a distance from the source, is given by

$$\Phi(d) = [dY(d)/dt] \times \Delta t$$

where Δt is the exposure time or measurement duration. We need both the quantities to calibrate a monitor in rate and integrate mode, respectively, though generally we refer to the measurements as fluence measurements, without distinguishing between the two quantities.

The emission of the source, however, is not always isotropic and so an anisotropy factor (i.e., relative variation of the emission rate with respect to the axis of the source) must be known to determine the source output in a particular orientation (say along the calibration direction). The anisotropy measurements are generally made in a low-scatter geometry using a precision long counter (PLC) to measure the output in steps of say 10°. For a chosen orientation (say along the calibration direction), the output remains constant with respect to time, when allowance is made for the radioactive decay. The National Physical Laboratory (NPL), UK, recommends that the ^{252}Cf sources be recalibrated at least every

TABLE 6.2

Radionuclide Neutron Source Characteristics

Source	Half Life	Mean Energy (MeV)	Specific Neutron Yield, S (#/MBq·s)	Specific Neutron DER Constant Γ_n (Sv m^2/MBq·h)
^{252}Cf	968 days	2.4	$(2.3 \times 10^{12})^a$	22
^{252}Cf + D_2O	968 days	0.54	2×10^{12}	5.2
^{241}Am–B(α, n)	432.7 years	2.8	16	1.8×10^{-10}
^{241}Am–Be(α, n)	432.7 years	4.4	66	7×10^{-10}

Note: ^{252}Cf + D_2O refers to a moderated source (source in a moderating sphere, shielded with 1 mm Cd).
^{a} For ^{152}cf, the yield and dose rate constant are per g and not per Bq.

two and a half years, and [241]Am-based sources at least every 10 years at the PSDL. Typical neutron calibration sources (in use at NPL, UK) are shown in Figure 6.6 [9].

A more accurate method of fluence standardization of the neutron field is by actually measuring the neutron fluence using the reference standard of neutron fluence, the PLC, or De Pangher long counter. The PLC mainly consists of a BF_3 proportional counter positioned along the axis of a cylindrical moderator phantom. The PLC is specially designed to show the energy independence of response (to within ±6%) from a few keV to about 5 MeV or higher, as shown in in Figures 6.7 and 6.8. (More details of fluence standards can be found in the literature [10,11].)

The fluence computed from the source output gives the primary neutron fluence from the source. If the source is not a collimated beam and is in open air geometry, the neutrons are also scattered by the room walls, floor, the ceiling, and the air inside the room. The actual neutron fluence at the calibrations point is then composed of both the primary and the scatter components of the neutron fluence. So, one should correct for the scatter contribution to the neutron monitor, at the calibration point, if one uses the inverse square law to determine the fluence at a distance from the source (which applies to primary fluence). The scatter contribution depends on the source to calibration distance, room size, and other parameters. The scatter neutron spectrum differs from the source spectrum and the accuracy of neutron monitoring very much depends on the flatness of the energy response characteristics of the reference standard used to standardize the calibration field. The energy dependence of neutron monitors is much more complex compared to the photon monitors.

The scatter is usually determined by the shadow cone experiment. The cone (almost completely) blocks the primary radiation and the instrument response is only due to scatter. From measurements made with and without cone, $M_1(r)$ and $M_2(r)$, respectively, the scatter can be determined.

Am-Be sources   [252]Cf sources

FIGURE 6.6
Typical neutron calibration sources.

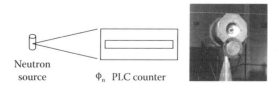

Neutron
source Φ_n PLC counter

FIGURE 6.7
Precision long counter for fluence measurement.

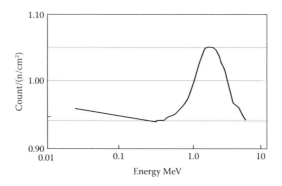

FIGURE 6.8
Response (sensitivity) of PLC with respect to energy.

FIGURE 6.9
Shadow cone experiment for scatter determination.

For instance, $M_1(r) = M_0(r) + M_s(r)$ without shadow cone

$M_2(r) = M_s(r)$ with shadow cone

$M_0(r) = M_1(r) - M_2(r)$

Figure 6.9 [12] shows the shadow cone intercepting the direct beam from the neutron source. Since the neutron monitor must be in the shadow of the cone, the cone size depends on the cone position with respect to the source and the monitor, and the size of the monitor. Since the cone must attenuate the neutrons and the neutron-produced gammas, the cone is a sandwich of stainless steel and borated polyethylene.

The calibration is generally performed in a low-scatter geometry (i.e., by placing the source in the middle of a large room, away from the walls and the floor, and the monitor mounted on a nonhydrogenous structure).

As in the case of photon fields, there are two steps involved in the characterization of the neutron fields:

1. Characterization of the field in terms of fluence, Φ.

2. Characterization of the field in terms of ADE, using appropriate ADE to fluence conversion coefficients, h_Φ, available in the literature [13,14].

TABLE 6.3

[ADE/Φ] Conversion Factors

Neutron source	Conversion Coefficient H*(10)/Φ pSv/cm^2
^{252}Cf	385
^{252}Cf+15cm D$_2$O moderator.	105
^{252}Am–B(α, n)	408
^{252}Am–Be(α, n)	391

Note: ADE, ambient dose equivalent.

The neutron sources are not monoenergetic but exhibit a spectrum so the mean conversion factors averaged over the neutron fluence spectrum is made use of.

$$h_\Phi = \int h_{\Phi,E}(10)\Phi_E dE / \int \Phi_E dE$$

The neutron spectrum typically covers eight to nine decades of energy ranging from 0.01 eV to a few MeV, to greater than 10 MeV in the case of accelerator sources.

The spectrum averaged conversion factors for the neutron sources are given in Table 6.3.

The following equation converts the fluence field into an ADE field for the purposes of calibration.

$$H^*(P) = h_\Phi(P) \times \Phi$$

$$[dN_{H^*}/dt]_{FI} = h_\Phi(P) \times d\Phi/dt$$

There are no pure neutron fields. Neutron fields of radionuclides are always accompanied by a background photon field and for the accurate assessment of neutron ADE values, the system must be able to discriminate against the photon response of the instrument.

The neutron sources have a half-life and output decreases with time. Hence, a decay correction (for ^{252}Cf decay) is necessary to refer the source output to the calibration date. There are other isotopes in the decaying of the ^{252}cf source. However, the contribution of other isotopes to the neutron yield is negligible during the useful life (about 15 years) of the neutron source. Even up to 25 years, the change in yield due to the competing emissions is insignificant ($\approx$ 1%) [15].

The calibration procedure is very similar to the calibration of photon monitors explained in the previous section. The details of neutron calibrations are based on the international standards ISO 8529, Part 1–3 [13–14] and IEC 61005 [16] that cover in detail the determination of the calibration factor, the linearity of response, and the photon sensitivity. The reader should consult these references for more details.

6.2.1.6 *ADE and ADER Calibration (Neutrons)*

As in the case of photon calibration factor, the neutron ADE calibration is given by

$$[N_{H^*}]_{FI} = [H^*(P)]/M\ K(T,P)$$

$$d[N_{H^*}/dt]_{FI} = [dH^*(P)/dt]/[(M/\Delta t)\ K(T,P)]$$

where M is the reading of the neutron survey meter, in integral mode, for an exposure time of Δt. M could be read in μSv from the display of the survey instrument (in which case, N becomes unitless). If M reads in counts it then displays the pulses counted. In this case,

FIGURE 6.10
A commercial neutron survey meter.

the units of N will be in Sv/count. The older neutron survey meters were reading dose equivalent (DE) in the old units of rem but the newer ones now read in the SI unit of Sv. The survey meter can also be calibrated directly in the rate mode when the display directly reads the dose rate (say in μSv/h). In that case, the rate quantity (M/Δt) in the above equation can be replaced by the display value.

More details on neutron survey meters (Figure 6.10) can be found in Chapter 7.

The following examples illustrate some of the concepts explained above.

Example 6.3

A shadow cone experiment was performed using a ^{252}Cf source. The source to calibration point distance was 1 m. H* (at 1 m) at the calibration point = 182 μSv. The survey meter readings were 188 μSv (without shadow cone, R_{wosc}) and 15.38 μSv (with shadow cone in place, R_{wsc}). The irradiation time was 360 seconds. Determine the survey meter calibration factor N_{H^*}.

Primary dose, R_0 (i.e., without scatter) = $R_{wosc} - R_{wsc}$ = 188 − 15.38 = 172.62 μSv

Calibration factor = N_{H^*} = 182/172.62 = 1.05

Example 6.4

A neutron survey meter was calibrated using a ^{252}Cf source. The calibration details are as follows:

The source strength on the calibration date = S = 1.7 × 10^7/s. Calibration distance (from the source) is 1 m. The survey meter reading at the calibration point, corrected for room scatter is 173.1 μSv. The exposure time is 360 seconds. The h(10) conversion coefficient is 385 × 10^{-12} Sv cm^{-2}. Determine the survey meter calibration factor.

$(d\Phi/dt)$ at calibration point = 1.7 × 10^7/s/(4 π d^2), where d = 100 cm

$$= 1.65 \times 10^7/(4 \times 3.14 \times 10^4) = 0.14 \times 10^3 \text{ (n/s)}$$

(dH^*/dt) at calibration point = h(10) × $(d\Phi/dt)$ = 385 × 10^{-6} × 0.14 × 10^3 μSv/s

$$= 53.9 \times 10^{-3} \text{ μSv/s}$$

For an exposure time of 360 seconds,

$$H^* = 53.9 \times 10^{-3} \times 3.6 \times 10^3 = 194 \text{ μSv}$$

$$N_{H^*} = 194/173.1 = 1.12$$

NOTE: The source decays with time. If the above source is used to calibrate a survey meter after 2 years from the above calibration date then the source strength on the new calibration date must be determined by applying a source decay correction (i.e., S in the above calibration × e$^{-(\lambda t)}$, where t = 730 days and λ = 0.693/$T_{1/2}$: $T_{1/2}$ for ^{252}Cf = 2.645 Y or 966.1 d).

6.2.2 Calibration of Personal Monitors

Since the personal monitoring devices (PMD) are worn on the body, they are also calibrated in the same geometry (i.e., with a back scattering phantom as the support) to include the body scatter in the calibration. So, unlike the workplace monitor, the personal monitor is placed on a back scattering phantom, for the personal monitor calibration. The calibration field geometry and the calibration method, however, remain the same.

PMDs are commonly used for monitoring the radiation dose received by the whole body or the body extremities (e.g., wrist, arm, leg, fingers). So, three types of phantoms have been defined (see Figure 6.11), in place of the theoretical International Commission on Radiation Units and Measurements (ICRU) slab phantom explained in Chapter 4, for the practical calibration of PMDs.

1. ISO water slab phantom of dimensions 30 cm × 30 cm × 15 cm simulating the trunk of the body
2. ISO water pillar phantom simulating the leg or the lower arm
3. ISO polymethylmethacrylate (PMMA) rod phantom simulating a finger

The front face of the slab phantom is 2.5 m thick while other sides are 10 mm thick. The pillar phantom is a hollow PMMA cylinder of diameter 73 mm and length 300 mm. The wall thickness is 2.5 mm and the end discs are 10 mm thick. The rod phantom is a PMMA rod of diameter 19 mm and length 300 mm. Figure 6.11 shows the phantoms used for PMD calibration.

The PMDs calibrated, as described above, will estimate the whole body doses (effective dose) or dose to wrist or ankles, or the fingers. Simple PMMA phantoms are also used by some laboratories, to simplify calibration procedures. However, Monte Carlo calculations of photon detection efficiency (PDE) suggest that the ISO phantom results are closer (very nearly the same) to the ICRU phantom (described in Chapter 4) than the PMMA phantom, particularly at low photon energies [17].

The AKR characterization of the photon field is carried out the same way as explained in a previous section. The PDE characterization of the calibration field requires the use of the conversion factor $[H_p(d)/K_a(P)]_{slab} = h_k$. These values are theoretically available [18]. The theoretically available conversion coefficients actually refer to the ICRU slab phantom computations and not to the much simpler PMMA calibration phantom. However, the conversion coefficients and the backscatter factors (BSF) are very close for the two phantoms and a small error this phantom may introduce may be ignored for protection-level calibrations where the calibration uncertainties far outweigh this error).

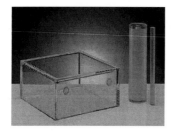 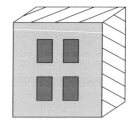

FIGURE 6.11
Phantoms for PM calibration.

The personal monitoring dosimeter being calibrated is then placed at the calibration distance (*mounted on the appropriate phantom*) and its reading M is noted, over an exposure interval Δt, for an integral value of the PDE, $H_p(d)$. $N_{Hp} = H_p(d)/M$ then gives the calibration factor in terms of the personal dose equivalent for the personal dosimeter. It must be noted that the calibration quantity does not refer to the value at the dosimeter position, but at depth d in the slab phantom.

6.3 Accuracies Required in Monitoring

The accuracy of calibration is a complex subject since the measurements are affected by the uncertainties of the various parameters that influence the measurements [19]. The uncertainties are generally classified as Type A and Type B uncertainties. The reproducibility of measurements (the reading of the reference standard and the monitor being calibrated) are referred to as Type A error and the errors associated with the parameters that influence these measurements act as systematic errors with respect to these measurements, and are referred to as Type B errors. Even if a Type A error is assumed to be 0, the measurement accuracy depends on the Type B error which cannot be eliminated. To give a simple example, if the length of a table is measured with a tape that is, say, short by 2 cm, no amount of repeated measurements of the table length can eliminate this error. This is a systematic error of the measurements, also known as bias, and is nonrandom in nature (it is now known as a Type B error instead of systematic error). A Type A error tells us how close our repeated readings are and is called precision of measurements. The Type B error decides whether what we measure is the true length of the table or not. In this case, it is easy to determine the error and actually correct for it (e.g., by using a measuring tape that is of the correct length). However, in most of the measurements there is no way of knowing the true value and limits have to be set around the measured value, taking into account both types of errors, so that it can be stated with certain degree of confidence that the actual value lies within the set limits. This is the accuracy of measurement.

The variations in the repeatability of measurements are random variables and usually follow a normal or near normal distribution (called t distribution) and the sample standard deviation can be determined and a confidence limit (or interval) set for the average of these readings. Usually, a 95% confidence limit is set for the mean of the readings. The Type B errors are also treated in a similar manner though they are not exactly random in nature. By assuming a distribution for the Type B errors, kM' can be treated as a standard deviation where M' is the maximum value of the distribution and the value of k depends on the nature of the distribution assumed. For a rectangular distribution, $k = 1/\sqrt{3}$, and $1/\sqrt{6}$ for a triangular distribution. Once the Type B errors are set as a standard deviation, the two types of errors can be added and a confidence limit set for the combined uncertainty. The accuracy is usually expressed as a percentage. If δH is the total uncertainty of the DE calibration factor of a monitor, set at 95% confidence level, the accuracy of the calibration factor, N_H is given by $[\delta H/N_H) \times 100\%]$.

A calibration factor issued by a calibration laboratory has an uncertainty associated with the calibration factor, usually expressed at 95% confidence limit. The conservative estimate of the accuracy of ADE and PDE calibration [20] is around 5% for kV X-rays and about 7–8% for gamma rays (at MeV). This accuracy applies only when the workplace monitoring conditions are identical to the calibration conditions, which is never the case. In a typical workplace situation, the survey meter sees radiations coming from all over the place and the spectral distributions are quite different from the well-defined beam qualities used for calibration. This brings into the picture the energy and angular dependence of the monitor which can be different for different types of monitors. Because of such influencing parameters, the monitoring process uncertainties are expected to be larger than the calibration uncertainty, at least by an order of magnitude [18]. According to ICRP 9 [20], for doses nearing the annual dose limits, the uncertainty of the monitored dose equivalent should not exceed a factor of 1.5 at the 95% confidence level. For doses less than 10 mSv, the uncertainty of the monitored dose equivalent should not exceed a factor of 2 at the 95% confidence level. For deriving the organ dose or the effective dose from the monitored DE values, additional computational uncertainties must be taken into account.

6.4 The Importance of Calibration

It is mandatory to calibrate workplace monitoring instruments (survey meters) and PDE monitors at regular intervals according to the country's regulations. The survey meter calibration services and personal monitoring services are generally offered by accredited external agencies but if electronic personal dosimeters (EPDs) are used, they too may require regular calibration. As of now, EPDs only serve as supplementary devices for monitoring purposes. Typically, survey meters are calibrated once in 2 or 3 years. Usually the protection instruments carry a sticker stating the calibration date and it is the responsibility of the institution to make sure that the protection instruments in use are never out of calibration.

Example 6.5

A typical calibration geometry for the calibration of thermoluminescence dosimeter (TLD) personal monitors is shown in Figure 6.12:
  The calibration details are as given below:
  Reference ionization chamber used for exposure measurement: 800 cc ionization chamber.

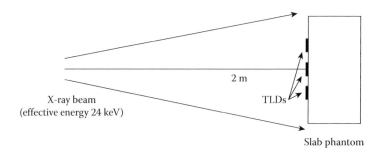

FIGURE 6.12
Calibration of TLD personal monitors.

Chamber calibration factor N_X = 4.44 mR/nC or 4.44 (mR/s)/(nC/s)
(at 22°C and 1013.5 mbar pressure)

Source to chamber distance = 2 m

Calibration room: temperature 23.7°C; pressure 1011.0 mbar

Chamber reading at calibration distance (2m): 5.664×10^{-12} A

Beam quality used for calibration: 24 keV (effective energy)

$[H_p(10)/k_a]$ = 0.79 mSv/mGy for this beam quality

Calculate the delivered $H_p(10)$ to the TLDs for an exposure of half an hour.
Steps:

1. Calculate exposure rate and air kerma rate at the calibration point
2. Calculate the PDE rate
3. Calculate the PDE for 0.5 hour exposure

$$\text{Chamber reading corrected for (T,P)} = 5.664 \times 10^{-12}\text{A} \times \frac{(273.15 + 23.7)}{(273.15 + 22)} \times \frac{1013.25}{1011.0}$$

$$= 5.664 \times 10^{-12} \times 1.0057 \times 1.0022 = 5.709 \times 10^{-12} \text{ A}$$

$$X(2m) = [4.44 \times 10^{-9} \times 10^{-3} \text{ R/s/A} \times 5.709 \times 10^{-12} \text{ A} \times 3.600 \times 10^{3} \text{ (s/h)}] \text{ R/h}$$

$$= 0.9125 \times 10^{-1} \text{ R/h}$$

Air kerma to exposure conversion factor = 8.76 mGy/R
This can be obtained from the equation $K_a \approx X \ (W/e)$

$$= X \text{ in R} \times 2.58 \times 10^{-4} \text{ (C/kg/R)} \times 33.97 \text{ J/kg}$$

and 1 J/kg = 1 Gy)

$$K_a(2m) = 0.9125 \times 10^{-1} \times 8.76 = 7.9935 \text{ mGy/h}$$

PDER in the slab phantom at 2 m distance from the source = $7.9935 \times 0.79 = 0.6315$ mSv/h

PDE or $H_p(10)$ at 2 m from the source for 0.5 h exposure = 0.3158 mSv = 315.8 μSv

Example 6.6

In Table 6.4, readings were obtained for the nine TLDs exposed to the above PDE on the slab phantom, as read from the TLD reader (marked in scale divisions). Calculate the individual calibration factor of the TLDs (all the TLD readings have been corrected for the background reading).

$$N_{TLD} = H_p(10)/\text{TLD readout}$$

For TLD5, N_{TLD5} = 315.8/292 = 1.08 μSv/Sc div.

Similarly, the calibration factors of other TLDs can also be determined.

TABLE 6.4

Readings of Exposed TLDs on the
PDE Calibration Phantom

TLD Identify	Reading (in Sc. Div.)
TLD1	287.4
TLD2	294.2
TLD3	283
TLD4	296
TLD5	292
TLD6	299.5
TLD7	291.7
TLD8	300.2
TLD9	289.6

References

1. The IAEA / WHO SSDL Network. http://www-naweb.iaea.org/nahu/dmrp/SSDL/ (accessed on October 2016).
2. A. Jennings, *Formulation of Quantities and units, their dissemination and traceability*, Consultant, NPL, *IRPA9 Conference*, April 19, 1996.
3. (ISO) International Organization for Standardization, *Reference Beta Radiations for Calibrating Dosimeters and Dose Rate Meters and for Determining their Response as a Function of Beta Radiation Energy*, ISO Report 6980, Geneva, 1983.
4. ISO (International Organization for Standardization), *X and Gamma Reference Radiation for Calibrating Dosemeters and Dose Rate Meters and for Determining their Response as a Function of Photon Energy – Part 1: Radiation Characteristics and Production Methods*, ISO Standard 4037-1, Geneva, 1995.
5. ISO (International Organization for Standardization), *X and Gamma Reference Radiation for Calibrating Dosemeters and Dose Rate Meters and for Determining their Response as a Function of Photon Energy – Part 3: Calibration of Area and Personal Dosimeters. Draft Standard*, ISO/DIS 4037-3, Geneva, 1996.
6. IAEA (International Atomic Energy Agency), *Calibration of Radiation Workplace monitoring Instruments*, Safety Reports Series No. 16, IAEA, Vienna, 2000.
7. Radiation Surveys. http://www.nrc.gov/reading-rm/doc-collections/cfr/part036/part036-0057.html (accessed on September 2016).
8. Activities of the neutron standardization at the Korea Research Institute of Standards and science, CCRI(III)/05-13, p. 86, Report of the 19th Meeting of BIPM, 2005.
9. Neutron Sources, Lecture Presentation. http://www.nrc.gov/docs/ML1122/ML11229A704.pdf (accessed on September 2016).
10. O.P. Massand, *Investigation of Some Properties of the Precision Long Counter*, Joint Nuclear Research Centre, Belgium (Report, EU4783e), 1972.
11. H. Park et al., Long counter and its application for the calibration of the neutron irradiators, *Radiation Protection Dosimetry*, 157, 1–5, 2013.
12. J.C. McClure et al., *Determination of Neutron Dose Rates for the Public Health England Neutron Facility*, Public Health England, Chilton, Oxon, UK, 2013.
13. ICRU Report 57, *Conversion Coefficients for Use in Radiological Protection Against External Radiation – International Commission on Radiation Units and Measurements*, Oxford University Press, UK, 1998.

14. ISO 8529-3, *Reference Neutron Radiations: Calibration of Area and Personal Dosimeters and Determination of their Response as a Function of Neutron Energy and Angle of Incidence.*

15. R. Radev and T. McLean, *Neutron Sources for Standard-Based Testing*, Los Alamos National Laboratory, Report LLNL-TR-664160, New Mexico, 2014.

16. IEC 61005, *Portable Neutron Ambient Dose Equivalent Rate Meters for Use in Radiation Protection*, Publisher IEC, Geneva, Switzerland.

17. M. Ginjaume and X. Ortega, *Influence of the PMMA and the ISO Slab Phantom for Calibrating Personal Dosemeters*, IRPA-10: 10th International Congress of the International Radiation Protection Association, Hiroshima, Japan, 2000.

18. G.F. Gualdrini and B. Morelli, *Air Kerma to Personal Dose Equivalent Conversion Factors for the ICRU and ISO Recommended Slab Phantoms for Photons from 20 keV to 1 MeV*, ENEA, Bologna, 1996.

19. Practical Radiation Monitoring. http://www.npl.co.uk/publications/good-practice-online-modules/radiation/uncertainties-in-radiation-measurement/assessment-of-uncertainty-in-a-measurement/ (accessed on September 2016).

20. ICRP 9, *General Principles of Monitoring for radiation protection of workers*, Pergamon Press, London, UK, 1982.

Review Questions

1. What is the meaning of traceability or dissemination?

2. What are the radiation quantities in terms of which the photon survey meter can be calibrated? Which quantity has been recommended by ICRU for the calibration of the photon survey meters?

3. How is the response or the sensitivity of a monitor defined? The AK calibration of a reference standard is 25.1 µGy/nC. What is its sensitivity?

4. Explain how a calibration beam is standardized in a calibration laboratory for offering ADE calibration for survey meters?

5. If the scales of two ADE monitors are marked in (a) µSv/h and (b) scale divisions, respectively, what will be the dimension of the calibration factor of the two monitors?

6. The ^{60}Co AK calibration factor of a spherical reference chamber is 24.5 µGy/nC at 20.0°C and 101. 325 k Pa pressure. The chamber is placed in the calibration beam and the beam is kept ON until the chamber electrometer collects a charge of 1.5 nC. The ^{60}Co calibration room temperature/pressure conditions are 22.5°C and 100.4 kPa pressure. The [ADE/K_a] conversion factor for the ^{60}Co beam is 1.16. How will you determine the ADE at the measurement point?

7. An electronic personal dosimeter (EPD) was placed on an ISO-recommended water slab phantom at a distance of 2 cm from a ^{137}Cs source. The field diameter at 2 m was 5 cm, enough to completely cover the EPD. The PDE at 2 m distance was measured as 2 mSv by the calibration laboratory, for a given irradiation time. The EPD reading was 1.87 mSv, for the same irradiation time. What is the calibration factor for the EPD? (The EPD detector is not an ion chamber and hence requires no temperature/pressure correction.)

7

Radiation Detectors for Area (Ambient) Monitoring

7.1 Introduction

A radiation detector generally produces a response (i.e., a signal) to the radiation incident on it which is measured by a suitable measuring system. The response depends on the nature of the detector. If the detector is a gas medium, the radiation interactions cause ionization, so the reader is a common current or charge-measuring system (electrometer). If the detector is a scintillation system, the interactions cause light scintillations so the reader is a light detector which may be further converted to a current or charge using a suitable transducer. If the detector is a film, the radiation interactions cause blackening of the medium, so the reader is an optical density (OD) reader that measures the amount of blackening of the film, in terms of light transmission. Here again, the transmitted light is converted to current using a photodiode. So when the detector signal is not a current or a charge, a transducer is used to convert the signal into a current or charge depending on whether the dose rate or the integral dose is being measured. The current or the charge signal is related to the measurement quantity (ambient dose equivalent rate [ADER] or ambient dose equivalent [ADE]) through a valid calibration as explained in Chapter 6. So in general, a radiation detector system is a detector plus a reader, as shown in Figure 7.1.

The detector is generally capable of registering the individual pulses produced by the particles or photons interacting in the sensitive volume of the detector. The readout system can be designed to count the pulses or the mean level output. Generally, in photon monitoring the current or the total charge is measured, produced in a detector for a predetermined time, in the workplace, which represents the radiation levels in exposure rate/air kerma rate or integral exposure/air kerma. When the radiation levels are high, the ionization events will occur at a high rate, and it is difficult to resolve the individual pulses. So the pulses are averaged over time and the dose rate or the integral doses are measured. In neutron monitoring which is always accompanied by photon background, the pulse mode of operation is generally made use of which would help to discriminate neutron pulses against photon pulses. At the reactor sites where the neutron fluences are high, it is necessary to resort to measuring the current.

The radiation detectors can be generally classified as

- Gas-filled detectors (GFDs)
- Solid-state detectors (semiconductor and scintillation detectors)

Under solid-state detectors, personal monitoring devices may be included such as

- Photographic film dosimeters (films)
- Thermoluminescence dosimeters (TLDs)

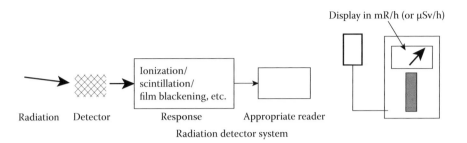

FIGURE 7.1
Principle of radiation detection.

and the less frequently used

- Radio photoluminescent dosimeters (RPLDs)
- Optically stimulated luminescence dosimeters (OSLDs)

and the personal monitoring devices that are now gaining popularity

- Electronic personal dosimeters (EPDs)

The personal monitoring devices (PMDs) will be dealt with in the next chapter. This chapter will discuss the detectors used for area monitoring.

7.2 Detector-Operation Regions in GFDs

Gas-filled detectors are generally cylindrical, spherical, or plane parallel geometry detectors with a wall and an enclosed gas. The detector may be sealed or open to the ambient. The advantage of a sealed detector is that its sensitivity can be increased by increasing the gas pressure. This would help in detecting radiations at very low dose rate levels. One disadvantage of this detector is that if the detector is leaking, its response will change and the monitoring will be erroneous. This problem does not arise for unsealed detectors but in this case, there will be changes in the mass of gas in the detector with changes in ambient temperature and pressure. This will also affect the sensitivity of the detector unless temperature–pressure corrections are applied to the detector readings. However, this is not a serious problem in radiation protection measurements, where the accuracy required is not as high as in clinical dosimetry.

The GFD detector needs a polarizing voltage to collect the ionization produced by the radiation interactions. The voltage separates the ions and the electrons (or negative ions) and collects the ionization on opposite ends at the respective (positive and negative) electrodes. Without the detector voltage, the ions will recombine and there will be no signal to detect. In other words, the height of the pulse will be zero. When the operating voltage of the GFD is gradually increased, the recombination gradually decreases and the ionization charge correspondingly increases. As the voltage is further increased, the detector sensitivity (or pulse height) increases due to various mechanisms occurring in the detectors, as

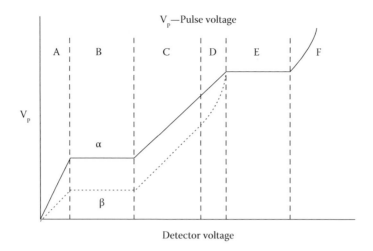

FIGURE 7.2
Different operating regions of a gas-filled detector.

a function of the detector voltage. As a result, the voltage versus the pulse height curve goes through different regions as shown in Figure 7.2. These regions are known by different names and the detectors carry these names corresponding to their regions of operation. Details of these regions and the detectors operating in the regions that are considered useful for particle identification or radiation monitoring will be discussed in this chapter.

The different regions of operation of the GFDs are known as follows:

A: Recombination region

B: Ionization region

C: Proportional region

D: Limited proportionality region

E: Geiger–Muller (GM) region

F: Continuous discharge region

7.2.1 Recombination Region

The recombination region and ionization chamber regions were explained in the earlier section. Since ion chamber based detectors operate in the least sensitive of the detector regions (in the saturation region), they are generally employed in high-radiation areas. The sensitivity of these detectors is generally increased by increasing the volume of the detector or by pressurizing the gas inside the detector so its use can be extended to lower radiation levels (e.g., for radiological survey in radiation oncology departments).

7.2.2 Ionization Region

Figure 7.3 shows the design principle of a basic gas-filled detector, usually known as the ionization chamber detector.

A point to consider in the ion chamber–based detector is the filling gas used in these detectors. In electronegative gases (e.g., oxygen), the ionized electrons are easily attached

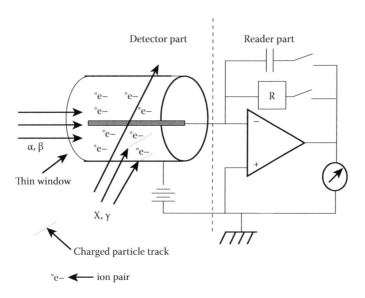

FIGURE 7.3
Principle of an ion chamber based gas-filled detector.

to the gas molecules, turning them into negative ions. On the other hand, gases like hydrogen, nitrogen, or noble gases have low electron attachment coefficients and do not attach to electrons. This becomes important in pulse-counting detectors because the collection time of electrons and ions vary by three orders of magnitude (due to the difference in masses) but when the detector is operated in current mode, electrons or negative ions can be collected. So, any filling gas can be used; the most common filling gas is air. Again, many ion chamber detectors are open to the atmosphere and may require a temperature–pressure correction. This is only possible when the filling gas is also air.

Another point to consider is the detector voltage. In order to collect the ionization (ion pairs produced) in the detector, a voltage is applied between the wall and the central electrode (separated by an insulator or insulator and guard to divert the leakage current away from the signal electrode). The function of the applied voltage is to provide an electric field which is necessary to sweep the electrons and ions across the two electrodes. Sufficient electric field (voltage) must be available for the detector to prevent recombination of the ionization. A plot of voltage versus ionization collected is shown in Figure 7.4.

The typical saturation voltage may fall in the range 50 to 250 volts for ion chamber–based detectors.

To determine the charge produced for an energy absorption of δε energy in the sensitive volume of the detector, it is necessary to know the W value, defined as the average energy required to produce an ion pair in the medium. This is about 35 eV in air. Since much of the energy is also going into excitation, W is always greater than the ionization energy.

Example 7.1

A charged particle deposits an energy of 105 keV in the sensitive volume of the detector. The detector is applied with a saturation voltage of 200 volts. The average energy required to produce an ion pair is about 35 eV.

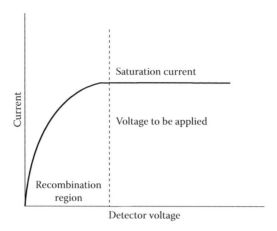

FIGURE 7.4
Determination of saturation voltage for an ion chamber detector.

1. What is the total charge produced in the detector? ($e = 1.6 \times 10^{-19}$C).
2. If this charge charges a capacitor of $C = 100$ pF, what is the voltage of the pulse produced?

$$\text{No. of ion pairs produced} = 105 \times 10^3 \text{ eV } / 35 = 3000 \text{ ip}$$

$$\text{Charge } \delta q \text{ produced} = 3000 \times 10^{-19} \text{ C} = 3 \times 10^{-16} \text{ C}$$

$$\delta V = \delta q / C = 3 \times 10^{-16} / 10^{-10} = 3 \text{ } \mu V$$

It can be seen from the above example that the ion chamber detector–produced pulses are very tiny and are not convenient for pulse measurements. In fact these detectors are not operated in pulse mode for monitoring X, γ, or radiation levels, but in current mode or charge mode that integrates the current over a period of time. The rate is a measure of the rate at which the particles are interacting with the detector. In the case of X or gamma rays, it can be related to the exposure or air kerma rate of the ambient radiation fields in the workplace. When the radiation fields are constant with respect to time, the detector shows a steady current subject to statistical variations due to the fluctuations in the photons or the reading electronics. Even for low radiation levels, the number of photons involved is quite large (about 10^5 or larger) and so the statistical effects on this count are negligible.

At radiation protection levels (say in the mR/h range), the currents produced are in the pA range or even lower. To measure currents of this order, the ion chamber detectors are designed with a guard to minimize the leakage current reaching the central electrode. Figure 7.5 illustrates the concept of guard (electrode) to divert leakage current from the central electrode.

If $V = 300$ volts is applied to the wall of the detector and R, the insulator resistance $= 10^{16}$ ohms, the leakage current, $I_L = 3 \times 10^{-14}$ A, which flows through the guard electrode and not the central electrode.

The survey meters used for monitoring the occupied areas in radiation oncology departments are the most well-known examples of ion chamber–based detectors. The detector

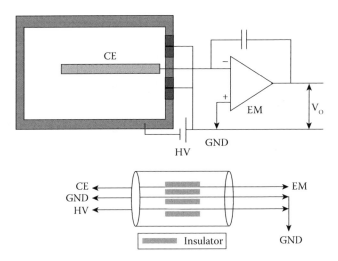

FIGURE 7.5
Chamber/electrometer/cable connections.

system is in rate mode when the detector current flows through the resistor in the feedback path (not shown in Figure 7.5) and in "integrate" mode when the detector current charges the capacitor in the feedback path (as shown in Figure 7.5). So, the reader system can be marked to read the dose rate (exposure rate or air kerma rate) in the rate mode, or the integral dose (air kerma or exposure) in the integrate mode when the charge is integrated over the capacitor for a fixed time. The integrate mode is very useful if the radiation levels near the door inside the linac room are required, during the operation of the linac.

Example 7.2

An ion chamber–based detector having a sealed volume of 100 cc and made of air equivalent wall reads the air kerma field outside the teletherapy room as 20 μGy/h. What is the approximate current produced in the detector? (The given conversion factors are 2.58×10^{-4} C/kg · R and 8.79 mGy/R).

An air kerma rate (AKR) of 20 μGy/h corresponds to 20 / 8.79 mR/h = 2.3 mR/h.

For an air equivalent–walled detector with equilibrium wall thickness, this would correspond to

$$2.58 \times 10^{-7} \text{ C/kg} \cdot \text{mR} \times 2.3 \text{ mR/h} = 5.93 \times 10^{-7} \text{ C/ kg} \cdot \text{h} = (5.93/3600) \times 10^{-7} \text{ A/kg}$$

100 cc corresponds to a mass of air = 220 cc × 1.293×10^{-6} kg/cc = 2.9×10^{-4} kg

So the detector current is $(5.93/3600) \times 2.9 \times 10^{-11}$ A = 4.7×10^{-14} A.

The above example refers to a typical survey meter volume. It may be noticed that the reader must be a high quality, low-current electrometer to measure this order of currents or the integral method is needed to determine the air kerma rate. One method of increasing the sensitivity of the ion chamber–based monitor is to increase the volume

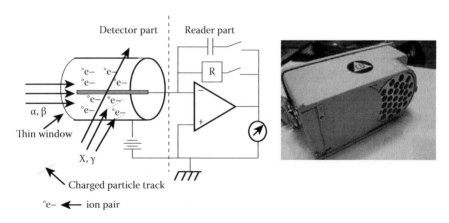

FIGURE 7.6
Ion chamber–based survey meter (β window side shown).

of the detector, or the pressure of the gas inside the detector, to cover a wide range (e.g., <1 mSv/h–1 Sv/h).

7.2.2.1 Ion Chamber Detector–Based Survey Meters

FIGURE 7.7
Cutie pie survey meter.

The most accurate and reliable GFD systems are ion chamber based and are the ones used in radiation oncology departments. They are also the ideal for pulsed beams. They operate in current and integral charge modes, as explained earlier. A basic ion chamber–based survey meter for β, γ monitoring is shown in Figure 7.6.

One of the earliest and the most popular ion chamber–based survey meter is the cutie pie survey meter. It is shown in Figure 7.7 along with its modern digital version.

7.2.3 Proportional Region

In the proportional region, the primary ionization produced in the detector is amplified by the secondary ionization produced by primary electrons in ionization by collision. If the detector can be designed with a very thin wire as the central electrode (Figure 7.8), the electric field near the central electrode exponentially increases, as can be seen from the electric field equation for a cylindrical detector (Figure 7.9).

$$E(r) = V \: / \: r \: \ln(b/a)$$

where V = applied to the detector, r = any distance from the central electrode wire, b = inner diameter of the detector, and a = outer diameter of the central electrode wire.

The electric field must be $>10^4$ V/cm or 10^6 V/m for the electron to gain sufficient energy to cause further ionization by collision. This is one reason for adopting the cylindrical configuration for the detectors. Figure 7.9 shows how the electric field varies with respect to the distance from the anode wire of a cylindrical detector, according to equation. Close to the wire ($\approx$ few tens of μm), the electric fields reach values required for causing ionization by collision.

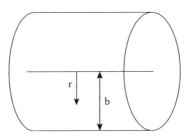

FIGURE 7.8
E(r) for a cylindrical detector.

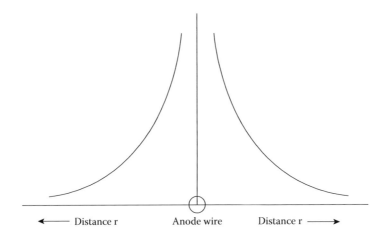

←——— Distance r   Anode wire   Distance r ———→

FIGURE 7.9
Variation of electric field as $(1/r)$.

This also implies that the gas multiplication region is confined to a very small envelope surrounding the central electrode wire. Irrespective of where primary ionizations occur, all the cascades of secondary ionizations are produced in the close vicinity of the anode wire, as shown in Figure 7.10 (the multiplication or primary avalanche region is greatly exaggerated in the figure).

The unit contains an air ion chamber coupled to a solid-state MOSFET input electrometer with built-in A-D converter. Readout is in mR/hr or mR. Rate range is 0.1 mR/hr – 9.999 R/hr in a single range. Dose range is 0.01 mR – 99.9 R in a single range. Accurate air equivalence is assured by 180 mg/cm^2 graphite-lined methacrylate walls. A thin (0.5 mg/cm^2) Mylar window allows high-sensitivity readings of alpha and low-energy beta such as C-14. Ion chamber survey meters are highly stable over time, making this instrument useful to reliably measure calibration sources, and radiation therapy around machines and sources. Because of their energy independent response, ion chamber survey meters are recommended for any dose rate measurements made for regulatory compliance (i.e., licensing, state regulations).

If N is the number of primary ion pairs produced and M is the amplification factor, the total ion pairs produced is given by M × N. M (and hence the pulse height) increases with the applied voltage, as can seen from the curve in the proportional region. So, a stable

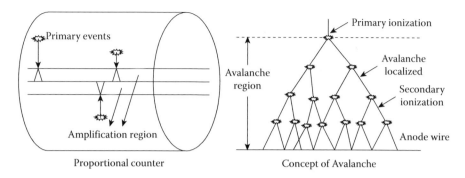

FIGURE 7.10
Pulse amplification in a proportional counter.

voltage source is an important requirement for detectors operating in this region. The detector is also known as a proportional counter (PC).

It may be noticed from Figure 7.10 that in the proportional region, the avalanche is localized—it is not spreading from one avalanche to the other across the entire length of the anode wire. So, if N ion pairs are produced in the detector, by energy absorption of δE, each is amplified by a factor M by the secondary ion pairs produced in the avalanche, that is, $\delta E = N \times M$. The proportionality between δE and N that existed in the ion chamber region is still preserved for energy discrimination. In other words, it can be assumed that $M = 1$ for the ion chamber–based detector where the electric field (or the applied voltage) was just sufficient to sweep the ions to the respective electrodes, but not sufficient for accelerating them to the extent they can cause further ionization.

$M \approx 10^3$ in PCs and it depends on the applied voltage V. As V increases, the avalanche starts farther from the wire and so the magnitude of the secondary ionization is larger. So, in the operation of a PC, stability of the applied voltage (power supply) is very important, making the system more expensive.

As explained above, in a PC the pulse height is proportional to δE. If Δx is the average dimension of the sensitive volume of the detector, along the particle track, the tracks of say an alpha and a beta particle can be compared.

$$(\delta E)_\alpha = (dE/dx)_\alpha \cdot (\Delta x)$$

and

$$(\delta E)_\beta = (dE/dx)_\beta \cdot (\Delta x)$$

$$\text{Since } (dE/dx)_\alpha > (dE/dx)_\beta \ (\delta E)_\alpha > (\delta E)_\beta$$

or alphas produce larger pulses compared to betas (see Figure 7.2). Similarly, when the PC is used for neutron monitoring, the neutron pulses can be easily discriminated against the gamma pulses. By adjusting the discriminator levels, the higher linear energy transfer (LET) particles can be counted easily. High LET radiations may create about 10^8 ion pairs and low LET radiations about 10^5 ion pairs in a PC.

Some of the well-known PCs used in monitoring are the BF_3 PC, and the ^3He PC used in neutron monitoring. Few proportional detector based β, γ survey meters used to exist in the market earlier and these days it is rare to come across these detectors for the purposes of gamma monitoring. This is mainly because GM detectors are very popular for monitoring low-radiation fields and ion chamber–based detectors are very popular in high-radiation fields.

7.2.3.1 Neutron Detectors Operating in the Proportional Region

PCs have a definite role to play in neutron surveys because neutron fields are always accompanied by a background gamma field, and PCs are ideal for rejecting the gamma pulses and counting only the neutron pulses.

Since neutrons are not charged particles, neutron detection must depend on the ionization produced indirectly by neutrons (as in the case of photons). A neutron field exhibits a wide energy spectrum. The fast neutron interactions have very low cross section. So, a more efficient detection method is to thermalize neutrons and detect these neutrons using thermal neutron capture reactions that have a high capture cross section. So, a neutron detector for area monitoring is always a moderated detector where the neutrons are moderated in a hydrogenous material to thermal energies and then detected using a thermal detector, placed along the axis of a cylindrical moderator, or along the diameter of a spherical moderator (Anderson & Braun design), the two geometries generally used in neutron detectors. Since a thermal neutron carries very little energy, a nuclear reaction of high Q value must be employed to produce reaction products of sufficient energies to get moderate signals. The pulse height depends on the deposited energy; maximum energy deposited in the detector $E_{dep} = E_n + Q$ and $E_n \approx 0$, so it does not depend on the neutron energy.

The most commonly used nuclear reaction for thermal neutron detection is

$$n_{th} + {}^3He \rightarrow p + {}^3H + 764 \text{ keV} \quad \text{(cross section 5330 barns)}$$

If the reaction occurs in a ^3He gas, the charged particles produced can ionize the gas and produce an electronic pulse. So, a sealed PC containing pressurized ^3He as the fill gas is the most commonly used thermal neutron detector.

Another common reaction for thermal neutron detection is

$$n_{th} + {}^{10}B \begin{cases} \nearrow \alpha + ({}^7Li)^* + 2.310 \text{ MeV} \ (94\%) \quad \text{(cross section 3840 barns)} \\ \searrow \alpha + {}^7Li + 2.792 \text{ MeV} \ (6\%) \end{cases}$$

A sealed PC containing BF_3 as the fill gas cum neutron detection gas—known as a BF_3 counter—is another well-known thermal neutron detector. This detector provides better gamma background discrimination compared to ^3He counters. However, BF_3 gas is toxic so the tubes are not pressurized, leading to lower detection efficiency. An attractive alternative is ^{10}B-lined counters where a coating of ^{10}B (about 0.3 or 0.4 mg/cm^2) is applied on the inside of the wall. They are also a good replacement for ^3He tubes; their design is identical to the ^3He tubes and has excellent gamma discrimination. The detection efficiency of the detector depends on the inner surface area available for coating. The detector has lower detection efficiency compared to BF_3 counters since the reaction products (α and Li ion) travel in opposite directions, and so only one of them enters the detection gas for detection. A fill gas of Ar at 0.25 atm. pressure and a little quantity of CO_2 is often employed in this detector. The detector can operate at voltages less than 1000 volts since the fill gas is not polyatomic. The lower gas pressure and lower voltage reduces the height of gamma pulses and makes neutron counting in very high gamma fields possible.

Figure 7.11 shows the thermal neutron detectors and a thermal neutron detector inserted into the moderator assembly along its axis. The thickness (or radius) of the moderator depends on the neutron spectrum in the workplace.

The neutron detectors are not very sensitive to gammas and can discriminate against smaller gamma pulses by adjusting the discriminator level. Because of the lower Q value of ^3He nuclear reaction, gamma discrimination is more difficult compared to BF_3 detectors. The sensitivity of the neutron detectors is given in terms of count rate per unit neutron fluence (cps/cm^{2}s^{-1}).

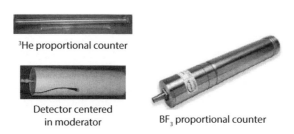

³He proportional counter

Detector centered
in moderator

BF_3 proportional counter

FIGURE 7.11
Principle of PC neutron detectors.

To increase the detection efficiency, ^{10}B can be enriched to near 95% in the gas (ordinary boron is 20% ^{10}B and 80% ^{11}B). Al, Br, Cu, and stainless steel is used as the cathode wall material of the counter. Al is typically used as wall material because of its small cross section for neutron capture. The anode is always a thin wire of about 0.04 mm thick tungsten running along the axis of the tube. The typical dimension of the tube is 1 to 2 inches in diameter, about 0.5 mm thick and about 8 to 10 inches or longer in length. The operating voltage of the detector is determined by plotting the characteristic curve, as explained earlier, and fixing the counter voltage beyond the knee portion of the curve. A typical detector voltage is around 2000 volts or less depending on the detector characteristics.

Another factor is the fill gas. For the multiplication to take place, the electrons must drift toward the multiplication region close to the vicinity of the anode wire. This means electronegative gases like oxygen (air) are not suitable as fill gas. Noble gases are generally used as fill gas which means the detectors have to be sealed or a continuous flow of gas through the detector must be maintained. (The continuous flow PCs are generally used in radioactivity standardization, and not in radiation survey). Ar is the most widely used noble gas since it is less expensive. Other noble gases like krypton and xenon are used for X or gamma ray monitoring due to their higher atomic number, which will enhance the gamma sensitivity of these detectors. Each primary ionization produces only one avalanche, retaining the proportionality between primary ionization and the total ionization. The most commonly used gases in neutron detectors are ^{3}He and ^{3}BF.

Example 7.3

In example 7.1, assume that the detector was a PC with a multiplication factor of 10^4. What will be the total pulse charge and pulse voltage?

The primary ionization in example 7.1 was 3000 ip.

The total ion pairs in the PC = 3000 $\times 10^4$ ips

So the charge will be 10^4 times greater, and so is the pulse voltage generated.

Example 7.4

An X-ray photon interacting in the sensitive volume of a PC produces 300 ion pairs. The amplification factor $M = 10^4$. The capacitance of the detector is 100 pF. Calculate the voltage of the pulse generated at the detector. ($e = 1.6 \times 10^{-19}$ C).

Total ip produced = $N \times M = 300 \times 10^4 = 3 \times 10^6$ ip

Total charge (of one polarity) produced $= N \times M \times e = 3 \times 10^6 \times 1.6 \times 10^{-19}$ C

$$= 4.8 \times 10^{-13} \text{ C}$$

Maximum pulse voltage $= Q/C = 4.8 \times 10^{-13}/10^{-10} = 4.8$ mV

It may be noticed that the pulse is still very small and may need further amplification and processing with a preamplifier and a main amplifier.

Example 7.5

A cylindrical detector is designed with the following dimensions: Inner diameter = 2 cm; outer diameter of the wire = 0.085 mm. The detector voltage is 2000 volts. The gas amplification is confined to what percentage of the detector volume? Assume 10^6 V/m as the electric field strength to cause ionization by collision. The length of the detector is L. The radius at which $E = 10^6$ V/cm is given by

$$r = 2000/10^6 \ln (10^{-2}/85 \times 10^{-6})$$

$$= 425 \text{ } \mu\text{m}$$

$$\begin{array}{l} \text{The amplification volume in the detector} \\ \text{(as a percentage of the detector volume)} \end{array} = \frac{\pi(425 \times 10^{-6})^2 \times L}{\pi(1 \times 10^{-2})^2 \times L} \times 100\% = 0.18\%$$

It can be seen from the above example that the critical factor in the detector operating in the proportional region is the extremely thin anode wire. So, it is not possible to just take a detector operating in the ionization detector region, apply a high voltage, and convert it to a proportional detector. These detectors operating in different regions are specially designed for those regions. Also, the geometry of the detector varies depending upon its application (e.g., for radioactivity measurements, or for alpha, beta, or gamma counting). In radiation oncology, the concern is mostly with monitoring X or gamma radiations, and in some special situations beta ray monitoring (e.g., for activated linac head monitoring). More details on these aspects can be found in the next chapter.

Example 7.6

Calculate the charge produced in a ^{3}He counter by a thermal neutron. The multiplication factor for the counter = 100. Assume the average energy required to produce an ion pair in the fill gas is 25 eV.

$$n_{th} + {}^3\text{He} \quad p + {}^3\text{H} + 764 \text{ keV}$$

The Q value of the reaction, 764, keV is shared by the recoiling daughter products proton and ^{3}H nuclei.

So, the charge of the pulse produced $= \{764 \times 10^3 \times 100 / 25\}$ ion pairs $\times 1.6 \times 10^{-19}$ C

$$= 3 \times 10^7 \times 1.6 \times 10^{-19} = 4.8 \times 10^{-12} = 4.8 \text{ pC}$$

7.2.3.2 *Proportional Detector–Based Neutron Survey Meters*

Two popular models of neutron survey meters are shown in Chapter 4 (Figure 4.27).

The $[H^*(10)/\Phi]$ response with respect to neutron energy was shown in Chapter (Figure). A neutron survey meter must have a similar energy dependence response—that is, its response per unit fluence, $[R_\Phi(E)/\Phi(E)]$, in units of cps/(n/cm^2·s^{-1}), or counts per unit

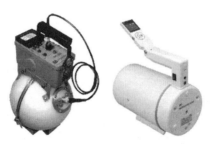

FIGURE 7.12
PC-based neutron survey meters.

fluence must follow the $[H^*(10)/\Phi]$ curve to be considered an ideal ADE or ADER monitor for measuring $H^*(10)$. In other words, the energy dependence curve $[r_\Phi(E)/h_\Phi(E)]$ for the neutron survey meter must be flat to within a certain percentage, which can be accounted for in the accuracy of measurement of the survey meter.

The above two designs are the most popular designs for a neutron survey meter. The cylindrical design, with CH_2 moderator and a perforated ^{10}B loaded layer, is by Andersson and Braun. It has a central BF_3 PC. The spherical design, with CH_2 moderator and perforated Cd layer, by Leake, is of spherical design with a centrally placed spherical ^3He gas PC (or thermal neutron detector). The design overresponds to the intermediate energy response, so a thermal neutron absorber with perforations to limit the thermal fluence to the required extent is placed at an intermediate depth in the moderator. The highest neutron energy that can be thermalized depends on the dimensions of the moderator. The commercially available survey meters have a radius of 8 to 10 inches which is adequate for surveying linac rooms for neutron levels.

In general, neutron survey meters are designed to exhibit an isotropic response. Since the introduction of the new radiation protection quantity ADE, which requires isotropic response by definition of the quantity, this assumes special significance. It is difficult to provide a perfect isotropic response to an instrument because of the detector cables, electronics, instrument handle, etc., which will influence the instrument response. Yet these two survey meters have an adequate isotropic response for survey purposes.

7.2.4 Region of Limited Proportionality

The proportionality between primary ionization N and energy deposition in the detector sensitive volume is maintained until M is about 10^4. As V is further increased, the avalanche starts occurring farther away from the anode wire when the proportionality becomes weaker and weaker. This is the region of limited proportionality. In this region, the gas amplifications become 10^5 to 10^7. This region is not of much use for particle detection.

7.2.5 Geiger–Muller Region

We saw in the earlier section of PCs that the avalanches were localized (see Figure 7.8). As the voltage on the detector goes beyond the region of limited proportionality, the avalanche extends across the entire length of the central wire. This happens because, on average, each avalanche produces another avalanche from the ultraviolet range photons emitted from the exited states of atoms, leading to a self-sustaining phenomenon, as shown Figure 7.13.

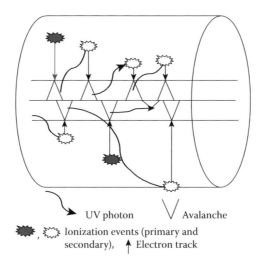

FIGURE 7.13
Spread of avalanche in GM detector.

This results in the production of 10^9 to 10^{10} ion pairs and a large pulse of several volts, even for a single primary event occurring in the GM detector. So, particles of different LETs cannot be distinguished in a GM detector. GM is thus not suitable for energy spectrometry but has maximum sensitivity to measure very low radiation levels. Also, because of the large voltage pulses, sensitive preamplifiers are not necessary for the GM-based systems.

7.2.5.1 Detector Quenching Gas

A distinction must be made between quenching gas and fill gas. Fill gas is responsible for primary ionization and pulse formation in the detectors. Quench gas is used as ultraviolet (UV) absorbers to prevent the production of spurious pulses in PCs by the UV photons–induced secondary avalanches. With many fill gases, not all the energy expended by the particle goes into ionization. Part of the energy goes into excitation and these molecules then de-excite by emitting photons. These photons can eject electrons from the gas molecules or from the wall by photoelectric or Compton interactions. The electrons can create spurious pulses in the counter through secondary avalanches occurring in a GM detector. This is highly undesirable in a PC since it can cause loss of proportionality between the primary and secondary ionizations. To prevent this from occurring, certain gases are added to the fill gas. These are polyatomic gases with lower ionization potential compared to the emitted photon energies. When these photons are absorbed by these molecules, they cause dissociation. Thus each avalanche event is terminated and not allowed to initiate another event, preventing the formation of spurious pulses. Often, methane is used as a quench gas. A popular fill gas is a mixture of 90% argon with 10% methane, generally known as the P-10 gas. This is generally used in gas-flow PCs. Since the quench gas gradually gets exhausted, these counters have a finite lifetime (typically about 10^8 counts). Neutron proportional detectors make use of polyatomic gases so they also serve the quenching function. Sometimes small quantities of CH_4 or CO_2 may be added for the quenching action.

GM detectors too require a quench gas but not for UV quenching. The UV photons are responsible for spreading the avalanches across the full length of the central wire in a GM detector. However, the avalanches initiated by a particle must terminate once all the avalanche electrons are collected. However, the positive ion sheath traveling to the wall of the GM tube are neutralized by attaching the wall electrons—ionization potential of ions are greater than the work function of the cathode material—and releasing the recombination energy (ionization potential–work function) in the process. This excess energy appears as a photon or as an ejected wall electron (if the excess energy exceeds the work function). This would reinitiate another ionization-avalanche phenomenon leading to a continuous discharge situation. In order to make pulse counting possible, the reinitiating of discharge is prevented by adding 5 to 10% of a quench gas.

There are two types of quenching—organic quenching and halogen quenching. In organic quenching the positive ions, before reaching the wall, collide with the organic gas molecules (e.g., ethyl alcohol) and neutralize them by electron transfer (due to the ionization potential differences between them, about 11 eV and about 16 eV for the primary gas molecule and quench gas molecule). Now the organic molecular ions reach the electrode instead of the fill gas molecular ions. Any excess energy released at the wall leads to the dissociation of the organic molecules instead of reinitiating any discharge. A typical filling of an organic quenched GM tube is 90% Ar and 10% ethyl alcohol. The organic gas becomes exhausted quickly and so these tubes have short lifetime (lasting 10^8 to 10^9 counts) and are not suitable for monitoring high radiation levels. To overcome this problem, halogen quenching extends its use in the GM tube. Neon with 0.1% of chlorine or bromine is a popular fill gas for halogen-quenched GM tubes. The halogen molecules also dissociate in quenching but they are able to recombine, thus extending the life of the tube.

7.2.5.2 GM Detector–Based Survey Meters

There are a few important points to consider in the design of various types of radiation survey meters:

- The penetrability of the particles or the photons to reach the sensitive volume of the detector
- The intended use (e.g., to monitor a surface for contamination or to standardize a radioactive source which requires almost 100% geometric efficiency)
- Energy dependence for monitoring photon fields
- Sensitivity

GM detectors are not ideal for detecting low energy αs or βs or very low energy β, γ emitting radionuclides like ^{13}C or ^{125}I, or very high-radiation fields. Since X and γs are very penetrating, the efficiency is low for these radiations. For low energy X-rays (in the lower keV region), both the wall and the gas contribute to ionization but in the gamma energy range, direct absorption of gammas in the gas is negligible and much of the response is due to the wall-generated electrons. Since the wall absorption amounts to a few percentage, that is the order of gamma efficiency that can be expected from these detector systems.

Figure 7.14 illustrates the different configurations of a GM-based radiation detector used for workplace monitoring. The detector part can be an internal (built-in) detector forming an integral part of the instrument, or it can be an external detector coupled to the reader from outside. Again, the detector could be just a gamma detector or a general purpose detector with a special window for detecting low-penetrating α and β radiations. The window wall thickness is typically about 1 to 2 mg/cm^2 mylar foil. The efficiency for αs and βs is 100% for the particles entering the sensitive volume, but due to small thickness of the

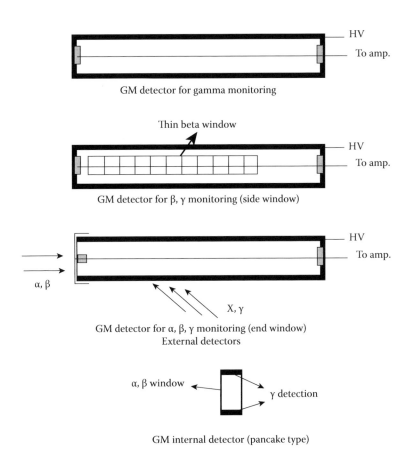

FIGURE 7.14
Different configurations of the GM-based detector.

mylar window there is a cutoff for αs for detection. In the case of β, γ survey meters, the β window mylar thickness can be larger making the system less fragile. When the window shield is closed for γ monitoring, it provides the electronic equilibrium thickness for the detector wall (about 500 mg/cm^2).

A pancake type β, γ GM survey meter with a sliding window for providing energy compensation is shown in Figure 7.15.

A GM-based survey meter with external detector attachments is shown in Figure 7.16.

FIGURE 7.15
Energy compensation filter for flattening the energy response.

Pancake type GM
detector and reader

End window type
GM detector

FIGURE 7.16
GM-based α, β, γ survey meter with external detectors.

With internal
GM detector

External detector
attachment

FIGURE 7.17
Analog and digital readout of GB-based detector systems.

Often there is an energy-compensated GM internal detector as part of the instrument apart from the capability to attach external detectors. So the system, without external probes, can be used for monitoring X or gamma radiation fields say around teletherapy or high-dose rate (HDR) rooms. The monitors with pancake-type external detectors are convenient for surveying surfaces for any contamination. They have higher counting efficiency since the detector is pointed toward the source with a thin window to allow particle entry. In some commercial models, like the one shown, the reader can be connected to any type of external detector, not necessarily just the GM detector.

The GM detector systems either read in the units of cpm (counts per minute) or in units of exposure rate or air kerma rate or ambient dose equivalent rate (see Figure 7.18). Most of the modern systems are digital systems and have digital displays.

With small GM tubes, miniature size GM-based survey meters are also available on the market with both internal detector and external detector attachment capability as shown in Figure 7.18.

7.2.5.3 Saturation and Energy Dependence of GM Detectors

A typical GM detector may become paralyzed at very high dose rates and may read zero instead of showing a maximum reading. This may give a false sense of safety, so ion chamber–based detectors are preferable for monitoring high-radiation fields. While using GM detectors in high radiation areas, the detector must be kept on before entering the radiation area so that a sudden kick in the needle momentarily showing a maximum reading can be seen before

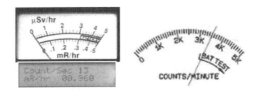

FIGURE 7.18
Pocket size GM-based survey meter.

going to zero. This would indicate a high-radiation field. Many GM detector systems have been designed with antisaturation circuits to prevent the detector from reading zero in high-radiation fields.

Also GM-based detectors are not suitable for monitoring pulsed radiation fields (e.g., linac room survey). The detector may indicate the pulse frequency instead of the actual dose rate, so ion chamber–based detectors must be used for linac room surveys. Both types of detectors can be used for the survey of a telecobalt room or an HDR room.

One of the serious drawbacks of a GM detector in gamma survey is its pronounced energy dependence in the keV region, due to the photoelectric interactions in the high Z wall (copper or stainless steel or Fe). The overresponse occurs in the energy region 30–200 keV. In the ^{137}Cs and ^{60}Co energy range (or beyond 300 keV), it shows a flat energy response like any other radiation dosimeter because the Compton interactions are independent of Z. A filtering material like tin or lead has high cross section for low-energy gamma interactions and by adding a filter material to the GM wall, the low-energy gammas will be preferentially attenuated, reducing the overresponse. By trial and error, energy-compensated GM detectors are designed to bring about a flat energy response to within say ±50% or as per the design considerations. The survey meters are usually calibrated at ^{137}Cs or at ^{60}Co energy, and used in workplaces where the photon spectrum is generally unknown. Without energy compensation, the detector would overrespond in the workplace containing low-energy photon components. Figure 7.19 shows the typical response of a GM detector before and after energy compensation.

7.2.5.4 Wall-Mounted Area Monitors

There are also wall-mounted units which can show changes in the radiation levels and can be set to give an audible alarm if the radiation level exceeds the preset value. One such commercially available unit is shown in Figure 7.20. There are also several light-emitting diode (LED) lights to visually indicate different radiation levels. These systems are mandatory in teletherapy installations. If the ^{60}Co source has not completely retracted to the parking position, the radiation levels are expected to be higher and this indicates an emergency situation. The monitor can be set to give an alarm to indicate an emergency situation. It is also very useful in brachytherapy departments where if a patient walks out with a radioactive source, the monitor can give a warning.

Multipurpose miniature GM–based survey meters are also available nowadays which can be used as a fixed wall-mounted area monitor, as a portable survey meter, and also as a personal dosimeter (which means it can read two different quantities—ADE and photon detection efficiency [PDE]).

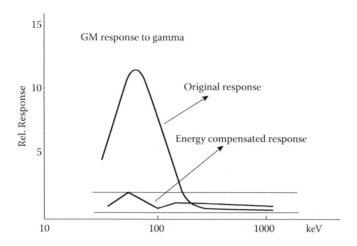

FIGURE 7.19
Energy response of GM detector system.

FIGURE 7.20
GM-based wall mounting area monitor.

7.2.6 Continuous Discharge Region

An increase in the detector voltage beyond the GM region takes the detector to the continuous discharge region. In this region, the detector spontaneously goes into continuous discharge from the stray electrons present in the detector. This will damage the detector and so no detector operates in this region; it is not useful for radiation detection.

7.3 Pulse Mode Operation of GFD

Basically any detector produces a group of ion pairs, or a charge of single polarity δQ, as a result of interaction of a particle or a photon, or a photon-produced charged particle in the sensitive volume of the detector. By applying a voltage to the detector, the charge is collected at the respective electrodes by the force of the electric field. The collection time depends on the mobility of the charge carriers and the distance that the charges have to travel before reaching the electrodes. This can vary from a few msec in the case of an ion chamber–based detector to few nsec in the case of a semiconductor-type detector. During the charge collection time, a transient current flows in the circuit. The time integral of the current over the charge collection time $\int_\tau i(t)dt$ gives the charge produced in the detector by one interaction event. The charge produced is proportional to the ionization produced, which is in turn proportional to the energy deposited by the charged particle in the sensitive volume of the detector. If the charged particle loses all its energy in the sensitive volume, the total charge of the pulse is proportional to the energy of the charged particle. The detectors are used for two purposes: for particle energy determination (energy spectrometry), or just for particle counting. In radiation survey, the interest is only in particle counting or the particle events, not energy spectrometry, but the energy discrimination becomes useful when one type of pulse needs to be rejected in favor of the other (e.g., to count the neutron pulses by rejecting the accompanying gamma background pulses). Here again, for neutron survey, the interest is only in the neutron counts, not the energy.

At the end of a pulse collection, the detector returns to the original state, and the detector is ready to register another particle. So, there is a finite time required by the detector and the associated electronics to register a pulse and when the radiation intensity is very high, the particles are incident on the detector in rapid succession. In such circumstances, it is difficult or almost impossible to operate the instrument in pulse mode. So, pulse counting systems are generally useful for low-radiation fields, and for high-radiation fields current or integral modes of operation are more suitable.

In the current mode, the detector current is proportional to the particle interaction rate, or the air kerma rate, or the ambient dose equivalent rate of the radiation field in the workplace. An electrometer connected to the detector can measure the average current—given by average number of interactions/sec × average charge per interaction—or the charge (integrated over a predetermined time) giving the ADER and ADE (for the integrated time) at the workplace. In this mode, the detector system (detector plus the circuit electronics) averages out the charges from the pulses over the response time of the system. At very low-intensity fields, the statistical fluctuations on the arrival of the particles in unit time would reflect in current fluctuations as can be seen from the needle fluctuations of the reader system. By increasing the response time, a steadier reading can be obtained but the detector system must be kept in the field for a longer time to get the steady readings. One drawback of the current mode of operation is the very low sensitivity of these systems. Even pA (i.e., 10^{-12} A) level of currents correspond to a large number of particle interactions (about 10^5) occurring per unit in the detector. Special types of electrometers are used for measuring currents of this magnitude, so counting systems are preferable at low-radiation fields (say μR/h level of fields). Ion chamber–based detectors are generally operated in current mode even at low-radiation fields, since the pulses produced by these detectors are very small and it is difficult to process and count these pulses with good noise discrimination. Here the signals are amplified by increasing the volume of the detector or pressurizing the gas inside the detector.

7.3.1 Pulse Formation and Pulse Shaping

Figure 7.21 illustrates the pulse formation in a GFD.

The pulse starts building up when charges are induced on the electrodes by the ions produced in the detector. To register a complete pulse, both electrons and positive ions must be collected at the respective electrodes. Since the velocity of the positive ions and electrons differ by three orders of magnitude (due to their mass differences), the initial part of the pulse is due to the electrons and the later part due to the ions.

There is a finite time for the detector to recover and reach its normal state following the occurrence of an event before it can register another particle. This is known as the dead time of the detector which is characteristic of the operation region. A pulse is defined by a dead time, resolving time, and a recovery time, as shown in Figure 7.22.

Dead time is the minimum time the detector would require to detect another pulse. Any event occurring during this time will produce no pulse since the positive ion sheath of the first pulse is still not completely collected. Recovery time is the time required for this to happen and the detector to reach its original state, and the pulse to reach its original height. The sheath of positive ions suppresses the field near the central electrode reducing the

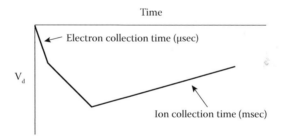

FIGURE 7.21
Typical pulse formation in a gas-filled detector.

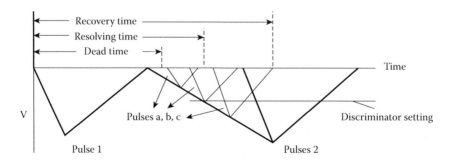

FIGURE 7.22
Dead time of a GM detector.

pulse height of the events occurring after the dead time. As the ion sheath moves toward the respective electrode, the electric field near the central electrode gradually recovers and the pulse height also recovers. (There is a minimum electric field necessary to collect charges at the electrode and form a pulse). When the pulse height is above the discriminator setting, it can be counted as a separate pulse. So, the resolving time depends on the discriminator level. For instance, in the above figure, pulse 1 and pulse "a" cannot be resolved since they fall within the resolving time. The pulses 1 and pulse "b" can be counted even though the later pulse has not fully recovered its height. This will not, however, matter for particle counting. The true counting rate and the observed counting rate will differ when particle interactions occur during the resolving time. The true counting rate, R_T, can be determined from the observed counting rate, R_0, using the following equation:

$$R_T = R_0/(1 - (R_0 \, \tau_r))$$

where τ_r is the resolving time.

Dead time is not a serious problem in a PC since the amplification is confined to a narrow region near the wire, resulting in a very short dead time, about 1 to 2 µsec. Much smaller dead times can be achieved for PCs, but there is a dead time not only for the detector but also for the associated electronics that take finite time to process the signals or pulses. The limitation for energy spectroscopy arises from the electronic circuits which require longer processing times, so an interval between pulses of the order of 100 µsec may be required for spectrometry purposes.

The large dead time of GM detectors limits its use for monitoring low-radiation levels and the lack of energy discrimination makes it unsuitable for energy spectrometry or for discriminating between beta and gamma pulses. On the other hand, the relatively large GM pulses make the counting system much simpler with no need for a preamplifier or a highly regulated power supply, or temperature/pressure corrections. Since particle counting in radiation survey is the key aspect and not particle energy, the pulses can be clipped using the circuit electronics soon following the fast rise time of the pulse by choosing a resistor–capacitor (RC) time constant smaller than the ion collection time.

This is followed by the decay of the pulse which is determined by the RC time constant of the circuit. There is a short burst of current in the detector during the pulse collection time, and when passed through a resistor this is converted into a voltage pulse, or when charging a capacitor it builds up a charge δQ on the capacitor, which leaks through the resistance R. The pulse height is a measure of the energy of the particle and the timing of the pulse registers a particle arrival. In radiation survey, the interest is in counting the events (or particles) and not energy determination. A long duration of the pulse is a hindrance to counting particles especially when the time gap between the arrival of particles is short (i.e., when the intensity of the radiation is high). So, in pulse counting instruments, the pulse is usually shaped into short pulses following its occurrence using a suitable time constant (RC) for the circuit.

7.3.2 Basic Principles of Pulse Counting Systems

The first stage of a pulse counting system is the preamplifier which is connected to the detector. The detector voltage signal at the input of the preamplifier depends on the effective capacitance at the input of the preamplifier, which is the sum of the detector capacitance, cable capacitance (which connects the detector to the preamplifier), and the input capacitance of the preamplifier. When the detector is close to the preamplifier (as in a portable survey meter with an internal detector), the cable capacitance is negligible. It may matter when the detector is not part of a portable instrument, but is externally connected to the reader system with a long cable.

A simple schematic of the first stage of the counting system is shown in Figure 7.23.

The charge pulse appears across the input capacitance of the preamplifier C. C is the effective capacitance taking into account the detector capacitance, the cable capacitance if any, and the input capacitance of the preamplifier. R is the input impedance. RC is known as the time constant of the circuit (T_C). The behavior of the circuit depends on the time constant value compared to the charge collection time of the detector. When RC << T_C, the transient current quickly flows through the resistance R developing a voltage. When RC >> T_C, the current charges the capacitor C and then leaks through the resistance with a time constant RC. Figure 7.24 illustrates the two cases.

Most of the counting systems are operated in the large time constant configuration. Here the pulse height depends—$V_{max} = Q_{max}/C$—on the capacitance value and the rise time to reach the maximum height (i.e., full charge collection) depends only on the charge collection time of the detector (see Figure 7.24) and not on the circuit electronics. The decay of the pulse, however, depends on the time required for the charge to leak through the resistance (i.e., on the RC time constant of the circuit).

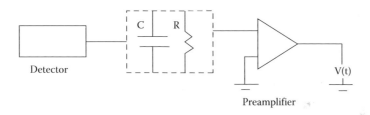

FIGURE 7.23
Preamplifier stage of a pulse counting system.

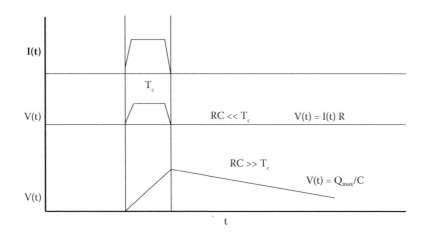

FIGURE 7.24
Pulse profile for short and long time constants.

For many detectors (e.g., GFDs) the detector capacitance is constant and $V \propto Q \propto$ particle energy. For some detectors (e.g., reverse-biased semiconductor detectors), the detector capacitance depends on the detector voltage and so the same charge Q would give rise to different V values. With such detectors, a charge-sensitive preamplifier is used instead of a voltage or capacitance-sensitive preamplifier. In a charge-sensitive preamplifier, the voltage corresponding to the charge Q is determined by the feedback capacitance in the feedback path and not the input capacitance. So any change in detector capacitance will not influence the output voltage. Figures 7.25 and 7.26 illustrate the concept of voltage-sensitive and charge-sensitive amplifiers.

The basic components of a pulse counting system are (1) preamplifier, (2) amplifier, (3) discriminator, (4) scalar/timer or a counter/speaker, and (5) microprocessor. These components will be discussed briefly in Figure 7.27.

For many detectors, the detector pulse is very small so a preamplifier is placed close to it to receive the signal without much degradation in signal-to-noise ratio. The characteristics of the preamplifier used in the system must match the characteristics of the detector in the circuit. The preamplifier also provides impedance matching between the detector and the main amplifier, by accepting the pulse across high input impedance and outputting a pulse across a low impedance. The amplifier intensifies and shapes these pulses as per the requirement of the counting system (for particle counting or particle energy measurement). The discriminator converts the pulse into a digital pulse and outputs the pulses according to the discriminator level set.

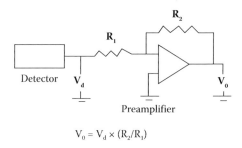

$$V_0 = V_d \times (R_2/R_1)$$

FIGURE 7.25
Voltage-sensitive preamplifier $V_0 = V_d \times (R_2/R_1)$.

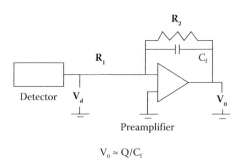

$$V_0 \approx Q/C_f$$

FIGURE 7.26
Charge-sensitive preamplifier $V_0 \approx Q/C_f$.

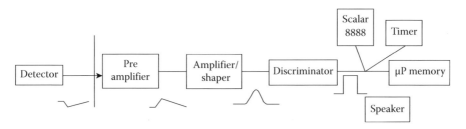

FIGURE 7.27
A generalized pulse counting circuit.

A scalar is a counter coupled to a timer which can be set for specific intervals. If the timer is set for a time interval δt, the scalar registers the counts that are output by the discriminator during that period. If δt is set for 1 second or 1 minute, it is possible to read cps and cpm directly. Otherwise, the count rate is given by $\delta N / \delta t$, where δN is the count registered by the scalar during the time interval δt. By increasing the time interval, better accuracy can be obtained in the cpm reading since the statistical fluctuations reduce with increasing δN. If there is a speaker provision in the detector system, it beeps for every pulse detected. The chattering of the speaker gives an idea of the count rate or the radiation level. The GM counter is actually not a counter but a rate meter that gives a needle deflection proportional to the rate at which the pulses are detected by the GM detector.

Since the pulse is digital, it can also be stored in a microprocessor memory and manipulated for further use.

Figure 7.28 illustrates pulse-shaping concept in simple terms. The preamplifier output shows the pileup of pulses but well resolved. If two pulses occur within the resolving time of the detector, the two will not be resolved and only one pulse or particle will be recorded. Thus there will be a count loss and the true count rate would differ from the observed count rate. In order to increase the pulse rate detection capability, the long tail of the pulse is replaced by a short-tailed pulse by the pulse-shaping amplifier.

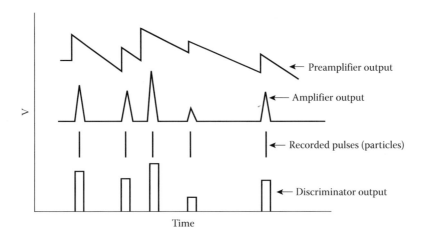

FIGURE 7.28
Pulses produced in a typical counting system.

The brief description of pulse electronics given above does not do any justice to the topic due to the author's lack of good knowledge in this area. Readers should refer to Knoll's book [1] for more information. It is the definitive text for learning about radiation detectors and reader systems.

Example 7.7

The resolving time of a GM detector is 200 μsec. The measured count rate was 9700 cpm. Calculate the true count rate.

$$R_0 \, \tau_r = (9700/60) \times 200 \times 10^{-6} = 0.032$$

$$(1 - R_0 \, \tau_r) = 0.9676$$

$$R_T = 9700/0.9676 = 10{,}024 \text{ cpm.}$$

From the value of $R_0 \, \tau_r$ (in the above example), it is easy to recognize that the error is about 3% or the true cpm would be 3% larger, or about 10,000 cpm. At 100,000 cpm, the measured count rate will be 75,000, an error of 25%. The error is small for small values of R_0 or τ_r. Microprocessor-controlled radiation monitors with dead-time correction circuits can correct for the resolving time losses and display the true count rates. The typical resolving time of GM detectors can vary from 50 μsec to 500 μs. With low dead-time counters and correction circuits, GM can also be employed at high count rates, but it is not a good detector for pulsed fields.

7.3.2.1 Characteristic Curve of the Pulse Counting Systems

A curve of count rate versus applied detector voltage is known as the characteristic curve. A typical curve for a PC is shown in Figure 7.29. There is a threshold voltage before the counter starts registering the pulses. As the voltage is further increased, there is a plateau

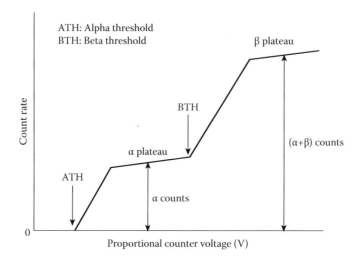

FIGURE 7.29
Characteristic curve of a PC.

region, where all the pulses are counted. Since at a given detector voltage, alpha pulses are larger than the beta pulses, when both the particles are present the alpha pulses pass the discriminator level and get counted first. The discriminator level decides the threshold detector voltage before the counter starts registering pulses. In order for the beta pulses to cross the same discriminator level, the pulse height must also reach the discriminator level. This is achieved by increasing the detector voltage to a much higher level. The characteristic curves therefore show two plateau regions—the first one counts all alphas, and the second both betas and alphas.

Thus, only αs can be counted in the presence of βs or both particles can be counted in a PC. Similarly, in a neutron PC, the neutrons can be counted in the presence of gamma background pulses. The plateau voltage is not the same for all detectors; it depends on the construction details of the detector. Larger tubes may require a larger voltage to produce the necessary electric field in the detector. The manufacturer usually mentions the correct voltage to be applied to the detector which is below the center point of the plateau region. So, the PCs are used in situations where there is a need to discriminate particles (e.g., αs against βs and neutrons against gammas).

The GM detector also exhibits a curve similar to the alpha plateau curve, but there is no energy discrimination in a GM detector and the curve goes into continuous discharge region beyond the plateau voltage. Ideally, the plateau must be flat but the typical curves exhibit a slight slope depending on the construction quality of the detector. A high-quality GM detector will have a slope of a few percentage per 100 volts.

7.4 Solid-State Detectors

Materials are classified as conductors, semiconductors, and insulators based on their band structure. When atoms come together and form a crystal or a solid, the energy levels of the individual atoms merge into a band of energy levels due to their mutual interactions. The gap between the discrete atomic energy levels become the band gap where electrons cannot stay since there are no energy levels available. The filled energy levels become filled bands of the solid and the partially filled levels become partially filled bands. The empty excited states of the atom become conduction bands. In materials like Al, Ag, or Ag, the valence and conduction bands overlap and so the electrons can easily move into the conduction band. That is the reason they are known as conductors. In materials like NaI, CsI, ZnS, there is a wide band gap and even at high temperatures electrons cannot gain sufficient energy to reach the conduction band. These materials are non-conductors and are known as insulators. Between these types of materials falls the semiconductor. A semiconductor is a material that has a small band gap between the valence band and conduction bands which can be less than couple of eV. These are materials like Si or Ge, the most well-known group 4 elements. Figure 7.30 illustrates the band structure of these three categories of materials.

Table 7.1 lists the band gap energies of important semiconductors and insulators used as radiation detectors.

The configuration in Table 7.1 is for ideal materials. In the practical world, all the materials have some defects (missing atom in the interstitial, etc.) or impurity atoms that have different energy levels compared to host atoms. These imperfections can have energy levels in the band gap region which can trap electrons and holes (e.g., produced by an

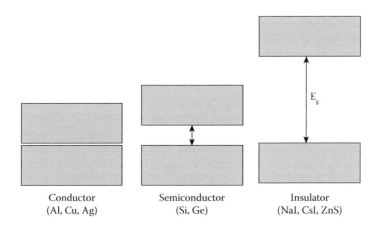

FIGURE 7.30
Band structure of solids.

TABLE 7.1

Band gap energies for various detector materials

Semiconductor	Band Gap in eV at 27°C	Insulator	Band Gap in eV at 27°C
Ge	0.66	NaI(T1)	5.9
Si	1.1	CsI(T1)	6.1
CdTe	1.44	ZnS(Ag)	3.6
GaAs	1.43		

ionizing particle) before they recombine emitting recombination energy in radiative or nonradiative de-excitations, depending on the energy level differences. Two types of radiation detectors used very much in radiation survey devices are the scintillator detectors and semiconductor detectors.

7.4.1 Intrinsic and Extrinsic Semiconductors

The most widely used semiconductors are Si and Ge. The band gap of germanium being very low can give rise to thermal carriers so even at room temperature T, some electrons can gain sufficient thermal energy to move into the conduction band, without any radiation interaction–caused ionizations. For Si, Eg > kT and so there are not many electrons in the conduction band at room temperature. Si then behaves more like an insulator and is the material of choice for radiation detectors. With a semiconductor material, say Si, two types of materials can be prepared—n-type Si and p-type Si. N-type will have excess of electrons compared to intrinsic Si and the p-type will have excess of holes compared to intrinsic Si. N and P types of Si can be prepared by doping intrinsic Si with a group V element (e.g., B) and group 3 element (e.g., P), respectively. In an intrinsic Si, each Si atom requires four more electrons to complete an inert octet configuration of the 3p subshell, so the Si atom can form covalent bonds with four neighbors to form the crystal lattice, which is the most simplified explanation for the Si crystal. So, a group 5 element will have an extra electron to contribute to the material and a group 3 element will require one more electron (or leave a hole) in the material which can move around by electron transfers. Figure 7.31 illustrates the concept of n- and p-type semiconductor creation by doping (the arrow points to the release of an electron or hole in the semiconductor).

While an electron is a physical particle, a hole represents a vacancy for an electron. When an electron moves from one vacancy to another, it is equivalent to a hole movement in the opposite direction. Since electron movement in conduction band and through hole vacancies (in valence band), are both real, the electron flow and the hole flow constitute currents. The hole however moves much slower compared to the electron since it requires the electron to find a vacancy to occupy that state, unlike electrons which can move freely in the conduction band. At room temperature, the electrons can acquire enough thermal energies to move to the conduction band, leaving holes behind. So, in an n-type semiconductor, holes are the minority carriers and in the p-type semiconductor the electrons are the minority carriers. In an electric field, the majority and minority carriers, being of opposite charge, always travel in opposite directions, as shown in Figure 7.32.

The doped semiconductors are also known as extrinsic semiconductors. The density of thermal carriers depends on the temperature and the band gap, being governed by the expression $e^{-Eg/kT}$.

7.4.1.1 P–N Junction Diode

P–N junction diodes are most widely used as radiation detectors. A p–n junction diode is formed by joining a p-type semiconductor and an n-type semiconductor. When the two are

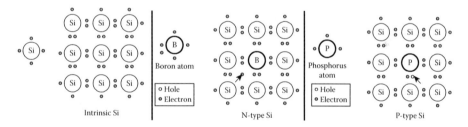

FIGURE 7.31
Concept of intrinsic and extrinsic semiconductors.

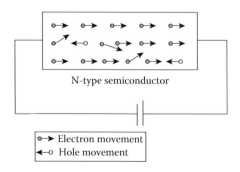

FIGURE 7.32
Majority and minority carrier movement in an extrinsic semiconductor.

joined together, the majority carriers diffuse across the junction leading to the recombination of electrons and holes. This leads to a junction region devoid of any majority carriers and hence named the depletion region. In the depletion region, there are exposed positive ions on the N side and exposed negative ions on the P side. Thus, a reverse bias appears across the junction, preventing all charge carriers from neutralizing one another. The depletion region now functions as a parallel plate capacitor with a voltage of 0.7 volts. The device is known as a p–n junction diode (see Figure 7.33).

With an intrinsic reverse bias, the diode does not conduct. There will be only a small minority carrier current across the junction since for minority carriers the reverse bias functions as a forward bias. If a forward bias is applied to the diode that exceeds the reverse bias, the diode will start conducting. The forward bias does not have much effect once the intrinsic junction voltage is neutralized. Only bulk resistance would then come into play, with a high-saturation current of majority carriers across the junction. So a forward biased p–n diode, in a conducting mode, is not useful for measuring the charged particle produced ionization current which will be a minute fraction of the majority carrier current.

7.4.1.2 Si P–N Junction Diode in Reverse Bias As a Radiation Detector

P–N junction capacitance is shown on the left side of Figure 7.34. The capacitance charges are immobile because they represent the lattice ions which cannot move, but there is an electric field connecting the two regions. The P and N regions with mobile carriers act

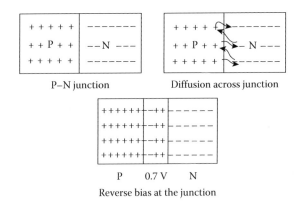

FIGURE 7.33
Concept of p–n junction diode.

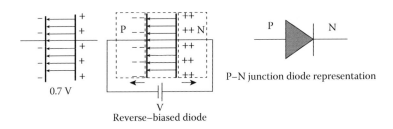

FIGURE 7.34
Depletion region as a capacitor, reverse bias increasing depletion region width, and p–n junction diode representation in electronics.

as the conducting plates of the capacitor. Since there are no majority charge carriers in the depletion region, it acts as the dielectric of the capacitor with high resistance. The P–N junction, even without an external bias voltage, can function as a radiation detector because of the intrinsic reverse bias of about 0.7 V. Though the voltage across the depletion layer is small ($\approx$1 V), the electric field is very high ($\approx 10^3$ V/ cm) because the depletion thickness is extremely small (in μM) and can sweep the ionization produced by a charged particle to the respective electrodes of the depletion layer to produce an electrical pulse.

If an external bias is applied in the reverse direction, the depletion length increases since the majority carriers move toward the battery exposing more layers of lattice ions. (C α A/W, where A is the area of the capacitor—N and P layers—and W the width of the depletion length). This is illustrated in Figure 7.34. Thus, the external reverse bias increases the sensitive volume (and hence the sensitivity) of the detector. The ionization produced by the interacting charged particle or quantum of radiation in this volume can be collected as in other detectors.

The external bias increases the minority carriers current but since their concentrations are small, the leakage current is still very small.

The current voltage characteristics of an Si junction diode is shown in Figure 7.35.

The approximate I–V characteristic of the diode is governed by the following equation:

$$I = I_0 (e^x - 1); \; x = (eV/k_BT)$$

I_0 is the reverse bias minority current (called the dark current) across the junction and Iis the injection current under a forward bias. The dark current is caused by the minority thermal carriers produced in the depletion layer and within a diffusion length on either side of the depletion region (which diffuse into the depletion region and are swept away by the internal electric field, because it constitutes forward bias for the minority carriers). I_0 is independent of the reverse voltage as Figure 7.35 shows, because they are thermally generated at the temperature of the ambient and have nothing to do with the voltage. The thermal carriers produced in the N and P regions do not contribute to this current and they recombine, as there is no electric field in this region. When the diode is forward biased, the reverse bias slowly gets reduced increasing the junction current, and once the reverse bias is overcome the diode offers very little resistance to current and the bulk resistance determines the current.

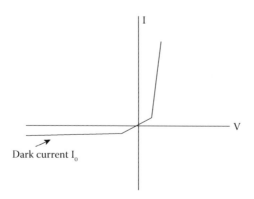

FIGURE 7.35
Si diode I–V characteristic.

When the diode is exposed to the radiation field, radiation causes additional e–h pairs in the diode. This current arises from the depletion region and within the diffusion length of the depletion region on either side of the junction. This is referred to as photocurrent I_p. The total current flowing through the junction is $I_0 + I_p$. The approximate I–V characteristic of the diode (exposed to radiation) is illustrated in Figure 7.36. It also shows the characteristics of the diode for two different intensities incident on the detector.

The above characteristic is governed given by the following equation:

$$I = I_0 (e^x - 1) - (I_p - I_0); x = (eV/k_BT)$$

$$(I_p - I_0) = \text{reduced produced current}$$

The diode output is proportional to the dose equivalent (DE) rate in its range of operation. The discriminator in the circuit cuts off the noise pulses, and the pulses are counted by the count rate meter. The detector counts one count for a certain design dose of say 1 μSv or 0.5 μSv; the counts accumulated and the count rate give the integral DE and the DER.

Figure 7.37 illustrates the principle of an Si diode–based area monitor.

Again, when forward biased, only the bulk resistance comes into the picture and the majority carriers flow across the junction from the three regions, causing the high current. This region is not used for radiation detection.

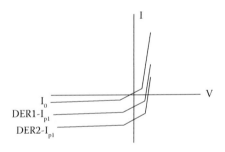

FIGURE 7.36
I–V characteristics of an Si diode exposed to radiation.

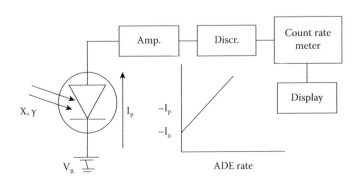

FIGURE 7.37
Si diode as a radiation monitor.

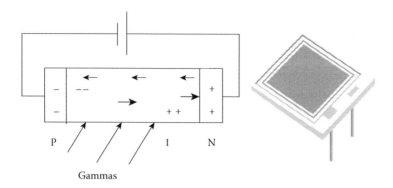

FIGURE 7.38
PIN diode with an intrinsic layer sandwiched.

7.4.1.3 PIN Diode in Reverse Bias As a Radiation Detector

The PIN diode is basically a refinement of the ordinary p–n junction diode. A p–n junction diode has a depletion region at the junction between the P and N layers. A PIN diode includes a layer of intrinsic material (slightly N doped) between the P and N layers (see Figure 7.38).

The capacitance of the PIN diode is proportional to (A/W) where A is the area of cross section of the detector and W the width of the depletion region. By increasing the sensitive area of the device, the detector sensitivity can be increased, but this will also increase the capacitance of the device which will reduce the amplitude of the output signal (V = Q/C). Commercially available devices use wider depletion regions to reduce the capacitance (C α 1/w). It also decreases the RC time constant.

Si diodes can also be used for beta or alpha survey if there is a thin window as in the case of GFDs or the particles are energetic enough to reach the depletion layer. Photodiodes are light sensitive and must be made opaque to light to use as gamma detectors. Otherwise the light current will overwhelm the gamma-produced ionization current. Semiconductor diode–based detectors are largely used in EPDs rather than in gamma ray survey meters. Many of the miniaturized, pocket-size survey meters do make use of GM tube or Si diodes.

7.5 Scintillation-Based Radiation Detectors

Scintillators are wide band gap materials (E_g >> hν of light photons are scintillations). Ionization in the material by dE/dx creates electron–hole pairs. The process can deliver sufficient energy to the electrons to promote them to the conduction band where they can move freely in the crystal. Electron energy above the bottom of the conduction band or below the top of the valence band are dissipated in lattice vibrations and they reach the lowest energy level of the bands. The probability of direct recombination of a hole and the electron is very low due to the wide energy gap and even if it occurs the transition quantum will not be in the visible range (i.e., not a light photon). Therefore, pure crystals are not scintillators. By doping the materials with suitable activators, energy levels can be created in the forbidden gap since their energy levels are different from the energy levels of the host atoms, and hence those levels need not fall in the allowed bands of the material. So, some levels appear in the band gap which makes light emission much more likely. The electron in the conduction band can now fall into this energy level, an excited state of the impurity activator atom. If the electron in turn transitions to a closely placed lower excited state, it will be a nonradiative energy transfer, the excess energy going into lattice vibrations. If the lower level corresponds to a light frequency (i.e., $E_1 - E_o = h\nu$), a scintillation is emitted. The crystal is transparent to these radiations (since band gap >> $h\nu$). The scintillation detector was perhaps the first radiation detector discovered, when Röntgen noticed some crystals fluorescing when the X-ray tube was activated. They are now widely used in X-ray cassettes, image intensifiers, gamma cameras, positron emission tomography (PET) scanners, etc.

The scintillating materials can be solid (organic or inorganic) or liquid. Inorganic solids are generally used for gamma ray survey meters while plastic scintillators are used for neutron detection. The most widely used inorganic scintillators for gamma survey are NaI(Tl), CsI(Tl) while ZnS(Ag) is used for alpha survey. Tl and Ag are the impurities added to the materials for scintillating. The mechanism of scintillator is illustrated in Figure 7.39.

Ionizing radiation interacting with the scintillator produces electron–hole pairs populating the conduction, and valence bands with electrons and holes. The electron may occupy a lower excited level of the impurity atom when it encounters one. It then transitions to the ground state, emitting a scintillation. In order to count the pulses a transducer is needed to convert the scintillations to an electric pulse; this function is performed by a photocathode. As the name implies, it is basically an alkali metal of low-work function (Sb, Rb, Cs) and when a quantum of light strikes it, it is absorbed by the metal and an electron is emitted in

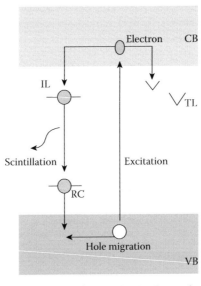

VB: Valence band   CB: Conduction band

IL: Impurity level   TL: Trap level

RC: (Electron–hole) recombination

FIGURE 7.39
Principle of scintillator.

a photoelectric interaction. (Electron energy = scintillating photon energy − photocathode work function). The electrical signal from the photocathode is amplified using a photomultiplier tube (PMT). Figures 7.40 and 7.41 illustrate the concept.

A PMT is basically a photocathode (often deposited on the inside of the input window of the PMT) plus a signal amplifier. The photocathode converts thermoluminescence (TL) light into electrons at the photocathode followed by a series of dynodes, at increasing voltages that produce more electrons than the incident number due to the increasing kinetic energies acquired by the electrons. They are finally collected at the anode as an output signal (current pulse). This amplifies the electrical signal to get a moderately larger pulse. The PMT gain (or amplification) can be 10^6 to 10^8. The gain depends on the PMT voltage which is typically around 1000 volts. The process is very fast and the pulse is generated in nanoseconds.

In Figure 7.41, the electrometer connected to the PMT measures the total charge (current integrated over a time period) which is proportional to the radiation dose (i.e., air kerma or ambient dose equivalent) or the rates of these quantities with correct calibration. Alternatively, the scintillator can also be operated in a pulse mode, as shown in Figure 7.42. For surveying low radiation levels, pulse counting is convenient and for high-radiation fields the system can be used in current mode. For high-radiation fields, simple photodiodes are more attractive alternatives. A PMT basically produces a current pulse which is converted to a voltage pulse by the amplifier. (If the PMT is voltage output type, an amplifier is also an integral part of the PMT module).

The size of the scintillator can vary from a few mm to few inches depending on the gamma energy. These scintillators are more popular for energy spectrometry because of

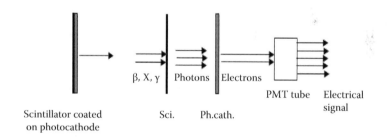

FIGURE 7.40
Signal generation in a scintillation detector.

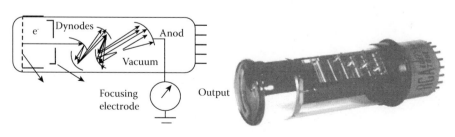

FIGURE 7.41
Principle of a photomultiplier.

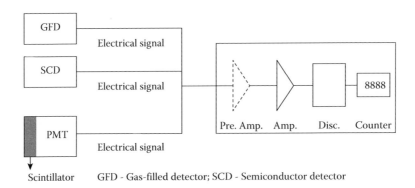

FIGURE 7.42
Pulse counting system for various detectors.

their higher absorption efficiency compared to GFDs. NaI is hygroscopic and hence must be enclosed in an airtight enclosure.

A ZnS(Ag) scintillator is generally used for alpha detection. A very thin coating of ZnS can be coated directly on the photocathode of the PMT and a thin mylar film (<1 mg/cm^2) window is provided for the entry of alpha particles.

Scintillation-based systems are more sensitive than GM counters because of higher conversion efficiency and dynode amplification. They are generally used for surveys at very low radiation levels (e.g., contamination monitoring and lost source-detection surveys like ^{125}I in brachytherapy). However, they can also be used at higher radiation levels, since their resolving time is quite low (a few microseconds or lower) compared with GM counters and can be used at high count rates.

Si semiconductor/CsI(Tl) scintillation-based survey meter and NaI(Tl) scintillation-based detector probe and survey meter are shown in Figure 7.43.

Typical specifications of the pocket-size survey meter are given in Table 7.2, as given by the manufacturer (Fuji Electric Company, Japan). It can be seen the compact device covers a fairly wide range and can be used in several workplaces for gamma monitoring.

Si and scintillation-based detectors are more commonly used in EPDs for personal monitoring. These devices, which are usually highly miniaturized for keeping in a

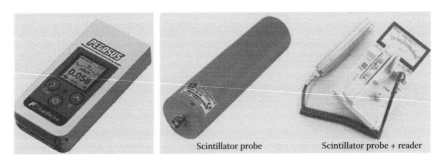

FIGURE 7.43
Si/CsI(Tl)/NaI(Tl)-based gamma survey meters.

TABLE 7.2

Characteristics of a Si-based and CsI(Tl)-based Compact Survey Meter

Radiation Detected	Gamma (X) Rays	Gamma (X) Rays
Detector	Silicon semiconductor detector	CsI(Tl) scintillator detector and Silicon semiconductor detector
Energy range	50 keV to 6 MeV	50 keV to 6 MeV
Measurement range		
Dose rate	0.01 μSv/h to 99.99 mSv/h	0.1 μR/h to 9.999 R/h 0.001 μSv/h to 99.99 mSv/h
Accumulated dose	0.001 μSv to 99.99 mSv	0.1 μR to 9.999 R 0.001 μSv to 99.99 mSv
Energy response	≤±30% (60 keV to 1.5 MeV, 137Cs reference)	
Accuracy	≤±10% (1 μSv/h to 99.99 mSv/h)	≤±10% 10 μR/h to 9.999 R/h) (0.1 μSv/h to 99 99 mSv/h,
Angular response	≤±25%	≤±25%
	(Horizontal Vertical ±90° : 137Cs except battery direction)	

pocket while working, make use of photodiodes instead of PMTs to convert the light pulse to an electrical pulse.

7.6 Neutron Detectors–Based on Semiconductors and Scintillators

Since neutrons do not interact with any of the well-known detectors (i.e., Si diode or NaI type of detectors) with appreciable cross section, neutrons need to be converted into charged particles in nuclear reactions, which can be revealed by these detectors. This can be done in three ways:

- Depositing a thin layer of neutron-sensitive element on the Si diode or other semiconductor detector
- Incorporating the neutron-sensitive atoms into the detector material
- Choosing a detector whose bulk material is sensitive to neutron interactions

Fast neutrons produce recoil protons in hydrogenous materials through elastic scattering. Organic scintillators (e.g., anthracene, stilbene), being largely made up of carbon and hydrogen, are widely used for the detection of fast neutrons. Organic scintillator detectors rely primarily on the scintillations produced by proton recoil.

At thermal and intermediate energies, neutrons have insufficient kinetic energy to transfer to the recoil nuclei. So, the detection of thermal neutrons relies on materials with large capture cross sections at low energy, leading to reactions producing charged particles with significant energy (due to the exothermic nature of the reactions, as the following equations show).

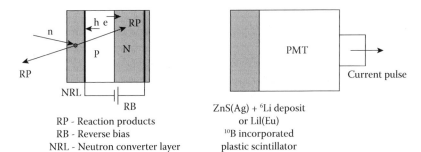

FIGURE 7.44
Principle of semiconductor/scintillation-based neutron detectors.

$$^3\text{He} + \text{n} \longrightarrow {}^3\text{H} + {}^1\text{H} + 765 \text{ keV}$$
$$^{10}\text{B} + \text{n} \longrightarrow {}^7\text{Li} + \alpha + 2.79 \text{ MeV (6\%)}$$
$$\longrightarrow {}^7\text{Li*} + \alpha + 2.31 \text{ MeV (94\%)}$$
$$^6\text{Li} + \text{n} \longrightarrow {}^3\text{H} + \alpha + 4.78 \text{ MeV}$$

The cross section for these capture reactions increases rapidly with decreasing neutron energy ($\approx \alpha \ 1/V$), making these reactions the most efficient for the detection of thermal neutrons. For thermal neutron detection, it is necessary to use the above materials as neutron converters or incorporate the atoms into the detector material (see Figure 7.44).

Some examples are the ^{10}B-coated Si diode or other semiconductor detector, or use of CdTe as a radiation detector. Cd has a high thermal neutron absorption cross section for thermal neutrons. Similarly for scintillator-based neutron detectors, ^{6,7}Li-deposited ZnS(Ag), ^{10}B incorporated plastic scintillator, and ^{6,7}Li doped NaI(Tl) have been used. Neutron capture isotopes have also been incorporated into a range of inorganic and glass scintillators (e.g., ^{6}Li-incorporated cerium-doped glass scintillator, ^{6}Li-enriched ^{6}LiI(EU) detector). Plastic scintillators, however, give reduced response to high LET radiations making gamma discrimination a real issue. Different methods are adopted for effective discrimination against gamma pulses.

Many of these detectors are used in neutron research rather than in routine neutron survey. A few EPDs based on semiconductor or scintillation are available for neutron personal monitoring, but not for area survey.

Question 7.8

The scintillation yield of NaI(Tl) is about 11%. The wavelength of light emitted is 410 nm (3 eV). How many light photons will be produced if a 1 MeV gamma is completely absorbed in the crystal?

The scintillation energy = 11% of 1 MeV = 11×10^4 eV

The number of light photons produced = $11 \times 10^4/3 = 3.6 \times 10^4$ = 36,000 photons are produced in the crystal.

Question 7.9

A scintillation-based survey meter operates in pulse mode. The sensitivity of the survey meter is given as 10^5 cpm/μSv/h. The detector in the workplace reads 5×10^5 cpm. What is the ambient dose equivalent at the monitored place ?

The ADE at the work place = $5 \times 10^5 / 10^5$ = 5 μSv/h

7.7 Main Features of a Survey Meter

Some commercially available models of survey meters were shown in the earlier figures but their unique features and properties were not discussed. The commonly available features of a modern survey meter and their important characteristics are:

- A "low battery" visual indicator
- Automatic zeroing, automatic ranging, and automatic back-illumination facilities
- A variable response time and memory to store the data
- The option of both *rate* and *integrate* modes of operation
- An analog, digital, or both displays, marked in conventional (exposure/air kerma) or ambient dose equivalent units
- An audio indication of radiation levels (through the chirp rate)
- A resettable/non-resettable alarm facility with adjustable alarm levels
- A visual indication of radiation with flashing LEDs
- Remote operation and display of readings

The important characteristics of area survey meters are

- Sensitivity

  This was discussed in earlier sections.
- Energy dependence

  The radiation spectrum and its angular distributions at the workplace are mostly unknown while the survey meters are generally calibrated at a few selective beam qualities in a well-defined beam direction. So, in order to obtain correct readings at the workplace, the survey meter must have a flat or low energy dependence and an isotropic response. In the past, the survey meters were designed to exhibit flat energy dependence for measuring exposure or air kerma, but are now expected to show a similar response for measuring the ADE.

  The ADE calibration of the monitor is given by

  $$(N_H)^* = [H^*(10)/M] = [H^*(10)/K_a] / [K_a/M]$$

  The instrument response must therefore vary as $[H^*(10)/(Ka)]$. The ion chamber detector used in survey meters is usually made of air equivalent walls of around 400 mg/cm^2 thickness but some higher Z material (e.g., aluminum foils) is added to the sides of the wall empirically to obtain the required ADE response for $H^*(10)$ measurements.

- Directional dependence

  By rotating the survey monitor about its vertical axis, the directional response of the instrument can be studied. A survey monitor usually exhibits an isotropic response, as required for measuring ambient dose equivalent, within ±60° to ±80° with respect to the reference direction of calibration, and typically has a much better response for higher photon energies (>80 keV).

- Dose range

  Survey meters may cover a range from nSv/h to Sv/h, but the typical range in use is mSv/h to μSv/h.

- Response time

  The response time (time to reach 10% to 90% of the steady state value) of the survey monitor is defined as the RC time constant of the measuring circuit, where R is the decade resistor used and C the capacitance of the circuit, which varies in different scale ranges of the survey meter. Low dose equivalent ranges (e.g., 0 to 50 μSv/h range) would have high R and hence high RC values, and so the indicator movement would be sluggish (say ≈ 8 to 10 seconds). It takes at least three to five time constants for the monitor reading to stabilize.

- Overload characteristics

  Survey meters must be subjected to dose rates of about 10 times the maximum scale range to ensure that they read full scale rather than near zero on saturation. Some survey meters, especially the older models, may read zero on overload (i.e., when the equivalent dose rate exceeds the scale range). Such survey meters should not be used for monitoring, since the worker may wrongly assume that there is no radiation in an area where the radiation field is actually very high.

- Long-term stability

  Survey meters must be calibrated in a standards dosimetry laboratory with the frequency prescribed in the regulatory requirements of the country, typically once every 3 years; they also need calibration immediately after repair or immediately upon detection of any sudden change in response. The long-term stability of survey meters must be checked at regular intervals using a long half-life source in a reproducible geometry.

- Particle discrimination

  Energy discrimination based on dE/dx differences was discussed earlier.

- Uncertainty of measurement

  The survey meter must be calibrated from an accredited calibration laboratory at regular intervals as per the country's regulations. The calibration certificate gives the accuracy of the calibration. There is an additional uncertainty due to the energy dependence, angular response, linearity, etc. These uncertainties must be added to the calibration uncertainty in quadrature and a coverage factor k = 2 corresponds to the total uncertainty of measurement at 95% confidence level.

The common controls and features on a very basic survey meter are shown in Figure 7.45 for a model of the survey meter that is commercially available.

Indicated use: Low-level gamma survey

(down to μR levels)

Detector: 2" × 1" (NaI) Tl scintillator

Sensitivity: Typically 700 cpm/(μR/hr) (*Cs*-137)

Energy response: Energy dependent

Meter dial: 0–5 μR/hr up

(maximum monitoring level 5 mR/h)

BAT test (*available*)

Multipliers: ×1, ×10, ×100, ×1000

Linearity: Linear to within ±10% of true value

Audio indication: Yes

Alarm indication: Yes

Reset button (for zeroing the meter): Yes

Response: Fast (4 s) and slow (22 s)

- to reach 10% to 90% of final reading

FIGURE 7.45
Basic controls and features of a survey meter.

7.8 General Procedures for Radiation Survey

Survey meters are required for surveying all workplaces where radioactive sources or radiation-generating equipment are handled. Considering the varied requirements in these situations, a whole range of survey meters are commercially available. In order to correctly survey a workplace three things are important:

- Workplace requirements
- Choice of survey meter
- Survey procedure

Familiarity with the workplace and critical areas to be surveyed are very important aspects. For instance, for surveying a linac room or an HDR room in radiation oncology, the critical areas to survey are the vicinity of entrance doors, vicinity of cable conduits, and vicinity of control console.

The workplace requirements dictate the choice of the survey meter. For instance, a gamma survey meter will not be useful for beta survey and a beta survey meter may not be able to detect alphas. A GM survey meter is not the right choice for surveying a linac installation or very high radiation fields. The dose range covered by the instrument is equally important. For surveying uncontrolled areas around a linac installation, the survey meter must be able to measure as low as 0.1 μSv/h. To cover emergency situations, the survey meter must be able to read up to or over 100 mSv/h. The survey meter (SM) must therefore be chosen based on the type and energy of radiation, and the levels of radiation.

The handling and operation of a survey meter are equally important for performing a proper survey. Important aspects regarding this will be briefly described here.

7.8.1 Pre-Survey Checks

Before starting the survey, there are some basic pre-survey checks that must be carried out, as listed below.

- Calibration check

  An SM out of calibration cannot be used for survey. The survey results have no legal validity. A sticker affixed to the side of the instrument may show the calibration due date and if not, it is advisable to apply such a sticker on the instrument.

- Visual inspection

  SMs are very delicate instruments. Any impact like dropping the instrument can affect the functioning or the calibration of the instrument. Any visible damage is an indication of such a situation.

- Battery check

  The indicator crossing into the "battery OK" range confirms the battery status is acceptable. Otherwise the battery needs to be changed, as the instrument readings are valid only when the battery check is okay.

- Source check

  Proper functioning of the SM may be checked with a check source for ensuring constancy of response in a fixed geometry, as per the manufacturer's recommendation. For example, in the Victoreen 451B ion chamber survey meter, by keeping the provided check source against the beta window (after removing the shield) the instrument must show a reading of 4 μSv/h.

- Zero check

  The meter must be zeroed using the knob provided on the meter before entering the radiation area. This will neutralize any background reading. If it cannot be zeroed, it must be deducted from the survey readings. In the case of scintillation-based SM, any light leak will give rise to a spurious reading. In many of the modern SMs, the zeroing is automatic (e.g., Victoreen 451 B ion chamber survey meter).

- Warm-up time

  Each SM requires some warm-up time (say about 5 minutes) and the manufacturer's recommended warm-up time should be heeded before using the instrument.

7.8.2 Basic Survey Procedures

- When surveying very low radiation levels, the response switch may be kept at F (fast response) so the meter is not too sluggish in registering the readings. Otherwise, the response switch is usually left in the S position.
- Low radiation levels can be surveyed in the integrate mode for a set time for more accurate measurement of the radiation level. (The modern SMs can show both the rate and integral values by using the mode button as a toggle switch).
- While surveying any area, the instrument must be kept close to the surface to be surveyed without touching the surface. This is because if the surface is contaminated, the instrument may pick this up, leading to the spread of contamination.
- Slightly higher radiation level than the design value may not be that important if those areas are inaccessible (e.g., cable conduit hole on the wall covered by the control console, or hot spots like cracks on the wall).
- The SM scale must be in the most sensitive range while entering the radiation area. (At very high dose ranges, the meter will read near zero for very low radiation levels, misleading the physicist. Many modern SMs have an automatic ranging facility).
- Before surveying a radiotherapy installation, a sketch of the layout must be made and the occupancies around all six sides (or five sides, in the case of a basement installation with no structures below) and the design values for these areas must be marked. The maximum values on all the walls and ceiling must be less than the design value for the maximum workload of the installation (more details of this survey will be described in the chapter on shielding).

7.8.3 Maintenance

No special maintenance is required if the instruments are stored in a dry place; in regions of high monsoon the instruments must be kept in desiccators to reduce leakage.

References

1. G.F. Knoll, *Radiation Detection and Measurement*, IV Edition, Wiley, New York, 2011.

Review Questions

1. What are the three operating regions of gas-filled detectors? Which detector is incapable of energy discrimination?
2. Name two scintillation-based detectors. Mention two ways of converting a light signal into an electrical signal.
3. Which is the ideal survey meter for the gamma ray survey of linac rooms, and why?

4. Name three thermal neutron detectors. What particles are involved in neutron detection? State one characteristic of each.

5. How does signal amplification occur in a proportional detector?

6. Why can pure NaI or CsI not produce scintillations?

7. Why can gamma ray survey meters not be used for α and β surveys as well?

8. Which radiation detectors can be used for measuring radiation levels at background radiation level, and moderately higher radiation fields (say few hundred mR/h)?

9. What are the advantages and disadvantages of GM-based survey meters?

10. What quench gas is added to the detector gas in GM detectors and what is its function?

11. In the scintillator NaI(Tl), what does Tl represent and what is its role?

12. Which detectors are used in the most common neutron survey meters? (a) gas-filled detectors (b) semiconductor detectors or (c) scintillator detectors.

13. What is the principle of a neutron survey meter construction?

14. Draw a block diagram of a pulse counting system. Briefly describe the function of each block.

15. What is an energy-compensated GM detector and what is the purpose of energy compensation?

16. What are the band gap values for Si, NaI(Tl), and CsI(Tl) materials?

17. Describe intrinsic and extrinsic semiconductors, and majority and minority carriers.

18. Describe the principle of an Si diode and how it can be used as a radiation detector.

19. What bias mode (forward or reverse bias) is applied to an Si diode to use it as a radiation detector?

20. How is a PIN diode different from a P–N junction diode?

21. Describe how a PMT converts its input light signal into an amplified electric pulse signal.

22. Explain the scintillation mechanism in an inorganic crystal.

23. What are the common features of a modern survey meter?

24. What are the important properties of a survey meter?

25. What are the important presurvey checks to be carried out on a routine survey meter?

8

Radiation Detectors for Individual Monitoring

8.1 Introduction

Individual monitoring, also known as personal monitoring, refers to the monitoring of doses received by people working with radiation. Those who work regularly in controlled areas or full time in supervised areas and most other workers exposed to radiation must be monitored as per the national regulations by wearing a personal dosimeter so that their doses can be regularly monitored and recorded. This will help to ensure that the doses received by the radiation workers are acceptable, and it also acts as an indicator of the effectiveness of radiation control practices in the workplace. It can also reveal any sudden changes in the radiation level in the workplace or provide information in the event of accidental exposures.

Occupational exposure may consist of either internal exposure or external exposure, or both—it is the total dose received by the individual that is of interest for regulatory compliance and for staff dose estimation. Since no internal radiation exposure is involved in radiation oncology, this chapter will deal with only external radiation exposure.

However, not all radiation workers need to be monitored; if the dose received by the individual is very low (about 2 mSv per year or less), then monitoring is not necessary.

Depending on the nature of work, three important parts of the body may require monitoring:

1. Whole body monitoring
2. Extremities monitoring (hands, feet, etc.)
3. Eye lens, thyroid, etc. monitoring (e.g., in interventional radiology)

This requires monitoring the affected radiation workers for $H_p(10)$, $H_p(0.07)$, and $H_p(3)$, with the personal dose equivalent (PDE) at 10 mm depth (for whole body monitoring), at 0.07 mm depth (for skin dose monitoring), and at 3 mm depth (for the eye lens dose monitoring). In radiation oncology, monitoring $H_p(10)$ is sufficient. In brachytherapy, the physicist handling ^{125}I source preparations or the surgeon performing ^{125}I plaque therapy or radioactive seed implant therapy may receive a small dose to the hands which can be monitored using a wrist dosimeter. However, experimental studies show a very small dose to the hands compared to the threshold dose, so there is no necessity to monitor extremities in radiation oncology. Eye lens and thyroid dose monitoring are particularly important in interventional radiology and interventional cardiology where the physician carries out fluoroscopy procedures inside the fluoroscopy room. However, this is not relevant in radiation oncology.

The personal monitoring of external exposure is carried out using a personal monitoring device (PMD) which is designed to measure, over a specified period of time, the radiation dose received by a person who is occupationally exposed to radiation. If neutron exposure is involved, neutron-monitoring devices (e.g., an albedo dosimeter or nuclear track emulsions, called neutron badges) must be worn by the person concerned. PMDs are generally passive and integrating detector systems, and the dose is known only after the period of its use. On the other hand, self-reading pocket dosimeters and electronic personal dosimeters (EPDs) are direct reading active dosimeters, and the EPDs show both the instantaneous dose rate and the accumulated dose at any point in time. They are the best devices for measuring the doses for a procedure (e.g., ^{60}Co source loading activity or an interventional cardiac procedure) so that the personal doses can be easily optimized. They are, however, not recognized as legal dosimeters for personal monitoring and maintenance of dose records and hence cannot be used in place of passive dosimeters like a thermoluminescence dosimeter (TLD), optically stimulated luminescence (OSL), or radio photoluminescent (RPL), depending on the device recognized by a country for this purpose.

The basic requirements of a personal monitoring system are

1. Ability to distinguish between the different types of radiations
2. Covering the required dose range
3. Less energy dependence
4. Minimum fading of signal during the monitoring period and till the processing of the monitor
5. Quick and simple readout system
6. Reusability
7. Good accuracy
8. Rugged, light weight, economical, and stable to changes in environmental conditions

Not all monitors satisfy all the conditions. In this chapter, the personal monitors that are commonly in use in various countries and their basic characteristics will be discussed.

In the previous chapter, under solid-state detectors, we listed the following detectors that are in use in individual monitoring:

1. Photographic film dosimeters (films)
2. Thermoluminescence dosimeters (TLDs),
   The less frequently used
3. Optically stimulated luminescence dosimeters (OSLDs)
4. Radio photoluminescent dosimeters (RPLDs)and the ones now gaining popularity, namely
5. Electronic personal dosimeters (EPDs)

The film- and luminescence-based dosimeters are passive dosimeters since they can be read only after the monitoring period by a service provider. They are not very useful for monitoring the dose received, say, in a day or during a particular operation. The EPDs belong to the category of direct reading personal dosimeters (DRPDs) or active dosimeters which are self-reading type and can give the dose rate or the integral dose. The working principles of these devices are discussed below.

8.2 Principles of Personal Monitoring Film

A personal monitoring film (PMF) is a photographic film which is a sheet of transparent plastic film base of polyester, about 100–200 μm thick, coated on one or both side(s) with a gelatin emulsion to about 5–10 or 20 μm thickness (Figure 8.1). The emulsion contains tiny light-sensitive silver bromide crystals well dispersed in the gelatin medium (approximately 50:50 ratio with a few percentage of AgI added). With two-side/single side film coatings, one is a fast film (for low doses) and the other is a slow film for measuring higher doses, giving a wide dose range of 0.1 mSv to 10 Sv to monitor. The shape, size, and orientation of the crystals, in addition to emulsion thickness, determine the sensitivity (and hence the speed) of the film to radiation, and the emulsion can be protected by a thin supercoat. The typical film size is 4 × 3 cm, similar to a dental X-ray film. The film is covered in a wrapper of thickness around 20–30 mg/cm^2. Any α's and low-energy βs (e.g., from ^{3}H or ^{14}C) will not be able to penetrate the wrapper and hence cannot be detected using the film.

8.2.1 Latent Image Formation

Sensitivity specks (regions) are formed in the crystal grains during the film emulsion production which are actually electron-trapping defects (perfect crystals, with no defect, show very little radiographic sensitivity). The speck is usually situated on the surface of the crystal. When radiation interacts with the film, some grains (not all) are exposed to radiation interactions. The interaction in the exposed crystal leads to the ionization of AgBr.

$$AgBr + h\nu \text{ or } \beta \rightarrow Ag^+ + Br^-$$

The electron released in the interaction diffuses and is trapped at the sensitivity speck, making the speck negatively charged. The speck now attracts the positive Ag ions in the neighborhood. The attraction makes the interstitial Ag ions migrate to the speck resulting in the neutralizing of the speck and leaving behind a metallic Ag atom. As more electrons become trapped the process is repeated, and the concentration of metallic silver at the speck increases. When there are 6–10 Ag atoms, it becomes a (developable) latent image.

8.2.2 Film Processing

Film processing is the film development process in a darkroom to convert the latent image on the film into a visible image. For a PMF, the image is just the blackening on the film

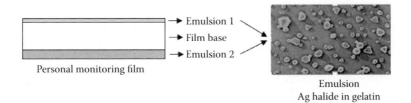

Emulsion 1
Film base
Emulsion 2

Personal monitoring film

Emulsion
Ag halide in gelatin

FIGURE 8.1
Structure of a PMF.

(due to the black metallic silver in the exposed Ag grains). The following steps are involved in film processing:

1. Developing

   The exposed film is dipped in the developer solution which transforms the latent image into a visible image in the exposed grain. For this development to happen, the free electrons in the developer solution must penetrate the grain and reduce all positive Ag ions to metallic silver, but because of the negatively charged Br ion repulsion, the electrons are not able to easily penetrate the undeveloped grain (Figure 8.2). However, in the case of exposed crystal grain, the sensitivity speck is neutralized and can attract an electron for trapping, thus allowing the entry of electrons into the grain. This leads to the complete development of the grain into metallic silver by neutralizing all the positively charged Ag ions. There is an optimal time (about 5 minutes) for the development of the film. If the film is left in the developer for too long or if the developer is hyperactive, the probability of electron entering the unexposed grain increases and this will lead to the development of unexposed grains as well. With increasing use, the developer becomes weaker, so it needs to be replenished at regular intervals. The number of Ag atoms reduced by the developer is about 10^{12} times the number of Ag atoms in the latent image, which gives an indication of why the latent image is not visible.

2. Rinsing

   After developing, the film must be rinsed in running water for 10–15 seconds. This procedure removes the excess developer solution from the film (so that unexposed grains will not also be developed) and all the soluble chemicals. The unexposed grains, however, are not water soluble and will not be washed away.

3. Fixing

   This procedure dissolves unexposed Ag halide crystals.

4. Washing

   This procedure removes the residual fixer solutions. The temperature of the developer, fixer, and water must be about the same to avoid any unequal expansion or contraction of the emulsion.

5. Drying

   This process removes water to dry the film. This can be done using a fan, or just hanging the film to dry in the surrounding air.

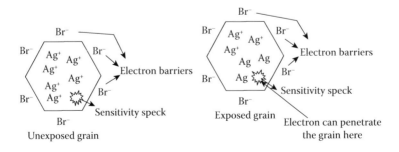

FIGURE 8.2
Silver halide grain development during film processing.

The chemicals used as developers are generally hydroquinone, phenidone, and dimezone; the ones used for fixing are ammonium thiosulfate and sodium thiosulfate. These chemicals are commercially available and must be used as recommended by the manufacturer for proper developing of the film. The developer, rinser, and fixer tanks are arranged left to right in the darkroom for processing.

The main development process is illustrated in Figure 8.3.

8.2.3 Film Reading

The transmission of light through a film depends on its opacity and can be measured in terms of the optical density (OD). OD or the radiographic density is defined as the logarithm of two measurements, $\log (I_0/I_t)$, where I_0 is the light intensity incident on the film and I_t is the transmitted intensity (i.e., the inverse of transmittance).

$$D = \log (I_0/I_t)$$

OD can be measured using a densitometer. The principle of a densitometer is illustrated in Figure 8.4.

The unexposed film would exhibit a background OD known as the fog density $(OD)_f$. The net OD is obtained by subtracting the fog density from the density of the exposed film.

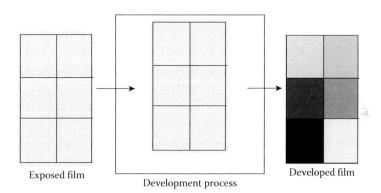

Exposed film   Development process   Developed film

FIGURE 8.3
Process of developing a latent image.

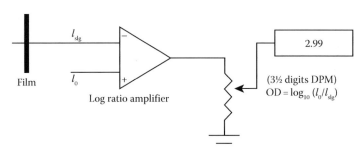

Film

Log ratio amplifier

2.99

(3½ digits DPM)
$OD = \log_{10} (I_0/I_{slg})$

FIGURE 8.4
Principle of a film densitometer.

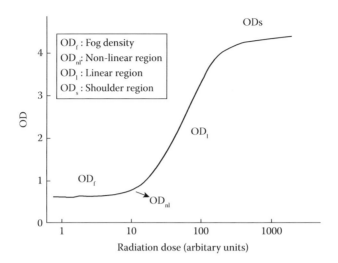

FIGURE 8.5
Typical dose response curve for a film.

Often, a single measurement would suffice as many densitometers are equipped with a zeroing facility to zero out the fog density.

The dose response of the film badge must be ideally linear so that calibration at any dose value can be used for all doses. But this is not always the case. Some emulsions are linear, some are linear over a limited range and some are nonlinear. So, the dose versus OD curve (known as the sensitometric curve or H–D curve) must be established for the film type before using it for any dosimetry work. A typical film response curve is shown in Figure 8.5.

The curve has three important regions, namely fog density (background [BG] exposure) region, a toe (nonlinear, underexposure) region, a linear (dosimetrically useful) region, and a shoulder and saturation (or overexposure) region where blackening is not a function of dose.

8.3 Personal Monitoring Film Badge Construction

The problem in personal monitoring is we may not know the spectrum of photons in the workplace or the type of radiations involved (X, β, γ, n), though it is much more straightforward in radiation oncology. The dose may also vary from very low levels (say in the μSv range) to very high levels in radiation emergencies (say in tens of mSv or even higher). In all such situations, the personal monitoring film badge (PMFB) must be useful to give us some important information. An added problem with the PMFB is the pronounced energy dependence at low photon energies (say <100 keV) due to the high Z of AgBr film. So, we need to design the film badge, not only to distinguish between the radiation types and their doses but also to reduce its response at lower photon energies compared to higher photon energies. So, a complex filter pattern is used in which the film is sandwiched, in a badge. So, a PMFB may be defined as a photographic film enclosed in a light-tight wrapper, sandwiched in the filter system and enclosed in a metallic frame. The filters used and the design of the film badges are not identical in all countries. However, the purpose of the badge is the same, namely to identify the components of the radiation, their energies and doses, and to give a fairly flat energy-dependent response. A typical design of the film badge is shown in Figure 8.6.

The badge has six windows—open window, window with a plastic filter (1 mm), window with a medium Z filter of small thickness (Cu 0.15 mm), window with a medium Z filter of larger thickness (Cu 1 mm), window with a high Z filter of large thickness (Pb 1 mm), and a cadmium window. The function of the various filters is given in Table 8.1.

The net OD under each filter must be converted to a dose with proper calibration before subtracting doses. Calibration curves are generated for the film badge, by exposing it to known doses of different types and energies of radiation and drawing calibration curves.

The double emulsion helps to cover a wide dose range for personal monitoring say from 0.1 mSv to 10 Sv for $H_p(10)$.

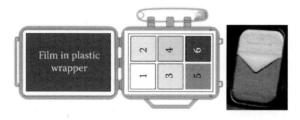

FIGURE 8.6
A typical film badge and monitoring film (on right).

TABLE 8.1

Design Features of a Film Badge

Window/Density	Filter Material	Purpose	
Window 1 D_1	None (open window)	Detects αs	Registers (β + X), β measured after correcting from other windows doses.
Window 2 D_2	1 mm perspex	Detects β	Registers X, filter cuts off β (<2 MeV) ($D_2 - D_1$) is a measure of β
Window 4 D_4	Thin (0.15 mm) Cu	Detects low energy X-rays	($D_2 - D_4$) is used to determine photon exposures for <150 keV energy.
Window 5 D_5	1 mm Cu	Detects high energy X-rays	Through proper choice of filter Z and thicknesses, D_4, D_5, and D_6 data can
Window 6 D_6	1 mm Pb	Detects γs	be used to determine photon energy information and photon dose over a wide energy range. The energy response curve at lower photon energies is also reduced for flatter energy response.
Window 3 D_3	1 mm Cd		($D_5 - D_3$) is a measure of thermal neutron dose.

8.4 Neutron Monitoring

Neutron monitoring is much more complicated compared to β or photon monitoring, for the following reasons: (1) Neutrons in the workplace exhibit the widest energy range, 0.025 eV to about 10 MeV, a variation of 9 orders of magnitude; (2) fluence to photon detection efficiency conversions vary by a factor of 40; (3) the quality by a factor of 10; and (4) the presence of high gamma field in the neutron environment [1]. It is difficult to find a single dosimeter that will have all the desirable properties (energy independence, wide dose range, etc.).

In addition, the monitoring device must have an isotropic response so that a device calibrated in a monodirectional beam can be used to monitor in a workplace where the radiations are generally from 4π directions. Generally, the neutron-monitoring devices do not show a flat energy response over a very wide range, nor adequate isotropic response, or a very low detection limit.

So, the accuracy of the neutron monitor mainly depends on the energy range of the neutrons in the workplace, the minimum detectable dose level and the device's angular dependence. The principles behind few neutron monitors that are in common use for personal monitoring will be described briefly in this chapter.

8.4.1 Film Dosimetry for Neutron Monitoring

In film dosimetry, for fast neutron (0.5–15 MeV) monitoring, nuclear track emulsions are used. The emulsions are usually one hundred to several hundred micrometer thick. The Nuclear tracks are produced by the fast neutron elastic collisions with hydrogen nuclei in the emulsion base, and the paper wrappings around the film. The proton recoils produced in the n–p reaction create tracks in the emulsion which can be counted using a microscope, with a magnification of 400X or 800X. In order to produce countable tracks, the emulsions are specially produced by reducing the grain size and increasing the grain density compared to the emulsion used for photon monitoring. One of the most widely used fast neutron-monitoring films is the Kodak type A (NTA) film. The number of tracks generated is proportional to the number of incident neutrons which can be correlated through calibration. Since the neutron dose equivalent (DE) per neutron fluence values are available, knowing the energy or the neutron spectrum in the workplace, one can relate the number of tracks to the DE. The films are calibrated using PuBe neutron sources (average energy 4 MeV). The film exhibits some thermal neutron sensitivity (through $^{14}N(n,p)^{14}C$ reaction with thermal neutrons) and so the film's thermal neutron response must be removed by using a thermal neutron absorber (e.g., ^{10}B carbide) as a filter in the fast neutron badge. The film exhibits fading with time, causing loss of tracks. Automatic counting methods give higher accuracy compared to manual counting.

The main disadvantages of nuclear track (NTA) film in neutron monitoring are [2] its high energy threshold (0.7–1 MeV, about 6 orders higher compared to thermal energy neutrons), significant energy dependence, high detection threshold ($\approx$0.5 mSv), signal fading with temperature, humidity and gamma ray sensitivity, and saturation at relatively low doses ($\approx$50 mSv).

8.4.2 Solid State Nuclear Track Detector for Neutron Monitoring

Many countries have replaced film dosimetry with Solid State Nuclear Track Detector (SSNTD)-based dosimetry because of their better characteristics compared to films. The dosimeter is also known as a plastic nuclear track detector or CR-39 (trade name) dosimeter since CR-39 is a type of plastic rather than a photographic film such as the NTA film used in the optical industry. It was adopted and widely used from the 1980s for intermediate and fast neutron monitoring, following the detailed studies by Tommasino et al. [3], Hankins and Westermark [4], Griffith et al. [5], and Dajko and Somogyi [6] on this detector.

The principle of monitoring is similar to the film dosimetry described earlier. The nuclear tracks are formed by the recoil nuclei in elastic neutron collisions with hydrogen, carbon, and oxygen atoms that form the components of the plastic. These tracks can be enlarged by suitable chemical or electrochemical etching processes or both combined, at elevated temperatures. Through proper calibration, the track density can be related to the neutron DE. The sensitivity of CR-39 dosimeter can be enhanced by keeping a hydrogenous material (e.g., polyethylene) in front of the dosimeter. The additional proton recoils generated in the proton radiator produce more tracks in the CR-39, increasing its sensitivity. This dosimeter can detect fast neutrons over a wide energy range (100 keV to 20 MeV), has a relatively flat energy response, and is insensitive to β or γ. The CR-39 dosimeters are commercially available and their individual sensitivities depend on the type of CR-39 material used, thickness of the radiator, conditions of the etching process, and so on, and must be used as per the etching recommendations of the manufacturer for correct results. One problem with the CR-39 dosimeter, however, is the high-background track density (due to natural background, material defects, etc.), which determines the lower dose limit for neutron monitoring. Since the BG track density increases with time, these dosimeters have a limited life.

The dosimeter is prepared by cutting a small piece (say 3 cm × 3 cm) from the CR-39 sheet which typically has a thickness of about 0.6 mm. A 1 mm thickness of polyethylene radiator of same dimensions is placed on top and sealed in an aluminized pouch, which protects the dosimeter from any contamination. The dosimeter is usually loaded into a plastic holder that has a clip, so it can be worn as a badge. The neutron monitor badge used in India is shown in Figure 8.7 [7].

Example 8.1

A CR-39 dosimeter worn in the workplace gave a reading of 1030 tracks/cm^2. The background track density, determined for an unexposed film, was 70 tracks/cm^2. The calibration factor of the CR-39 neutron monitor is 2.088 μSv/(tracks/cm^2). Calculate the PDE received by the user.

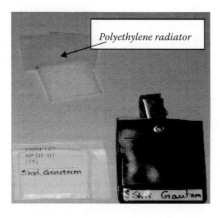

FIGURE 8.7
CR-39 neutron badge in polyethylene cover used in India.

The BG reading of the CR-39 dosimeter is subtracted: $1030 - 70 = 960$ tracks/cm^2.

The dose received by the user $= 2.088 \times 960 = 2$ mSv

The minimum detectable dose is determined by the background tracks and the fluctuations in the tracks determined from a set of unexposed dosimeters. Assuming that the standard deviation in the background tracks in the above example is 16, the maximum BG counts expected at the 3 σ limit are $70 \times 3 \times 16 = 118$ tracks/cm^2. This corresponds to a background dose of 0.25 mSv, which can be considered as the minimum reliable detection limit of the dosimeter. This dosimeter is very useful for neutron energies in the range of 200 keV to 20 MeV [8].

The response of the CR-39 to lower energy neutrons (intermediate and thermal neutrons) can be increased by substituting the proton irradiator by a thin ^{10}B incorporated layer (few tens of mg/cm^2 thickness); ^{10}B is one of the best thermal neutron absorbers and this absorption cross section varies as $(1/V)$ as the neutron energy increases. The CR-39 will then register the α tracks produced in the ^{10}B(n,α)^{7}Li reactions by thermal and intermediate energy neutrons. Figure 8.8 illustrates the concept of monitoring neutrons of all energies.

The proton radiator layer may also extend to the full length of the CR-39 film. Then part of the film will register fast neutrons and part will register fast as well as the lower energy neutrons. Proton ranges being much longer, compared to α ranges, there is very little attenuation of the protons in the ^{10}B layer. For instance, the Landauer Corporation (Illinois) company offers commercially two different neutron-monitoring badges, one for fast neutron monitoring (Neutrak-J) and one for monitoring all energy neutrons (Neutrak-T), as shown in Figure 8.8. An idea regarding the neutron spectrum in the workplace is important for accurate neutron monitoring. The CR-39 detector has been used in radiation oncology in the control room of an 18 MV therapy linac for personal monitoring purposes [9].

8.4.3 Bubble Detectors for Neutron Personal Monitoring

A bubble detector is a simple tube filled with an elastic polymer gel that contains very tiny droplets (about 0.025 mm diameter) of superheated liquid (i.e., liquid at a temperature above its boiling point) dispersed throughout the transparent medium. The neutron interacting with the liquid droplets or with the polymer in the vicinity of the droplet produces charged particles. The passage of the charged particle through the droplet vaporizes, forming a visible gas bubble (about 1 mm diameter) that gets trapped inside the gel (see Figure 8.9).

The number of bubbles can be counted which is directly proportional to the neutron PDE (NPDE), thus it gives visible response to the presence of neutrons. The counting can also

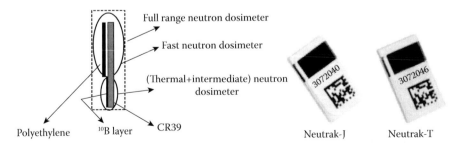

FIGURE 8.8
Working principle of CR-39 film for neutron monitoring.

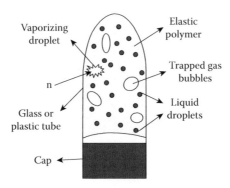

FIGURE 8.9
Principle of a bubble detector.

FIGURE 8.10
BD-PND neutron bubble dosimeter and reader.

be automated. The detector can be reset by simply screwing down a piston cap—the increased pressure causes the bubbles to recondense which can be reheated by another neutron exposure, thus it is a reusable detector. After a certain number of cycles though, the bubbles may no longer recondense.

Some important properties of the bubble neutron dosimeter are (1) zero gamma sensitivity, (2) isotropic angular response, (3) dose rate independence, (4) energy independence in the range 200 keV–25 MeV, (5) convenient size like a pocket dosimeter, and (6) reusability (say, 50–100 times depending on the manufacturer's specifications). One disadvantage of the dosimeter is the temperature-dependent sensitivity. For best accuracy, the dosimeter must be used as per the manufacturer's recommendations.

One of the most widely used neutron personal monitoring dosimeters is the BD-PND bubble dosimeter from Bubble Technology Industries, Canada (see Figure 8.10).

The dosimeter is reset by screwing down the piston on the cap to clear the gel of the bubbles; the increased piston pressure causes the recondensation of the bubbles. The monitoring PDE range is about 1 μSv to about 5 mSv, depending on the dosimeter sensitivity. The detector has been used in the control room of an 18 MV therapy linac for personal monitoring purposes [9].

Example 8.2

The average sensitivity of the neutron bubble dosimeter is 30 bubbles/mrem. Express the sensitivity in bubbles/μSv. If the dose received by the user is 6 μSv, how many bubbles are expected in the dosimeter?

$$1 \text{ mSv} = 10 \text{ μSv. So sensitivity} = 3 \text{ bubbles} \div \text{μSv.}$$

For 6 μSv, about 18 bubbles are expected in the bubble dosimeter.

8.5 Modern Personal Monitoring Dosimeters

The film badges are rarely being used these days. They have been almost completely replaced by other types of solid state dosimeters that will be discussed in the following sections. These are

1. Thermoluminescent dosimeters (TLDs) or TLD badges (as they are more familiarly known)
2. Optically stimulated luminescent dosimeters (OSLDs)
3. Radiophotoluminescent dosimeters (RPLDs)

All these materials are generally broad band gap insulator materials which cannot give rise to any light emissions since the band gap is much larger than the energy levels involved in light emissions. However, none of the materials exist in a pure state, they always contain some defects and impurities which may introduce trap centers for electrons (below the conduction band) and for holes (near the valence band), known as electron traps and hole traps. The trapping centers are empty before irradiation. In order to have desirable luminescence properties, these materials are deliberately doped with specific impurity atoms which, based on luminescence research, have been known to produce suitable energy levels in the forbidden gap to produce luminescence.

The luminescence involves two steps. In the first step, radiation exposure ionizes the atoms, creating electron–hole pairs. The free electrons move to the conduction band and the holes are left behind in the valence band. They migrate in the respective bands until they recombine or come across some trap levels (i.e., defects or impurity atoms) and become trapped. Now the radiation energy is partly stored by the crystal, so it is a storage phosphor. In the second step, by a suitable stimulating mechanism the trapped electrons can be re-excited to the conduction band. These electrons again migrate in the crystal until they are trapped in the excited states of the luminescent centers. The spontaneous transition to the ground state then gives rise to the luminescence. Through proper calibration, the luminescence can be related to the radiation quantity, which can be exposure, air kerma, or PDE. These days, PM badges are calibrated in terms of PDE.

If the stimulating mechanism is heat, then the dosimeter is known as a thermoluminescent dosimeter. If the mechanism is optical stimulation, then the dosimeter is known as an optically stimulated dosimeter. In the case of photoluminescence, the luminescent centers are created following the irradiation, where the electrons can stay in the ground level until they are vacated by annealing the material at high temperatures. Here, the electron is raised to the excited level of the luminescent center (not to the conduction band) by optical means and hence these dosimeters are known as photoluminescent dosimeters or radiophotoluminescent (RPL) dosimeters. The luminescence phenomenon is very complex but the basic principle behind the phenomenon can be seen in Figure 8.11.

PMFDs were the only personal monitoring system in use until the 1980s when the thermoluminescence dosimeters or the TLDs started coming into the picture. Over the next 30 years, most of the users switched to TLDs for personal monitoring. Since the development of sensitive OSL phosphor $Al_2O_3{:}C$, some users started experimenting with OSL dosimeters for personal monitoring purposes. These dosimeters became more popular when Launder developed the Luxel personal dosimeter system and made them commercially available. Japan developed radiophotoluminescent glass dosimeters for personal dosimetry in the

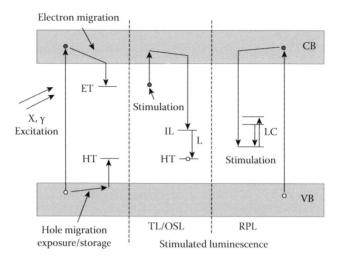

ET: Electron trap; HT: Hole trap; IL: Impurity level;
LC: Luminescent centre

FIGURE 8.11
Basic principle of luminescence.

1970s at the same time TLD development work was going on, but it did not become very popular due to its low sensitivity. The introduction of a pulsed ultraviolet (UV) source for excitation increased the sensitivity of the dosimeter. In 2001, Japan introduced an Ag-activated RPL glass dosimetry system for personal monitoring which became commercially available. Now, many centers in Japan, Germany, and few other countries use RPLDs for personal monitoring. These systems will be described in brief in the following sections.

8.5.1 TLD Dosimetry

Most countries have moved to the use of TLDs because of their several advantages over film badges. There is quite a large number of TLD materials in the field. The most widely used TLDs for personal monitoring are LiF:Mg, Cu, P, liF:Mg, Ti (United States, Europe), CaSO$_4$:Dy (India, Korea, Australia, and Brazil). A few countries use Li$_2$B$_4$O$_7$:Cu or other dopants like Mn or Si. The elements mentioned after the TLD are the dopants. The dopants are chosen to produce deeper traps—not close to the conduction or valence bands—to prevent thermal excitation of traps at room temperature. Otherwise, there would be significant fading of the signal with time and it would not be useful for reproducible results in monitoring. The color (frequency) of the light emitted depends on the specific dopants used (e.g., Ti emits in the blue, ultraviolet region). Defects and impurities are responsible for both the traps and luminescent centers.

TLDs are available in various forms such as powder, chips, rods, and ribbons depending on their use in dosimetry Figure 8.12.

8.5.1.1 Thermoluminescence Emission Phenomenon

The probability for a charge carrier to escape from the trap level is proportional to $e^{-E/kT}$, where E is the energy of the trap level, and T is the temperature. The equation shows that the probability of escape (and hence the escape rate or the thermoluminescence [TL] output) increases with the TLD temperature. The escape rate reaches a maximum at a specific temperature and then decreases rapidly. This occurs for each temperature that corresponds to a trap depth. As an example, Figure 8.13 shows three trap levels in a TLD where $E_1 < E_2 < E_3$, the traps corresponding to temperatures $T_1 < T_2 < T_3$. The TL peaks occur in the same order, while heating the TLD. Each TL peak can be related to a specific temperature and corresponding trap depth. The number of TL peaks exhibited by a TLD depends on the number of impurities and defects present in the crystal.

A plot of TL output against temperature is called the glow curve, which can be generated by a TLD reader. A TLD reader heats up the exposed TLD as a function of time and records the TL output.

If the traps are shallow, many of them may be excited by room temperature and this signal will fade fast, so it is not suitable for monitoring. Deeper traps need heating to higher temperatures and are therefore stable against decay at room temperatures. These peaks are used for dosimetry. Their lifetime at moderate room temperatures is relatively long and little signal is lost between the exposure time and the reading time of the TLD. So, the signal fading is less troublesome in personal monitoring compared to PMFs. The dosimetry peaks used in the PMTLDs occur around 200°C, as shown in Table 8.2 [10].

Traps that are too deep are also not very useful since they will leave a residual signal during annealing, for example, annealing at 400°C for 1 hour. Ideally, the TLD must have a single dosimetry glow curve with no interfering glow curves nearby.

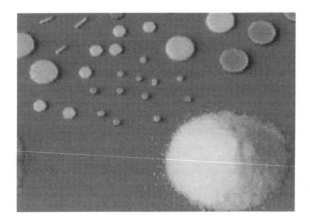

FIGURE 8.12
TLD available in various forms.

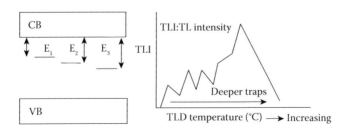

FIGURE 8.13
Relation between trap depth and excitation temperature.

TABLE 8.2

Dosimetry Peak of Common TLDs

TLD	Z_{eff}	Main Peak (°C)	Max. Emission λ (nm)
LiF:Mg,Ti	8.2	200	400
LiF:Mg,Cu,P	8.2	210	400
$Li_2B_4O_7$:Cu	7.4	205	368
$CaSO_4$:Dy	15.3	220	480, 570

8.5.1.2 Principle of a TLD Reader

A basic TLD reader system consists of a planchet for placing the TLD, a heater system to heat the TLD, a photomultiplier tube (PMT) to convert the TLD signal into an electric signal and an electrometer giving the signal current or charge (giving the TL intensity or the total TL output under a glow peak). The TLD reader can also record the glow curve. Figure 8.14 shows the basic components of a TLD reader.

The planchet in the reader is heated with a coil or with hot nitrogen gas, which gives more uniform heating. A thermocouple measures the temperature of the heating cycle. The TLDs are heated to 100°C for 10 minutes to remove the low TL peaks in the glow curve. On further heating the planchet, the dosimetry peak will be read by the reader, giving the TL output. This is followed by post-readout annealing, according to the established annealing procedure of the TLD, to ensure the signal has been completely erased and the TLD is ready for reuse. Pre-use annealing is important to erase the residual signal. The heating rate must be kept constant for a reproducible glow peak intensity. A reproducible readout cycle is very important for accurate estimate of TLD dose. Figure 8.15 shows a typical heating cycle and TL readout for an automatic TLD reader [11].

The TL output as read from the TLD reader, after proper calibration, gives the TL dose. If the heating rate is kept constant, then the glow curve represents the TL intensity (α measured current).

8.5.1.3 TLD Badge Design

The requirements of the PMTLD badge are the same as for the PMFs, so a filter system is part of a TLD badge. There are many different TLD materials used for personal monitoring.

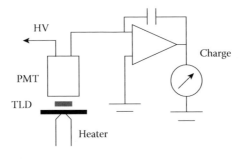

FIGURE 8.14
The components of a TLD reader.

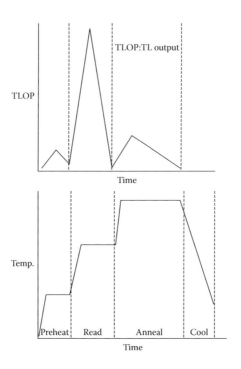

FIGURE 8.15
Heating cycle and TL readout for TLD.

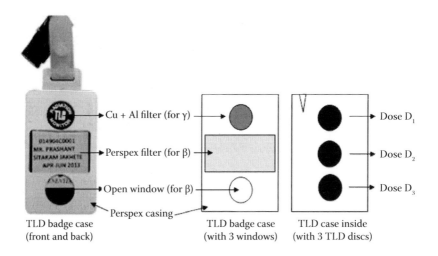

FIGURE 8.16
A CaSo$_4$:Dy TLD badge for personal monitoring.

A CaSO$_4$:Dy TLD badge will be discussed in some detail here, as used in India. Figure 8.16 shows the TLD badge, the three windows with filters (in the plastic casing on both sides), and the TLD card with three CaSO$_4$:Dy discs imbedded in a Teflon matrix.

The TLD card is Ni coated Al (52.5 mm L × 29.9 mm W × 1 mm H) with a V cut to ensure proper orientation of the TLD card in the cassette. The three TLD discs are fixed in the TLD

TABLE 8.3

The $CaSO_4$:Dy PMTLD Badge Details

Window/TLOP	Filter Material	Purpose	Detector
Window 1 D_1	0.9 mm Cu + 1 mm Al	Registers γ	$CaSO_4$:Dy-Teflon disc (0.8 mm × 13.2 mm dia.)
Window 2 D_2	1.5 mm perspex	Registers X	$CaSO_4$:Dy-Teflon disc (0.8 mm × 13.2 mm dia.)
Window 4 D_4	Open Window	Registers (X + β)	$CaSO_4$:Dy-Teflon disc (0.8 mm × 13.2 mm dia.)

card and are sandwiched between the three filters on the casing of the badge. The badge detail is given in Table 8.3.

The card is enclosed in a paper wrapper with the identity of the user and the period of use printed on it. This wrapper also protects the card from any kind of contamination.

The doses of β, X, and γ radiation registered by the TLD are evaluated, as in the case of PMFs, by measuring the TLD output under different filters. The results are then compared with calibration curves established for the TLD badge that had been exposed to known doses under well-defined conditions.

A control badge is always used in conjunction with the user TLD, which is stored in a low-background area. While reading the TLD, the control badge reading is subtracted which removes the background radiation dose, and any dose received during transit.

The control badge is used to measure the normal background radiation level or doses received in transit and must be stored in a low-background area.

8.5.2 Other Popular Personal Monitoring TLDs

The other very popular PMTLD is LiF:Mg,Ti. The LiF TLDs come in three different versions known as TLD-100, TLD-600, and TLD-700. TLD-100 is natural LiF (7.5% ^{6}Li and 92.5% ^{7}Li). TLD-600 is LiF enriched with ^{6}Li (to near 96%) and TLD-700 is an LiF enriched in ^{7}Li (with a very small quantity of ^{6}Li). They all have about the same sensitivity for gammas but with thermal neutrons, TLD-600 will be about 10 times more sensitive compared to natural LiF. TLD-700 with very minute fraction of ^{6}Li will not significantly respond to neutrons. So, TLD-600 can be used for neutron dosimetry.

Other common PMTLDs are also known by similar nomenclatures:

CaF_2:Dy (TLD 200), about 30 times more sensitive than TLD-100; CaF:Mn (TLD-400);

LiF:Mg, Cu, P (TLD-200A or GR-200) or MCP type TLD, more sensitive than TLD-100;

$CaSO_4$:Dy (TLD-900), about 20 times more sensitive compared to TLD-100; $Li_2B_4O_7$.

The TLD badges based on these TLDs are also designed on the same principles to monitor β, X, and γ with a fairly energy-independent response. Low Z TLDs are near tissue-equivalent and so are fairly energy-independent (see Figure 8.26) and do not require complex filter systems like the high Z TLD badges. The dose computations are also simpler.

One of the most widely used TLD card configurations for TLD monitoring is from Harshaw, which is used in many European countries. The typical configuration of this TLD card is shown in Figure 8.17 [12]. Harshaw also offers TLD materials, filters, etc., for designing the TLD badge.

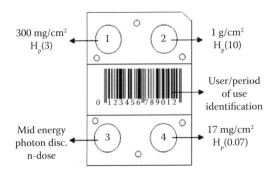

FIGURE 8.17
Harshaw TLD card configuration.

Two TLD discs, one usually thicker (with a thicker filter) and one thinner are used in these badges for monitoring $H_p(10)$ and $H_p(0.07)$, respectively; a 3-mm filter is used for $H_p(3)$ monitoring. TLD-600 (LiF which is ^{6}Li enriched) is used in the element 4 position for neutron monitoring. Each card is uniquely barcoded for identification.

8.5.3 TLD Albedo Neutron Monitors

Most of the dosimeters are insensitive to fast and intermediate energy neutrons, while many are sensitive to thermal and slow neutrons. When neutrons with a wide energy range are incident on the body (of a radiation worker), they slow down and are reflected by the body. Monitoring of the reflected slow neutrons is an indirect way of monitoring the incident neutrons. An albedo dosimeter can do this job; it can measure neutrons in the wide energy range and also discriminate against the gamma field dose. A personal neutron albedo dosimeter (PNAD) is worn on the body of the person like any other personal monitoring badge, and it responds to thermal and slow neutrons reflected (albedo) by the body which acts as a moderator. The most popular PNAD used today is based on a pair of TLD-600 and TLD-700 dosimeters though OSL-based PNADs are also now available. The TLD-600 registers the alpha and triton dose from the thermal and slow neutron reaction ^{6}Li(n, α)^{3}H, and also registers gamma. TLD-700 is insensitive to neutrons but has the same sensitivity to gammas as the TLD-600. So the two are used as a pair, and the difference in reading gives the dose due to thermal neutrons leaving the body. If the detector is covered on the front side by a thermal neutron absorber (Cd or B), they will absorb 95% of thermal neutrons and the detector will give only the backscattered thermal neutron dose. If high-energy neutrons are significant, track detectors are more suitable due to the pronounced energy dependence of albedo neutrons [13]. Several designs of albedo dosimeters are in use [2]. PDE > 0.05 mSv can be measured with this dosimeter with 30% accuracy.

8.5.4 Monitoring Extremities for $H_p(0.07)$

Unlike film badges, TLDS, because they come in different shapes and sizes, are very convenient dosimeters for monitoring eyes or the extremities (arms, wrist, fingers, etc.) and are widely used for these purposes (Figure 8.18).

Since skin doses are required, tissue-equivalent (TE) dosimeters are generally made use of for extremity dosimeters (e.g., LiF:Mg, Ti disc, or chip). The dosimeter can be easily

FIGURE 8.18
Extremity dosimeters. (Courtesy of Control Atom, Vilvoorde, Belgium and P T Sinergy Lintas Persada, Tangerang Selatan, Indonesia.)

wrapped around a finger or wrist, or placed near the eye using a convenient device such as a head band). These dosimeters are generally used by interventionists who carry out fluoroscopy procedures on patients, or by the physicists handling radioactive sources, or by researchers working with low-energy X-rays (e.g., X-ray diffraction units). The dosimeters have lower sensitivity compared to high Z TLDs but that is not of much of a disadvantage since the lower detectable limit is much higher. This is because the extremity dose limits are nearly 25 times greater than the whole body dose (500/20). These dosimeters are generally not required in radiation oncology departments.

8.6 Optically Stimulated Luminescence Systems

Al_2O_3:C is the most sensitive OSLD, developed in the 1990s, that is used for personal dosimetry. It is estimated that about 25% of the PM badge users make use of OSLD badges for personal monitoring [10]. The most popular commercial OSL dosimeter is from Luxor. A Luxor OSL badge (Landauer Corporation) is shown in Figure 8.19. It uses two thin Al_2O_3 strips for photon, β, and (if necessary) neutron monitoring (by adding a Cr-39 film).

The luminescence from the OSL detector is read by stimulating the dosimeter with select frequencies of pulsed laser light, such as green. The stimulating frequency depends on trap depth which can vary from infrared to ultraviolet. The luminous emission (blue in Al_2O_3:C) is proportional to the PDE. The emission λ is characteristic of the OSL material; the rate of luminescence emission is proportional to the stimulating laser light intensity, and the reader integrates the photons over the period of simulation. The OSL reader is also commercially available from Landauer Corp.

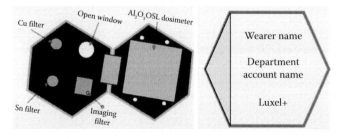

FIGURE 8.19
Luxor OSL badge design. (Courtesy of Landauer Corporation, Illinois, USA.)

The luminescence response ratios under different filters are used to discriminate between β, X, and γ radiations, and a dose algorithm determines the doses. The measurement can be repeated since the electrons drop to the ground level of the luminescence center, and so can be restimulated. Using a CR-39 strip, the neutron dose can also be estimated using the same badge. The dosimeter covers a wide dose range—10 μSv to 10 Sv for photons, 100 μSv to 10 Sv for β's, and 100 μSv to 50 mSv for neutrons. The dosimeter can measure $H_p(10)$ and $H_p(0.07)$ for photons (>20 keV) and β's (>200 keV), with ±15% accuracy.

The OSL reader is similar to the TLD reader, described earlier. Instead of thermal heating, a light source is substituted in the reader. It illuminates the dosimeter in a light-tight chamber. The detector-emitted luminescence falls on the PMT tube which converts it into an electrical signal.

8.7 Radiophotoluminescent Glass Dosimetry Systems

Radiophotoluminescent (RPL) glass dosimeters are the integrating-type, solid state dosimeters based on the radiophotoluminecence phenomenon to measure the radiation dose. It is estimated that about 8% of badge users make use of RPLD glasses for personal monitoring. Though many materials are capable of RPL, the most widely used system, following extensive research, development, and use in Japan, is the silver-activated phosphate glass from Chiyoda Technol Corporation, Tokyo, Japan. The dosimeters come in the shape of a small glass rod which can be enclosed in a holder (see Figure 8.20).

When silver-activated phosphate glass is exposed to radiation, stable luminescence centers are created in silver ions, Ag^0 and Ag++, by trapping the ionized free-moving electrons. The readout technique uses pulsed ultraviolet laser excitation. A PMT registers the orange fluorescence emitted by the glass. Figure 8.21 illustrates the general readout concept of luminescent dosimeters.

The RPL signal is not erased during the readout, thus the dosimeter can be read repeatedly. Commercially available RPL dosimeters typically cover the dose range of 30 μSv to 10 Sv. They have a flat energy response within 12 keV to 8 MeV for $H_p(10)$. The RPL signal exhibits very low fading and is not sensitive to the environmental temperature, making it convenient in individual monitoring.

The characteristics of the Chiyoda Technol Corporation RPL personal monitoring badge are given on the company's website [14].

Table 8.4 compares the characteristics of the various PM dosimeters discussed.

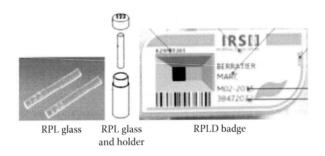

RPL glass   RPL glass   RPLD badge
             and holder

FIGURE 8.20
RPL dosimeter badge from Chiyoda Technol. (Courtesy of Chiyoda Technol. Corporation, Tokyo, Japan.)

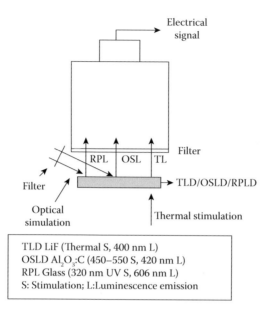

FIGURE 8.21
Readout principle of luminescent dosimeters.

TABLE 8.4

Comparison of the Characteristics of Passive Personal Monitoring Badges

PM Characteristics	Film	TLD	OSLD	RPLD
Material	Radiographic film	Crystal	Crystal	Glass
Discrimination against radiation type	Yes	Yes	Yes	Yes
Measurement principle	Optical density	Luminescence signal	Luminescence signal	Luminescence signal
Source of excitation (stimulation)	Light transmission	Heat	Optical laser	UV laser
Sensitivity	Moderate	Good	Good	Good
Repeated reading	Yes	No	Yes, sensitivity going down	Yes, with no change in sensitivity
Repeated use	No	Yes	No	Yes
Typical measurement range	100 μSv–10 Sv	10 μSv–10 Sv	10 μSv–10 Sv	30 μSv–10 Sv

8.8 Direct Reading Personal Monitors

In addition to passive dosimetry badges, direct reading (or active) personal dosimeters (APDs) are also widely used, for example:

1. To provide direct readout of the dose at any time
2. For tracking the doses received in day-to-day activities

3. In special operations (e.g., source loading survey, or handling of any radiation incidents or emergencies)

4. As alarm dosimeters to warn against high doses or dose rates

Passive dosimeters are not capable of these functionalities.
DRPDs fall into two categories:

1. Self-reading pocket dosimeters
2. Electronic personal dosimeters

8.8.1 Self-Reading Pocket Dosimeter

A self-reading pocket dosimeter resembles a pen and was one of the earliest self-reading dosimeters developed for personal monitoring. It consists of three parts: a reader, an ion chamber, and a capacitor. The ion chamber is hermetically sealed and also functions as a capacitor. Figure 8.22 illustrates the working principle of a pocket dosimeter.

The chamber is made of a metallic or plastic wall with a graphite coating inside for conducting. When it is made of metal, and also because of the bronze electrode, it shows pronounced energy dependence below 100 keV, like other high Z dosimeters. There is a movable fiber in the ion chamber which moves from the fixed bronze electrode when the dosimeter is charged.

A commercially available pocket dosimeter and charger are shown in Figure 8.23.

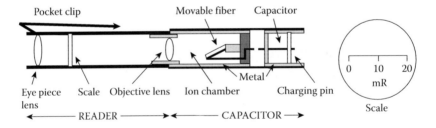

FIGURE 8.22
Working principle of a pocket dosimeter.

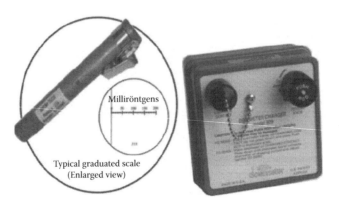

FIGURE 8.23
A pocket dosimeter and charger.

When the capacitor is fully charged, the fiber position can be adjusted (using the zeroing knob) to read zero on the scale. When the pocket dosimeter is exposed to radiation, the ionization produced in the chamber discharges the capacitor. This makes the fiber move on the scale, which is proportional to the discharge, which in turn is proportional to the dose. The scale is commonly calibrated in mR or R, depending on the dosimeter sensitivity (i.e., the value of the capacitor). The dosimeter can easily be read against light and no special external reader is required for this purpose. So, only the initial charging requires a charger unit. It can be used several times without recharging and the dose increment will give an idea of the dose for subsequent uses. The accuracy of measurement is around ±10%; the energy dependence is about ±15%, in the energy range 80 keV to ^{60}Co energy. The typical leakage of the pocket dosimeter is about 1–2% of full scale per day. They can measure as low as 2 mR full scale to few hundred R full scale, and they need to be calibrated annually since they are more prone to shock and rough handling (e.g., if they are accidentally dropped). However, the use of pocket dosimeters has declined in recent years because of their poor useful range, charge leakage problems, and poor sensitivity compared to EPDs.

8.8.2 Electronic Personal Dosimeters

Electronic personal dosimeters (EPDs) are commonly based on a miniature GM counter, silicon diode detector, or scintillation detector. They are available with the measurement range down to 20 keV photon energy. The modern EPDs are calibrated in PDE, that is, in terms of $H_p(10)$ or $H_p(0.07)$ for both photons and beta radiation. They are much more sensitive compared to passive dosimeters, and can read as low as 0.1 μSv. Two types of EPDs are in common use—the digital pocket dosimeter (replacing the analog self-reading type discussed previously) and pocket-type EPDs that can be carried in the chest pocket, as shown in Figure 8.24.

Since the detectors are very small in size it is not difficult to miniaturize EPDs. They can monitor and display both the instantaneous PDE rate and also the accumulated PDE, with a single display either toggling the two values or displaying both values simultaneously. They also have auto-ranging facilities like the digital survey meters, and audio indication by flashing and chirping at a frequency proportional to the dose rate, in addition to a visual display, (e.g., for every 0.5 μSv of dose or the set design value, the monitor will give one beep). With this audio facility, increase in radiation field levels can be easily

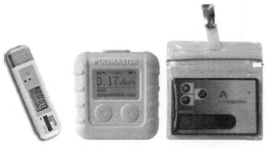

Dosicard

FIGURE 8.24
Pocket-type EPDs.

recognized immediately. Thus they are extremely useful in emergency handling situations. Some of the EPDs can also be used with preset dose alarms. Their use as legal dosimeters is already established in a few countries, and this trend may grow in the near future.

The EPDs are generally used for photon monitoring of PDE, $H_p(10)$, in the energy range of around 30 keV to 1.25 MeV. Microprocessor-controlled EPDs with data acquisition and storage facilities are also commercially available (e.g., the Dosicard in Figure 8.24). The EPD circuit is not very different from the GM- or Si-based area monitor circuit shown in the previous chapter, but needs to be calibrated in terms of $H_p(10)$ instead of $H^*(10)$.

A comparison of the commercially available EPDs was carried out by the International Atomic Energy Agency (IAEA) in 2006, which led to the following conclusions [15]:

- The performance of EPDs (or APDs) were comparable with, and even better than the passive PM systems in certain parameters.
- Very few EPDs gave satisfactory performance in pulsed radiation fields (requiring further improvement).
- Not all EPDs were designed for any radiation field, so the user must know the requirements in order to choose an appropriate EDP.

This study is now more than 10 years old and there are now many more models of EPDs on the market which can be used for personal monitoring purposes. For radiation oncology departments, the requirements are not highly demanding though the suitability of the chosen EPD for monitoring pulsed radiation fields must be ensured before its purchase. In the case of EPDs meant for neutron monitoring, sensitivity for thermal and slow (<500 keV) neutrons must be checked for monitoring in the control areas.

8.9 Properties of Personal Monitors

The basic characteristics of personal monitors are summarized in this section.

8.9.1 Sensitivity and Linearity

Film and TLD badges can measure PDEs in the range 0.1 mSv to 10 Sv. The RPL dosimeters are more sensitive, with a lower detection limit of 10–30 μSv. Personal dosimeters are generally linear in the dose range of interest in radiation protection.

8.9.2 Energy Dependence

This is the most important characteristic of any dosimeter because they are usually calibrated at a few representative beam qualities. However, they are used in workplaces where the multiple scattering and interactions of the radiation (photons or neutrons) give rise to a much wider spectrum of photons. So, the PM devices must be designed to have a fairly uniform energy response over a wide energy range. Earlier, the dosimeters were measuring exposure or air kerma and so were designed to exhibit a flat response with respect to these quantities. Non-air equivalent or non-TE dosimeters (high Z dosimeters) usually over-respond at low photon energies (<100 keV). This response is minimized with some filter systems to preferentially absorb the low-energy photons, to produce a flatter response with respect to exposure or air kerma. Later, the same dosimeters were adapted to measure the PDE. So, these dosimeters must exhibit a flat energy response with respect to the PDE.

$$N_{Hp} = [H_p(10)/M] = [H_p(10)/K_a]/[K_a/M],$$

where N_{Hp} is the PDE calibration of the dosimeter at an energy E and M is the reading of the personal dosimeter.

Since the $[K_a/M]$ is constant within a certain percentage, depending on the design, the dosimeters response with respect to energy must vary as $[H_p(10)/K_a]$. The $[H_p(10)/K_a]$ conversion factor curve with respect to energy is given in the last chapter (Figure 8.27).

In the case of neuron monitors, they usually measure the neutron fluence Φ, and the conversion factor $[Hp(10)/\Phi]_E$ is used to determine the PDE for each energy in the spectrum or the effective energy of the spectrum. Ideally one would like to have a flat energy response for the quantity $[Hp(10)/\Phi]_E$ so that the PDE calibration of a personal monitor at some representative energies can be used to measure the PDE in the workplace with reasonable accuracy. For this reason, the PMDs must be designed such that the response per unit fluence closely follows the fluence to PDE conversion factor.

$$[Hp(10)/M]_E = [H_p(10)/\Phi]/[M/\Phi] = \text{constant}.$$

8.9.2.1 Energy Dependence of PMFs

Film exhibits strong energy dependence and film badges are empirically designed to reduce their energy response to within ±20%.

The approximate energy dependence of the film with and without a filter is shown in Figure 8.25. It can be seen that the over-response occurs only at low energies due to the photoelectric interactions which are highly Z dependent. Preferential absorption of the low-energy photons in a high Z filter flattens the response, so that a Cs-137 or Co-60 calibration can be used for monitoring photon energy radiations in the keV region as well.

Thus, film badges are suitable for monitoring doses received from photons (15 keV to 3 MeV), beta ($E_{max} > 500$ keV) radiations, and thermal neutrons. For instance, the film

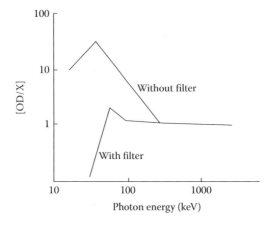

FIGURE 8.25
Energy-dependence response of the film with and without filter

badge cannot detect alphas or very low-energy betas (e.g., ^{14}C) due to their absorption in the wrapper used with the film to protect against light.

8.9.2.2 Energy Dependence of TLDs

The low Z TLDs (LiF) are TE while the high Z TLDs (CaSO$_4$) are highly sensitive and may be ideal for personal monitoring purposes but, like the film, they will exhibit pronounced energy dependence at low photon energies (see Figure 8.26).

The energy dependence curve is simply the ratio $[(_m\mu_{en})_{Zeff} : (_m\mu_{en})_{air}]$ since the dosimeter response is proportional to the energy absorption in the sensitive volume, which depends on $[(_m\mu_{en})_{Zeff}$ for the effective Z (Z_{eff}) of the dosimeter. The term $(_m\mu_{en})_{air}$ appears in the denominator because exposure (measured in röntgens) is defined for air, and so the relative exposure response depends on this term. This applies not only to TLDs, but to any clinical dosimeter because it depends only on the effective Z of the dosimeter. CaSO$_4$:Dy, for instance, shows significant energy dependence, and its energy response is reduced by empirical adjustments in the badge design. An LiF TLD is nearly TE and exhibits good energy-dependence characteristics.

8.9.2.3 Energy Dependence of RPLD and OSLD

Commercially available RPL dosimeters (e.g., Asahi, PTW, and Toshiba) have a flat energy response from 12 keV to 8 MeV, while commercially available OSL dosimeters (e.g., Landauer) have a flat energy response from 5 keV to 40 MeV. For direct reading pocket dosimeters, the energy dependence is within ±20% over the range 40 keV to 2 MeV. For EPDs containing energy-compensated detectors, the energy dependence is within ±20% over the energy range 30 keV to 1.3 MeV. The energy response values quoted here can vary in energy range and in the degree of flatness, depending on the individual monitor material and construction details.

8.9.2.4 Energy Dependence of Neutron Monitors

The trend in the variation of the conversion factor with neutron energy is shown in Figure 8.27 (the actual curve is not piecewise linear but varies more smoothly). Generally, the

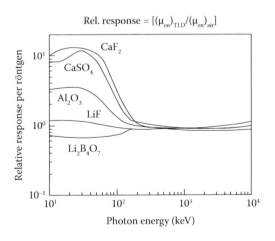

FIGURE 8.26
Energy dependence of TLDs.

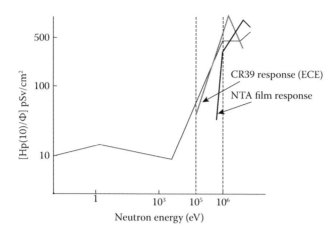

FIGURE 8.27
Neutron PDE per unit effluence versus energy.

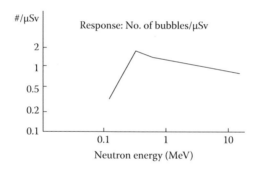

FIGURE 8.28
Energy response of neutron bubble dosimeter (BD-PND).

monitoring devices do not show a flat energy response. For instance, Figure 8.27 also shows the energy-dependence response of two neutron monitors whose responses do not cover the full energy spectrum. This is typical of all neutron monitors—strong energy dependence, and only partial coverage of the neutron energy range [16].

The response of a popular bubble neutron dosimeter (BD-PND from bubbletech, Chalk River, ON, Canada) is shown in Figure 8.28 [9].

8.9.3 PDE Range

Personal monitors must have as wide a dose range as possible, so they can cover both radiation protection and accidental situations (typically from 10 mSv to about 10 Sv). The dose range normally covered by film and TLDs is from about 100 mSv to 10 Sv, and that by optically stimulated luminescent and radiophotoluminescent dosimeters is 10 mSv to 10 Sv.

Self-reading pocket dosimeters can measure down to about 50 mSv; the upper dose limit of the available pocket dosimeters is around 200 mSv. EPDs measure in the range 0.1 mSv to 10 Sv.

8.9.4 Directional Dependence

According to the International Commission on Radiation Units and Measurements (ICRU), an individual dosimeter must be iso-directional, that is, its angular response relative to normal incidence must vary as the ICRU directional DE quantity $H'(10, \Omega)$. The directional dependence must be evaluated and the appropriate corrections derived.

8.9.5 Discrimination between Different Types of Radiation

Film dosimeters can identify and estimate doses of β, X, γ, and thermal neutrons. TLDs, OSLDs, and RPLDs generally monitor doses of β, X, and γ radiations.

8.9.6 Uncertainties in Personal Monitoring Measurements

The ICRU has stated that, in practice, it is usually possible to achieve an uncertainty of about 10% at the 95% confidence level ($k = 2$) for measurements of radiation fields in laboratory conditions. However, in the workplace, the energy spectrum and orientation of the radiation field are generally not well known and the user conditions and calibration conditions influence other properties of dosimeters. This includes energy dependence and angular dependence, and the uncertainties in a measurement made with an individual dosimeter will be significantly greater. This may be a factor of one for photons and still greater for neutrons and electrons. The uncertainty in measurements with EPDs is about 10% for low-dose rates (2 mSv/h) and increases to 20% for higher dose rates (<100 mSv/h) in laboratory conditions [17].

8.10 Personal Monitoring Regulations and Recommendations

The regulations and recommendations may differ slightly from country to country, but the following are the basic requirements in personal monitoring. The respective national regulations or recommendations for country-specific details should always be consulted.

8.10.1 Personal Radiation Monitoring Services

Personal radiation monitoring services are provided by agencies approved by the government or the regulatory body or any recognized body in a country (e.g., the Environmental Protection Agency).

8.10.2 Eligibility for Monitoring

Some recommendations say that if the annual dose is unlikely to exceed one-tenth of the dose limit, area monitoring is sufficient and personal monitoring is not necessary. However, the need for monitoring arises not only from the magnitude of the dose but also the potential for high-radiation exposure (e.g., radiotherapy room and high-radiation area). Also, monitoring helps to ensure the effectiveness of radiation protection and assure the individual regarding good protection in the workplace. It is helpful for documenting the wearer's radiation exposure, especially in an accidental or emergency situation.

Therefore many regulatory bodies require most radiation workers to wear badges except in mammography or dental radiography, where the area of exposure is restricted or the radiation is not highly penetrating, and to restricted areas (mammography), or where the doses are very low. In radiation oncology, all the radiation workers must wear a personal monitoring badge (PMB). According the regulatory body in India, monitoring of anyone who is likely to be exposed to radiation over a long period of time in their work should be compulsory.

Neutron area monitoring is required in radiation oncology department if the linear accelerator (linac) energy is >10 Mv. Individual monitoring is necessary if the monthly dose is, or is likely to be, >0.2 mSv.

If extremities are at risk then extremity dosimeters must be worn to monitor the doses (e.g., monitoring eye lens dose and wrist dose may be necessary for an interventional cardiologist). The dosimeter must be worn in such a way that it faces the incident radiation.

In addition to a personal dosimeter, an alarm dosimeter or an alarming survey meter must be used in certain situations, such as for industrial radiography workers or teletherapy source-loading operations.

8.10.3 Frequency of Monitoring

The frequency of monitoring depends mainly on the potential of exposure. Staff in radiation oncology and radiology are monitored on a quarterly basis, while industrial radiographers are monitored on a monthly basis.

8.10.4 Badge-Wearing Position

The badge is usually worn at chest level and at the waist level—the position must be representative of the whole body. For monitoring pregnant women, the badge must be worn near the fetus position since the fetal dose is of interest. In interventional cardiology, the person performing fluoroscopic procedures on the patient generally wears an apron. Here, the whole body monitor is worn inside the apron and another badge is worn near the collar level to monitor doses to the exposed parts of the body such as the eyes or thyroid. Here, dose estimation is more complex and special procedures have been developed for this purpose. In addition, special dosimeters such as a wrist dosimeter may be worn to monitor doses to the extremities. The situation is much simpler in radiation oncology.

8.10.5 Monitoring a Pregnant Radiation Worker

Pregnant radiation workers, on declaration of pregnancy, must be provided with a badge to be worn at waist level, to monitor the fetal dose. The staff will be monitored until the end of the pregnancy.

8.10.6 Control Badge

The personal radiation monitoring service (PRMS) issues a control badge along with the staff badges. The control badge must be kept in a background radiation area, away from the radiation rooms. This is used for correcting for the background dose, and the transit dose of the staff badges at the end of the monitoring period.

8.10.7 Badge Rotation

At the end of each monitoring period, all the staff badges must be collected by the RSO and along with the control badge these are sent to PRMSs for processing and dose estimation. TLD badges are issued on a monthly, bi-monthly, or quarterly basis depending on occupation and desired specifications. The PRMS offering the dosimetry service sends a dose report to the clients 1 or 2 weeks after receipt of used badges.

8.10.8 Responsibility for Monitoring a Radiation Facility

Each radiation facility must have a personal monitoring program to record and assess the occupational exposures received by the individuals covered under the program. The RSO is responsible for the monitoring of all the radiation workers in the hospital. The RSO is responsible not only for issuing the badges and collecting them at the end of the monitoring period, but also for ensuring (by randomly inspecting the imaging departments) that all radiation workers wear the badges while working with radiation and that they wear them correctly.

8.10.9 Proper Use of Monitoring Badges

- Badges must be worn only while working with radiation. Other times they must be placed in a background radiation area (e.g., the administrative office of the department) along with the control badge. The badge should never be left in the radiation room, regardless of the reason).
- PMBs must be kept away from extreme heat, light, or chemicals.
- It is the responsibility of the radiation worker to wear the badge correctly and hand it over to the RSO at the end of the monitoring period for mailing to the PRMS.
- The badge cannot be worn by any staff member to whom it has not been assigned. (This will erroneously show the dose in the wrong dose records).
- The badge cannot be worn in another facility, if the staff member works in multiple facilities, to which it has not been allotted.

8.11 Dose Records Issues and Maintenance

- The PRMS regularly mails the dose records to the user's facilities.
- It is mandatory for the employer to keep records of all monitored radiation workers, and also share the dose records with the departments concerned.
- The radiation workers must be informed about the doses they have been receiving in their work, and the dose records should not be treated as a confidential matter.
- The records must be kept even after the radiation worker ceases to work at the facility, for a minimum period of 30 years or until the worker reaches the age of 75 (IAEA).
- Many countries have a national occupational dose records registry where all the dose records are kept. (In India, PRMSs enter the dose records of all the monitored individuals into the National Occupational Dose Registry which is maintained by the Bhabha Atomic Research Center, an atomic energy research facility.)

References

1. J. Chwei-jeng et al. *The development, characterization, and performance evaluation of a new combination type personnel neutron dosimeter*. Oak Ridge, TN: ORNL, 1989.
2. D. E. Hankins. Progress in personal neutron dosimetry. Paper presented at IRPA 3 Conference, Washington, DC, 1973. http://www.irpa.net/irpa3/cdrom/VOL.3A/W3A_77.PDF (accessed on November 2016).
3. L. Tommasino, G. Zapparoli and R. V. Griffith. Electrochemical etching mechanisms. *Nuclear Tracks* 4, 191–196, 1981.
4. D. E. Hankins and J. Westermark. Preliminary study on the use of track size distribution on electrochemically etched CR-39 foils to infer neutron spectra. *Radiation Protection Dosimetry* 20, 109–112, 1987.
5. R. V. Griffith, J. H. ThornGate, K. J. Davidson, D. W. Rueppel and J. C. Fisher. Mono energetic neutron response of selected etched plastics for personnel neutron dosimetry. *Radiation Protection Dosimetry* 1(1), 61–71, 1981.
6. G. Dajko and G. Somogyi. Study of spot development around track and electric-tree induced perforations through an aluminized track detector. *Nuclear Tracks and Radiation Measurement* 8, 125–128, 1984.
7. R. Pal et al. *Present status of neutron personal dosimetry system based on CR-39 solid state nuclear track detectors*. BARC/2011/E/015. BARC, Mumbai, India.
8. E. V. Benton, R. A. Oswald, A. L. Frank and R. V. Wheeler. Proton-recoil neutron dosimeter for personnel monitoring. *Health Physics* 40, 801–809, 1981.
9. D. Muca. Estimation of neutron dose contributions to personnel working around high-energy medical linear accelerators for radiation therapy. MS Thesis, Lund University, 2006.
10. B. C. Bhatt. Thermoluminescence, optically stimulated luminescence and radiophotoluminescence dosimetry: An overall perspective. *Radiation Protection and Environment* 34(1), 6–16, 2016.
11. A. Savva. Personnel TLD monitors, their calibration and response. MS Thesis, University of Surrey, UK, 2010.
12. TLD badges and monitoring materials, ADM Nuclear Technologies, Victoria, Australia. http://www.admnucleartechnologies.com.au/tld-badges-and-monitoring-materials (accessed on November 2016).
13. H. Q. Tuan et al. *The procedure for evaluation of neutron personal dose using thermoluminescence (TL) dosimeter type Harshaw 8806, Dosimetry Lab, Institute for Nuclear Science and Technology*. Vietnam Atomic Energy Institute. http://www.iaea.org/inis/collection/NCLCollectionStore/_Public/45/058/45058920.pdf (accessed on November 2016).
14. Personal Monitoring, Chiyoda Technol. Corporation, Tokyo, Japan. http://www.c-technol.co.jp/eng/e-small (accessed on November 2016).
15. IAEA. *Intercomparison of personal dose equivalent measurements by active personal dosimeters*. IAEA-TECDOC-1564. Vienna: IAEA, 2007.
16. Assessment of occupational exposures due to external radiation sources and intake of radionuclides. Personal dosimeters, IAEA talk. http://slideplayer.com/slide/6626518/ (accessed on November 2016).
17. G. Rajan and J. Izewska. *Chapter 4: Radiation monitoring instruments, radiation oncology physics: A handbook for teachers and students*. Vienna: IAEA, 2005.

Review Questions

1. What is the purpose of personal monitoring?
2. What are the basic requirements of a personal monitor?

3. What is the presently recommended dosimetric quantity and its unit for personal monitoring? What quantities monitor doses received by the whole body, the extremities, and the eye lens, respectively?

4. What radiation types are generally monitored by a PM badge? Name personal monitoring devices and the one most widely used among them. What is the typical dose range covered by the PM badges?

5. What are the advantages and disadvantages of a film badge?

6. What are the purposes of using filters in the design of a PM badge?

7. Explain the principle of TLD, OSLD, and RPLD dosimeters used in personal monitoring.

8. What are the advantages and disadvantages of a PMTLD badge?

9. How is a light signal converted to an electrical signal in luminescent dosimeters?

10. What is meant by passive and active personal dosimeters (APDs)? What are the advantages of APDs over passive personal dosimeters?

11. Name a TE personal dosimeter. What is the disadvantage of a high Z personal dosimeter over a low Z personal dosimeter?

12. What is the most widely used phosphate glass dosimeter?

13. What are the detectors commonly used in EPDs?

14. How is an Si diode used for personal monitoring purposes?

15. What is the principle of neutron monitoring?

16. Name two neutron badges used in neutron monitoring. What is the principle behind their use?

17. What is a bubble detector and how is it used for neutron monitoring?

18. Which TLD is sensitive to thermal neutrons and why is it sensitive?

9

Transport of Radioactive Materials in Radiation Oncology

9.1 Introduction

The ^{60}Co teletherapy sources that are used in telecobalt units, in radiation oncology departments, are replaced every 7–8 years due to decay. Similarly, ^{192}Ir sources are procured by the department for use in the ^{192}Ir high-dose rate (HDR) afterloading units. Discrete radioactive sources are also used in brachytherapy departments for interstitial and intracavitary treatments of cancer. The decayed sources are sent back to the suppliers, or to a national or regional facility for safe disposal. So, an understanding of source ordering, transport, and disposal (or return) is important for radiation safety. Around 10–15 million packages are transported annually throughout the world. Nearly 75% of these packages are for medical use (diagnosis and therapy) and almost 75% of those are transported by air, with around 25% by road/sea/rail and a small percentage transported by mail. The number of accidents involving radioactive consignments over all these years of transportation is extremely small, with no adverse health effects. This should be compared with the accidents in the case of other forms of dangerous goods, which shows the radioactive transport regulations are highly effective in the transport of radioactive materials (RAMs).

Radioactive materials are classified as dangerous goods. Hence, unlike the ordinary materials, the transport of RAMs (TRAMs) must conform to safety regulations. This will ensure that RAMs do not cause any harm to the public or the environment. However, the RAMs used in medicine (or in industry, agriculture, or research) do not have additional dangerous properties like explosiveness, inflammability, corrosiveness, or toxicity. Under normal or even under accident conditions of transport, the packages containing RAM used in medicine must not be capable of causing an explosion or a fire.

9.2 Transport Regulations

Most of the countries have their own regulations for the TRAMs and almost all of them are based on the International Atomic Energy Agency (IAEA) Safety Standards [1,2], shown in Figure 9.1. The IAEA issued regulations for the safe transport of RAMs, which have been adopted by the professional bodies in many countries (e.g., by the Nuclear Regulatory Commission, USA). In some countries, the competent authority for enforcing the transport regulations could be the licensing authority for radiation protection. For instance, in India, the Atomic Energy Regulatory Board (AERB) is the competent

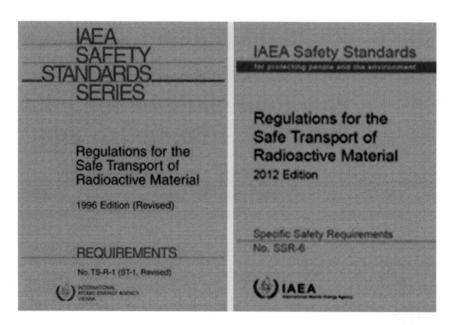

FIGURE 9.1
IAEA Specific Safety Requirements for the transport of RAMs.

authority for enforcing the regulations. The TRAM code developed by AERB, based on the IAEA Safety Standards, is the safety code AERB/SC/TR-1. Other agencies like the Department of Transportation, (USA), or even multiple agencies may be involved in regulating the TRAMs. The regulations cover all modes of transport (air, land, and sea), all aspects of transport (dispatch, transport and delivery), and all forms of RAMs (in special or nonspecial form).

9.3 Classification of Packages

The RAMs must be properly packaged before being transported. The most important requirements in the packaging of RAMs used in radiation oncology are (1) containment of RAMs and (2) shielding, to control radiation levels around the package. It is also important for the public to recognize packages containing RAM so they can keep away from them, and seek expert assistance in case they come across such packages in any transport accident or abandoned in a public place. In order to achieve these objectives, there are approved designs for the packages carrying RAMs, and approved labeling for the packages to recognize a radioactive package.

There is no direct correlation between the physical size of the material and the radioactivity content. For instance, contaminated materials may be in large quantities with a small amount of activity. A pound of ^{238}U contains about 150 µCi activity while a small ^{60}Co teletherapy source has thousands of curies of activity.

Since the activity handled varies over a very wide range (µCi to tens of thousands of curies) they do not cause the same level of harm if there is any loss of package integrity. It is not cost effective to design all packages to the same stringent specifications. The sturdiness of the package is obviously related to the risk of exposure in case of any accident. Unsealed sources of small activity used in medical diagnosis are permitted to be transported in packages of simple design (Type A package). On the other hand, very high activity sources like the ^{60}Co teletherapy sources (exceeding 10,000 curies) warrant a very sturdy, accident-resistant package (Type B package). Some sources are of extremely low activity and they cause negligible exposure. Such sources may be transported as excepted packages (e.g., radioimmunoassay (RIA) kits).

9.4 Classification of Radioactive Materials

The RAMs can be in a special form or other (nonspecial) form. The RAMs designed to retain the radioactivity by encapsulating them are said to be in a special form. The special form radioactive material is required to be subjected to stringent regulatory tests. Once compliance with the test requirements has been demonstrated, the manufacturer of the radioactive source has to obtain a special form certificate from the competent authority. Since these sources are required to meet the ISO standards for sealed sources, they already satisfy most of the requirements for special form RAM. The isotopes supplied as unsealed sources, the radioactive waste, and the RAM in the form of gas or liquid, or in powder form are examples of RAMs in normal form and can disperse into the environment, if there is a loss of integrity of the package in an accident. Release of any RAMs in dispersible form entails higher risk, as it can lead to intake and internal exposure. So, the amount of radioactivity that can be transported in a package depends not only on the source activity but also on the nature of the material. The maximum activity that can be transported in a package in special form or other form is referred to, respectively, as A_1 and A_2 values. The above classification of RAMs is shown in Table 9.1.

9.5 Classification of Package Types

Depending on the nature and activity content of the RAMs, four different packages have been defined for the transport of RAMs:

1. *Excepted.* Extremely low activities are contained in some devices used in the public domain such as smoke detectors. Since the individual risk is negligible with such

TABLE 9.1

Classification of RAMs in Different Forms

RAMs	Examples	Maximum Activity Permitted
Special form	Neutron source, ^{192}Ir HDR source, ^{60}Co teletherapy source	A_1
Non-special (normal) form	Liquid, powder or gas form in container, waste in bag, etc.	A_2

Note: HDR, high-dose rate; RAM, radioactive material.

materials, the transport of packages containing these materials is subject to minimal regulatory requirements. Such packages are known as excepted packages. The activity permitted in excepted packages is typically one thousandth to one ten thousandth of the activity permitted in Type A packages. Postal packages used to carrying radiopharmaceuticals such as RIA kits are permitted to contain one-tenth the quantity permitted in excepted packages.

2. *Industrial.* Low specific activity materials and contaminated materials transported in the nuclear industry come under this category. In radiation oncology, we are not concerned with the above two categories.

3. *Type A.* Type A packages are designed to carry moderate activities of RAM and so they are required to withstand (i.e., retain their integrity) only in normal conditions of transport. The integrity of the package may be breached in accidental conditions, but the activities transported in these packages are limited to A_1 and A_2 according to whether the material is in special form or other form. So, if some activity is released in an accident, the risk from external exposure or contamination is very low. Radioiostopes used in agriculture, medical diagnostic, or nuclear medicine (e.g., Tc generators) and some very low activity brachytherapy sources (e.g., ^{125}I, ^{192}Ir wires) are transported in this type of package. The allowed activities for various materials are tabulated in the transport regulations [1].

   The Type A package can be a cardboard box, a wooden box, or a metal drum of any shape (Figure 9.2).

4. *Type B.* When activities exceeding the limits for Type A packages are required to be transported, Type B packages used. Type B packages are required to withstand (i.e., retain their integrity) in severe accidental conditions of transport. What constitutes normal and accidental conditions of transport has been defined in the regulations, and the packages must pass the tests performed on them under simulated accident conditions in order to be accepted by the national approving authority for TRAM. A Type B package developed in India for the transport of telecobalt sources is shown in Figure 9.3 which also shows a Type B package being surveyed for radiation levels before being labeled and dispatched [3].

FIGURE 9.2
A Type A package.

FIGURE 9.3
A Type B container and a Type B package survey.

For transporting very high activity gamma sources (e.g., ^{60}Co teletherapy source) the package must have substantial gamma shielding and multiple containers with thermal dissipation capability, as shown in Figure 9.3.

9.6 Classification of Package Categories

The packages are classified into three categories (I, II, and III) depending on the radiation levels around the package, as described below:

- Category I—White label indicates that the radiation levels outside the package are very low
- Category II—Yellow label indicates that the radiation levels outside the package are slightly higher
- Category III—Yellow label indicates that the radiation levels outside the package are higher

The maximum permissible radiation levels at the surface and at 1 m from the package are given in Table 9.2 for different categories of packages. The radiation level at 1 m from the package is known as the transport index (TI). The TI is defined as a number that expresses the maximum radiation level at 1 m from the surface of the package, expressed in mrem/h, or in mSv/h divided by 100.

1 mrem/h = 0.01 mSv/h, so the TI can also be expressed in SI units. Table 9.2 shows the permitted radiation levels around various package categories expressed in both the SI and non-SI units. The radiation levels at 1 m from the surface of a radioactive package are illustrated in Figure 9.4.

When either the maximum radiation level at the surface of the package or the TI exceeds the limits for the specified category, the package is assigned to the next category.

The radiation level at the surface of a package and the transport index can be measured by surveying the package with a radiation protection monitor, as shown in Figure 9.5.

TABLE 9.2

Permitted Radiation Levels for Various
Package Categories

Category	Max. Radiation Level at the Surface (mSv/h)	TI
I white	0.005	0.0
II yellow	0.5	1.0
III yellow	2	10.0

Note: 1 mrem/h = 0.01 mSv/h ≈ 1 mGy/h
≈ 1 mR/h (for photons); TI, transport
index.

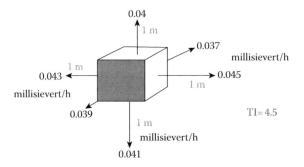

FIGURE 9.4
Radiation level at 1 m from the surface of a radioactive package.

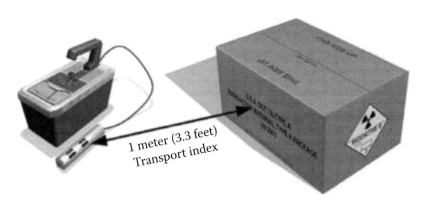

FIGURE 9.5
Determination of transport index. (Courtesy REMM, US Department of Health & Human Services.)

9.7 Labeling Radioactive Packages and Placarding Transport Carriers

As far as the public safety is concerned, it is the label that gives all the relevant information regarding the contents of the consignment. The label indicates (through symbols) the type of hazard and the level of hazard associated with a package. The label, in addition to the visual indication, provides the following information:

- Dangerous goods classification (Class 7)
- Nature of material (radioactive)
- Source name (e.g., ^{60}Co source)
- Source strength
- Type of package
- Category of package
- Transport index

This is described in Figure 9.6.

According to the UN classification, RAM falls under Class 7 of dangerous goods. The different UN numbers identify the different packages. For example, UN 3332 refers to RAM, Type A package, special form, and UN2916 refers to RAM, Type B(U) package, special form.

It is a mandatory requirement to affix the label on two opposite sides of the exterior of the package. Before fixing the label on the package, the sender must survey the package and determine the radiation level at the surface and also the TI, which helps in the choice of the proper label category (i.e., I, II, or III).

In addition to the label, the package must carry the following additional and relevant information:

1. Address of the consignor and consignee
2. Gross weight of the package
3. Type of certificate issued to the package (e.g., Type B)

The Type B package may carry a unilateral certificate (BU) or multilateral certificate (BM). A BU-certified package design is approved by the country of origin while a BM

FIGURE 9.6
Category I, II, and III package labels.

FIGURE 9.7
Placard for vehicle-transporting RAMs. (Courtesy REMM, US Department of Health & Human Services.)

certified package design requires approval from the competent authority of every country it has to pass through.

The symbol helps the public to be cautious about the nature of the package (from the symbol) even if they cannot read what is inscribed on the package. Anyone finding a package with a radioactivity symbol should stay away and should not try to touch it, or pick it up, or let anyone else do so. The appropriate authorities such as the police or the fire brigade must be contacted and informed about the package. The label also helps inform the transportation workers and emergency response personnel about the radioactive contents of the consignment. The information on the radiation levels will also help in assessing the integrity of a package following an accident, although if there are many packages at the scene their presence will interfere with the survey of any single package.

The vehicle carrying packages containing RAM must be suitably placarded. The placards must be large compared to the labels (about 10″ × 10″) and generally contain the information on the nature of hazard. The placard usually says RADIOACTIVE and also mentions the UN Class 7. An example of a placard for fixing on the vehicle is shown in Figure 9.7.

The one on the right in Figure 9.7 (with a black border), is used in the US for shipments containing large quantities of radioactivity and the highway routes are controlled, to ensure public safety. The placards must be displayed on all four sides of the vehicle. Unlike the different categories of labels, there is only one type of placard and no other information which is on the label (like the source identification, TI) needs to be written on the placard. The details of the contents of the packages can be known only from the labels and the shipping papers carried by the driver.

9.8 Carrier Responsibilities

Some of the responsibilities of the carrier are:

- Handling the package with care
- Properly securing heavy Type B packages to the vehicle
- Placarding the vehicle

- Carrying the package to the destination without undue delay
- Not storing the package with other dangerous goods (e.g., explosives, inflammables) or radiographic films
- Not stacking the packages in one place that results in a total TI value of the packages exceeding 50.

According to transport regulations, the sum of TI of all the radioactive packages in the vehicle, or in any single group in storage, cannot exceed 50, which can be determined by adding the TI values given on the package labels.

9.9 Providing Transport Documents

It is the responsibility of the consignor to provide the carrier with the transport documents:

- Consignor declaration (certifying the contents details)
- Special form certificate
- Instructions for carriers
- Transport emergency card (TREMCARD)
- Emergency handling instructions

The TREMCARD lists the basic actions to be taken in case of an emergency. The card also includes the details of the sender and receiver of the package, and the regulatory agency that must be contacted in case of any emergency. In the absence of proper shipping documents, the authorities can view the movement of any RAM as unauthorized and suspicious.

9.10 Transport of RAM

The RAMs must be transported only by the authorized persons, the institutions, or companies. In radiation oncology, the source suppliers transport the sources (e.g., brachytherapy or teletherapy sources) on receiving an order from the user institution. Radiation oncology departments in turn transport the decayed brachytherapy or teletherapy sources back to the supplier or to any authorized waste management facility for safe disposal.

9.10.1 Transport Under Exclusive Use

When the activity of the source in the package is large it is desirable, from the safety point of view, to transport the package directly from the sender to the receiver in a specially chartered vehicle. This type of shipment is termed as under exclusive use. In India, for instance, all teletherapy sources and RAM used in irradiation facilities are transported under exclusive use. If the radiation level at the external surface of the package exceeds 2 mSv/h but does not exceed 10 mSv/h, and if the TI of the package exceeds 10 it can be transported in an exclusive use (contract carrier) closed vehicle, provided the dose equivalent at 2 m from the vehicle does not exceed 0.1 mSv/h. (If it is open, even under exclusive use, it cannot exceed the TI limit of 10, meant for common carriers.)

9.10.2 Prerequisites for Ordering a Radioactive Source by a Radiation Oncology Department

Prior to ordering a radioactive source (e.g., ^{192}Ir HDR source), the user must apply to the regulatory body (RB) for license/authorization for source procurement. Based on the request, the RB issues the authorization. Normally the authorization is issued for procuring a single source at a time. However, in the case of relatively short-lived sources (e.g., the ^{192}Ir unit needs source replacement every quarter) the user, to avoid repetitive work, can seek authorization for the procurement of four sources over a period of 1 year and this could be granted by the RB. The user can then order the source as per the specifications in the authorization. However, only one source can be ordered at a time.

9.10.3 Transport of RAMs by the Source Supplier

On receiving the order, the supplier ships the source to the user institution. Before handing over the package for shipment, the consignor (e.g., supplier of RAM to the user institution) must ensure that the package to be used meets the design requirements. The consignor must also make sure that the consignee is authorized and prepared to receive the RAM, before forwarding the package for transport.

The supplier/shipper/consignor prepares the package, labels it, prepares the shipping papers, signs the papers, and hands it over to the carrier. The shipper is primarily responsible for the safe transport of his sources. The carrier then examines certification papers, checks packages for proper labeling, placards the vehicle, and properly secures the package before accepting it for transport.

9.10.4 Source Transport Documents

At the time of booking the package for transport, the consignor declares it as RAM under UN Class 7 in the transport document. The radioactive package is booked as an item of cargo and cannot be carried in buses or local trains, or in passenger cabins (unless it is Category I White). They must be carried in placarded vehicles accompanied by the relevant transport documents.

9.10.5 Source Certifications

A certificate of source must accompany the brachytherapy source showing the following:

- *Integrity against leak.* According to ISO recommendations, the wipe test of the source must show a contamination <200 Bq (5 nCi). So, the certificate would say that the source has been leak tested by wipe test and was found to have say 0.4 nCi removable contamination. (The leak testing instrument must be sensitive to detect at least 50% of the limit).

  For example, when the radiation oncology department procures a brachytherapy source, the certificate assures the department that the source is not leaking, has the order of activity stated therein and the output in terms of air kerma strength or the reference air kerma rate, for example, is as mentioned in the certificate.

- *Output.* For a brachytherapy source, the output may be specified in terms of air kerma strength, or reference air kerma rate at a measurement date; for a teletherapy source, the output may be specified in terms of RMM (i.e., output at 1 m in R/min).
- *Activity.* The nominal activity of the source is shown as certified by the source supplier.
- *Warranty.* A guarantee, regarding the source workmanship and integrity for a certain number of years (say 15 years for a teletherapy source) should be shown. This implies that the source has undergone the safety performance tests regarding temperature, external pressure, impact, vibration, puncture, etc., as prescribed by ISO for teletherapy sources.

9.10.6 Transporting RAMs by the Institution

Authorization is required for the user institution, not only for the procurement of sources but also for transporting it outside the institution. As in the case of procurement, the user must, in some countries, obtain prior authorization for transporting the source. On receiving authorization, the user prepares the package for transport as explained above.

During source exchange, a wipe test is carried out near the collimator part of a telecobalt unit or on the applicators in the case of a ^{192}Ir HDR unit. Any radioactive contamination would indicate a breach of source integrity or a leaking source. A source that fails the leakage test should not be transported without obtaining approval from the competent authority.

References

1. International Atomic Energy Agency, *Specific Safety Requirements No. SSR-6: Regulations for the Safe Transport of Radioactive Material*, IAEA, Vienna, Austria, 2012.
2. *Safe Transport of Radioactive Materials*, AERB draft safety code No. AERB/NRF-TS/SC-1, 2005, Atomic Energy Regulatory Board, Mumbai, India.
3. D.C. Karr, et al., *Development of Type B(U) Package for Cobalt-60 Teletherapy Source Transportation*, Bhabha Atomic Research Centre, Mumbai, India, 2009. http://www.barc.gov.in/publications/nl/2011/pdf/DAE%20EA/paper%2013.pdf

Review Questions

1. Describe "special" and "other" forms of RAM. To which of these classifications do the radioactive sources used in radiation oncology department belong?
2. State four types of packages defined in the regulations for the transport of RAMs.
3. What types of packages are used for transporting brachytherapy and teletherapy sources to a radiation oncology department? What type must be used for transporting ^{192}Ir HDR source?
4. How many package categories have been defined?
5. What information is provided on the label displayed on the package? How does it differ from the placard displayed on the transport vehicle?

6. The label on a package reads a TI of 0.2. What should be the maximum radiation level at 1 m from the surface of the package?

7. What is the maximum permitted radiation level on the surface and at 1 m from the surface of a Type B radioactive package? If the radiation level exceeds these values how can the package be transported? What is the difference between BU and BM type packages?

8. A survey of a Type B package at 1 m from the surface yields the following values in μSv/h: 5.1, 8.2, 7.4, 9.6, and 8.3. What is the TI of the package?

9. Define transport index. Is there any restriction on the transport index if a large number of radioactive packages are transported in a vehicle?

10. What are the transport documents that should be provided to the driver transporting RAMs?

11. What are the main carrier responsibilities?

10

Radiation Protection in External Beam Therapy

10.1 Introduction

The success of radiotherapy treatment is the joint responsibility of the radiation oncology team, which comprises professionals from various disciplines such as radiation oncology, medical physics, medical dosimetry, and X-ray treatment technology. Over the years, the radiation oncology practice has become quite complex. Developments in 3D imaging, 3D treatment planning, and onboard imaging techniques, in the last decade, have led to new developments in treatment delivery and verification technology. This has made it possible to deliver radiation dose to the clinical target volume with unprecedented geometric and dosimetric accuracy. There is a high level of complexity in the modern treatment techniques of intensity-modulated radiotherapy (IMRT), image guided radiotherapy, stereotactic radiotherapy and stereotactic radio surgery (SRS), and tomotherapy. Considering the very high dose rates being delivered by the treatment delivery equipment (TDE) compared to the diagnostic X-ray machines, there is great potential for overexposure to the patients or staff, and also to the general public, if proper safety and quality audit (QA) procedures are not adopted, or if the regulatory requirements in radiotherapy are not adhered to.

The employer delegates responsibility at various levels so they have collective responsibility for meeting the same objective, as stated in Chapter 5. Those having subsidiary responsibilities are listed below:

- Equipment supplier
- Radiation oncologists
- Radiation safety officer (RSO)
- Ethics committee
- Any party or person with specific delegated responsibilities (e.g., technician)

By his qualification, the medical physicist becomes mainly responsible for radiation safety and regulatory compliance though the licensee, and the party operating the radiation oncology facility is legally responsible for the safety and security of the practice.

10.2 Safety in Treatment Delivery Equipment Design

According to the basic safety standards of the International Atomic Energy Agency (IAEA) [1], the equipment used in medical exposure shall be so designed that:

1. Failure of a single component of the system is promptly detectable so that any unplanned medical exposure of patients is minimized
2. The incidence of human error in the delivery of unplanned medical exposure is minimized

The International Electrotechnical Commission (IEC) has introduced standards to ensure the electrical, mechanical, fire, and environmental safety of radiotherapy equipment. The IEC Standards applicable to radiotherapy are [2–6]:

- IEC 601-2-1, for medical electron accelerators
- IEC 60601-2-11, for external beam radiotherapy
- IEC 60601-2-17, for remote afterloading brachytherapy
- IEC 601-2-8, for superficial therapy with X-rays
- IEC 60601-2-29, for therapy simulators

Many national regulatory bodies demand that the equipment meets the requirements of IEC or equivalent national or international standards during type approval testing of the equipment, or as demonstrated through the manufacturer's records of compliance testing. More detailed guidance on this aspect can be found in the IAEA document [7]. IEC also specifies different grades of testing for the equipment.

In the European Union, medical devices are subject to European Directive 93/42/EEC that specifies the safety requirements for the equipment [8]. The "CE" marking indicates compliance with those directives. The equipment must have the CE marking for marketing in EU and in several other countries outside the EU. The accessories that influence the treatment, apart from the medical equipment, must also have CE marking, indicating safety compliance. These accessories include Treatment Planning Systems and Record and Verify Systems, and virtual simulation software.

In the United States, the equipment must have 510(k) premarket clearance notification from the Food and Drug Administration (FDA), which ensures that the equipment is safe and is substantially equivalent to the devices already on the market. Many national regulatory bodies insist on 510(k) certification for allowing the equipment for clinical use.

The performance specifications and operation, maintenance, and safety instructions must be available in a major world language understandable to the users, and also in the local language when appropriate. Many national regulations require that the documents are available in the local language. The display on the operating console must be in a major world language acceptable to the users (e.g., English).

The high-energy radiotherapy equipment must be provided with:

- Two independent fail-safe systems for terminating the exposure
- Safety interlocks or other means to prevent any unauthorized use

The equipment design must allow interruption of exposure from the control console; on interruption the exposure must be reset and treatment resumed only from the control console. In the case of TDE containing a source (e.g., ^{60}Co teletherapy equipment), a suitable device is to be provided for manually returning the source to the parking position in the event of an emergency. For a gamma knife (GK) unit, it must be possible to close the shielding door manually. The external beam therapy equipment containing one or more sources (e.g., teletherapy equipment, ViewRay MRI-guided cobalt radiotherapy system, GK system) must display a permanent "radiation present" sign indicating the existence of radioactive material.

The above requirements are met by all the radiation equipment used in radiation oncology practice. In addition, the equipment installation must meet the regulatory requirements of the country in which it is being installed. A few of the important requirements, in this respect, are:

- The TDE and accessories must be purchased only from authorized suppliers.
- Only type-approved equipment must be installed in the hospital (the type approval details are given in Chapter 5).
- In countries that do not have a national protocol for type testing or acceptance testing of the equipment, an established national or international protocol (e.g., The American Association of Physicists in Medicine (AAPM), The Institute of Physics and Engineering in Medicine should be adopted for testing.
- It is mandatory to carry out acceptance testing of every unit installed at any hospital to ensure that the installed unit meets the requirements of the protocol.
- It is the responsibility of the supplier and licensee to ensure that the equipment and its installation meet all the regulatory safety requirements.

10.2.1 Safety in Equipment Leakage

Radioisotope machines (e.g., teletherapy cobalt equipment, high-dose rate (HDR) brachytherapy equipment, GK equipment) show leakage both when the machine is ON or OFF. This is because the radioisotope emits radiations continuously and cannot be switched off. On the other hand, a linac produces radiation and shows leakage only when the machine is ON. For radioisotope machines, the leakage characteristics are different in machine ON and OFF conditions because the source positions are different for the two conditions, as can be seen from Figure 10.1, for a telecobalt head.

It is important to restrict the machine head leakage to reduce the risk to both the staff and the patient. Head leakage outside the useful beam in the patient plane must be restricted to reduce the risk of secondary carcinoma to the patient from leakage radiation dose. Head leakage in other directions, around the head, also must be restricted since it is undesirable from a radiotherapy point of view and also from the room shielding point of view. Leakage around the teleisotope machine gives unnecessary dose to staff during patient setup or other measurements with the machine, and hence must be regulated.

IEC regulations, and other national and international protocols require that the leakage from the head of a linac must not exceed an average of 0.1% and a maximum of 0.2% of the primary beam (output at the isocenter, for a 10 cm × 10 cm field) over a 2 m radius in the patient plane (Figure 10.2).

Apart from the patient plane recommendation, the leakage at any point 1 m from the target (or source for a telecobalt unit) must also not exceed 0.5% of the open beam output at the isocenter for a 10 cm × 10 field. For the linac, this recommendation also applies to points 1 m from the electron path in the waveguide (since some X-rays are produced in

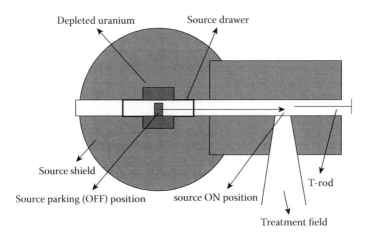

FIGURE10.1
Principle of a telecobalt head.

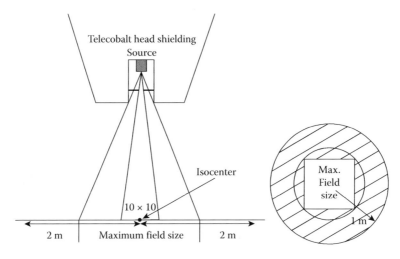

FIGURE10.2
Leakage measurement in patient plane in beam-on condition.

the waveguide as well, due to interaction of scattered electrons with the waveguide). More details regarding this can be found in the section on radiation survey.

The manufacturers generally design the equipment with leakage much less than 0.1%. Hence the National Council on Radiation Protection and Measurements (NCRP) 151 assumption of 0.1% leakage [9] for shielding calculations includes a safety factor.

In the case of telecobalt units, leakage in the beam-off condition also must be restricted (by regulation) not to exceed 20 µGy/h (or 2 mR/h) at 1 m from the source and 200 µSv/h (20 mR/h), at 5 cm from the surface of the head for the maximum head capacity [3], so that staff would receive a very small dose while working on patient positioning. If the leakage during beam-on condition also satisfies this criterion (as in the case of some commercial telecobalt units) then it need not be taken into account for shielding calculations. When the beam-on condition leakage is higher, may have to be considered for shielding calculations.

Both leakage tests are important because in the beam-on condition, the source is in a different position where the shielding is less than in the parking position. The on-position leakage must be carried out outside the maximum useful beam area, with the collimators completely closed. If the collimators do not close completely (as used to happen with some cobalt machines) the open area must be completely blocked with three tenth-value layer (TVL) thickness of lead. In the useful beam area, the dose rates can be higher since the allowed leakage for the collimators are higher.

For beams >10 MV, neutron leakage may also have to be regulated. In the patient plane, according to the recommendations, neutron leakage must not exceed a maximum of 0.05% and an average of 0.02% of the maximum absorbed dose in a 10 cm × 10 cm at the center of the patient plane. For other points 1 m from the target or the electron path, neutron leakage must not exceed 0.5% of the primary beam dose rate at the isocenter, for a 10 cm × 10 cm field.

10.3 General Concepts in Radiation Hazard Control

The purpose of radiation protection in radiation oncology is to protect the patient, the staff, and the public from unwanted radiation exposure. This is realized through the concept of justification/optimization and dose limitation/dose constraints as explained in Chapter 5. In optimizing staff exposures, the time, distance, and shielding (T/D/S) concept plays a significant role. This simple concept can be explained from Figures 10.3 through 10.5, where an exposed individual is depicted standing at a distance from the source of radiation. Dose rate (DR) represents the dose rate produced by the source at the position of the exposed individual. If the individual spends certain time in the radiation field, the dose received is given by DR × Time (duration of exposure).

Since the same amount of radiation is spread over a larger and larger area as the distance increases, the energy fluence per unit area falls as $1/d^2$ and so $DR_2 < DR_1$. In Figures 10.3 and 10.5, the distance between the source and the individual remains the same. Yet, the shielding attenuates the radiation (in Figure 10.5) and so $DR_3 < DR_1$.

Example 10.1

The radiation field (in air kerma rate) at a workplace is 50 μSv/h. If you are not to exceed a dose (air kerma) of 10 μSv, how long you can remain in that field?

Permitted duration of stay = 10 μSv/(50 μSv/h) = 0.2 h or 12 min.

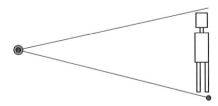

FIGURE 10.3
Dose received = DR_1 × Time of exposure.

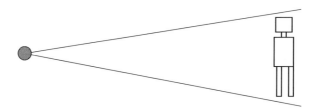

FIGURE 10.4
Dose received = DR_2 × Time ($DR_2 < DR_1$).

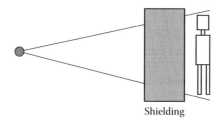

Shielding

FIGURE 10.5
Dose received = DR_3 × Time ($DR_3 < DR_1$).

TABLE 10.1

Concept of time, distance, and shielding in radiation exposure control

Parameter	Recommendations	Comments
Time	Keep as short as possible	Dose received is directly proportional to time (duration of exposure)
Distance	Keep as large as possible	Inverse square law
Shielding	Use shielding and protective barriers or devices wherever applicable	High density and high atomic materials are good attenuators for photons

Example 10.2

The air kerma rate (AKR) at 1 m distance from the source is 40 μSv/h. What is the AKR at 2 m distance?

The dose rate or the AKR falls as $1/d^2$. So, the AKR at 2 m = 40 μSv/h/4 = 10 μSv/h

Table 10.1 shows how the T/D/S concept can be used to control radiation exposure and hence the radiation hazards.

All three factors play a role in the protection of the interventional cardiologist who performs cardiac procedures on a patient. How much time he takes to perform the procedure, how far from the patient he is standing (which is decided only by the convenience in doing the procedure and so is usually around 0.3 m) and how many protective devices he is using (e.g., lead apron, leaded eye glasses, thyroid shield) will decide the dose received by him.

In radiation oncology, time and distance play an important role in reducing staff exposures while handling low-dose rate (LDR) brachytherapy sources, or in reducing public

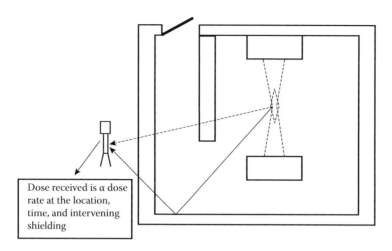

Dose received is α dose rate at the location, time, and intervening shielding

FIGURE 10.6
Parameters controlling individual's radiation exposure.

exposures while admitting visitors to spend time with a brachytherapy patient. In external beam therapy, where the source output is very high, it is not practical to reduce the dose rate at the position of the individual using the distance parameter alone. Shielding is the principal method to reduce the dose rates at the position of the individuals to the required levels. Shielding also plays a primary role in storage and transport of brachytherapy sources, and in use of HDR brachytherapy equipment or external radiation TDE in radiotherapy treatment. By manipulating the shielding thickness, the radiation dose rate outside the shielding barrier can be controlled. The TDE used in external beam therapy or remote afterloading HDR brachytherapy is controlled remotely from the control console situated outside the shielded treatment rooms.

Figure 10.6 illustrates a typical linac room used for radiotherapy treatments. The dose received by an individual standing just outside the shielded room is governed primarily by the room shielding, though distance and time play a part.

10.4 General Aspects of Treatment Delivery Room Planning

Radiotherapy facilities fall into three groups:

1. External beam therapy (telecobalt, linac, particle therapy, tomotherapy) facilities
2. LDR brachytherapy facility
3. HDR brachytherapy facility

The basics in the design of radiotherapy facilities are:

1. Protection of patient, staff, visitors, and the public
2. Geographical and functional integration of activities and facilities related to radiation oncology, to reduce unnecessary public exposures to as low as reasonably achievable (ALARA) levels.

Several factors must be taken into account in the planning of a radiotherapy facility. For instance, the related facilities of radiation oncology must be close to one another and must be located in the area of minimum occupancy in the hospital (i.e., in a corner away from the busy departments).

A convenient room size for installing a linac is about 7 m × 7 m × 4 m (height) with the machine isocenter placed approximately at the room center. This allows adequate space around the machine for attending a patient, for the machine installation, and for the maintenance. (The simulator room must also have these dimensions as it is similar in size to a linac). The maze width must be 2 m minimum for equipment movement and for easy movement of the patient on a trolley or wheelchair. The maze must be provided with the door on the side of the maze wall with a maze length ≥ 10 to 11 feet. Such a layout may not need much radiation (X or neutron) shielding for the door, for ≤15 MV linacs. The door may still be required for physical safety and for door interlock provisions. The door shielding requirement, however, needs to be verified for each installation, and if required the door may have to be provided with the necessary radiation shielding. The space occupied by the wall-mounted lasers must be considered when computing the shielding thickness for the walls. Provision for dimming the room light must be provided for patient alignment in the treatment position. The room must have wall cabinets to store treatment devices like the immobilization devices and QA equipment.

Adequate shielding must be provided for all the walls, door, the ceiling (and the floor if there is any occupancy below) so that the doses do not exceed the design limits. This important topic will be discussed in a separate section. For linacs operating beyond 10 MV, adequate door shielding for neutrons and capture gammas must be provided. Prior to the construction of the radiation treatment room, the shielding plan must be submitted to the competent authority for approval. The approving agency scrutinizes the shielding calculations and approves the plan if it complies with the applicable rules.

If more than one linac is to be installed or if there is any future plans of adding more linacs to the facility, it is highly desirable to plan the two rooms adjacent to each other so that one of the walls can be common for both rooms. This will reduce the cost of construction.

Proper air conditioning (to keep temperatures in the range of 20 °C to 25 °C and humidity in the range of 30% to 70%) and ventilation must be provided so that any toxic gas production (e.g., ozone) does not build up but is efficiently discharged by the ventilation system.

The treatment room must be provided with safety interlocks and warning lights to prevent any inadvertent entry of persons during treatment. In the case of a cobalt teletherapy room, a zone monitor must be provided at the entrance of the treatment room to monitor the radiation level continuously. In the case of a "source getting stuck" the zone monitor will give an alarm or indicate a radiation level above the background level pointing to an emergency situation. An appropriate patient viewing system and two-way audio communication system must be available at all times so the technician can observe the patient from the control room during treatment and can also communicate with them.

Apart from the emergency off switch on the console, there should also be one provided on the couch and on the wall in the treatment room (at a convenient height), and at maze entry, to shut off the machine in case of any emergency. However, no emergency shut off should switch off the room lights, safety interlocks, and the motorized treatment room door, if any. The switch must be activated if there is any danger to the patient (e.g., any medical emergency, likelihood of gantry collision with the patient, a fire in the room). The emergency switch will be activated even if there is any danger to the staff or member of the public (e.g., any member of staff or the public remain in the room when the treatment is initiated).

Radiation symbols must be posted prominently at the entrances of the treatment room and control room. A notice in English and the local language mentioning radiation hazard and restricted entry must also be posted below the radiation symbol display.

10.5 Shielding Concepts in Radiation Oncology

The NCRP (USA) has published several reports on this topic over the years [10–12] and the most recent publication, NCRP 151 [9] is the most widely followed report for the purposes of shielding (see Figure 10.7). The shielding of the linac room must be designed as per the recommendations of NCRP 151 taking into account the country's regulatory requirements. There are a few other reports and presentations that deal with this topic in detail [13–25]. The reader must consult these publications for a thorough understanding of the subject. This section deals with the basic shielding design concepts, which will give some grounding to do the shielding calculations.

The shielding design consists of three steps:

1. Setting design values for the effective dose in occupied areas
2. Computation of dose in the occupied area (in the absence of any room shielding)

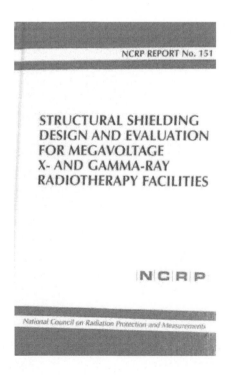

FIGURE 10.7
NCRP 151 shielding report for radiotherapy facilities.

3. Estimation of the dose reduction factor (RF) to reduce the unattenuated effective dose in (2) to the attenuated effective dose in (1)

We must shield against three types of radiation, given below in decreasing order of importance:

1. Primary radiation (actual treatment beam)
2. Leakage
3. Scatter radiation

In addition, for linacs with energies >10 MV, consideration must also be given to the shielding of photoneutrons produced by the linacs. The three components of the radiation are shown in Figure 10.8.

10.5.1 Primary Radiation and Primary Walls

The primary beam is the beam emerging from the collimator and incident on the patient for treatment. The transmitted part of this beam falls on the wall known as the primary wall. The maximum field size of the linac at 1 m from the target (40 cm × 40 cm, the maximum collimator opening) decides the beam size on the primary wall. To this width a margin is added, for a more conservative design, and this part of the wall will be much thicker than the remaining part of the wall since it receives the primary radiation, which is the most penetrating. The remaining part of the wall receives only the scatter and leakage radiation of lower energy and hence lesser penetration. Since the machine is isocentric mounted

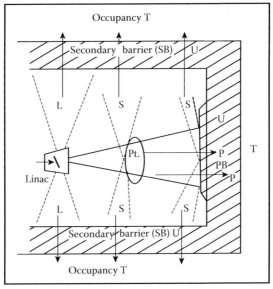

P: Primary; S: Scatter; L: Leakage; U: use factor; Pt: patient

FIGURE 10.8
Three kinds of radiation in a radiotherapy treatment room.

and the head rotates in a transverse plane, only the lateral walls, the floor, and the ceiling receive the primary radiation.

One way of reducing the primary barrier thickness is to incorporate a beam stopper in the machine, behind the patient so that the beam reaching the wall is significantly attenuated. (A beam stopper attenuates the radiation by a factor of 10^3, reducing primary walls to secondary walls.) The machine with a beam stopper was a popular design for the ^{60}Co teletherapy machines in earlier times, but with the introduction of megavoltage imaging systems, in place of beam stopper, this design was no longer possible and the required thickness had to be provided only in the room walls. If a lower energy TDE is replaced by a higher energy TDE in the room, only the walls and ceiling have to be modified to provide additional shielding.

10.5.2 Secondary Radiation and Secondary Walls

Scatter and leakage radiations are known as secondary radiations (see Figure 10.8). All the surfaces of an external beam radiotherapy room see the secondary radiations (apart from the primary barriers) and must be shielded against these. Two main sources of scatter of the primary beam are the patient and the primary barriers. The scatter intensity is a small fraction of the incident intensity, and the scatter radiation is scattered at least once before reaching the secondary barrier. Other scatterings are second- and third- order scatterings with very low intensities and can be disregarded in shielding calculations. Most of the scatter arises from the patient where the incident beam (treatment beam) intensity is the highest, at the patient skin.

Primary radiations are more penetrating than secondary radiations, so primary barriers automatically shield for any scatter or leakage incident on them; no separate shielding calculations are necessary. When the treatment beam is directed toward a primary barrier, all the remaining portions of the walls (beyond the primary beam area), ceiling, and the floor constitute secondary barriers and receive scatter and leakage radiations. These barriers must be thick enough to shield against only these radiation components. Since scatter and leakage radiations are of lower energy and intensity, secondary barriers will be less thick (about half) compared to primary barriers.

Leakage radiation is the radiation transmitted by the machine head. The leakage from the head contains not only the extra-focal transmitted X-rays but also the multiply scattered X-rays in the linac head shield, so leakage radiation has an effective energy much lower than the primary energy. Since the linac head can be at different angles with respect to the patient, on an average the head leakage can be assumed to be emanating from the isocenter position. The effective energy of the leakage radiation is much less than the primary radiation, due to the multiple scatterings occurring in the head. For >10 MV linacs, the neutron leakage also must be considered. There are strict regulations restricting the leakage radiation levels of the TDE.

In radiation oncology, the main mode of interaction in the photon energy range of interest is the Compton effect. The primary beam interacting with the primary barrier or the patient gives rise to scatter radiation in the treatment room. The scattered radiations are of much lower energy than the primary or the leakage radiations.

The secondary walls must shield both scatter and leakage radiations. Since the effective energies of scatter and leakage components are different, separate shielding calculations are necessary for scatter and leakage radiations, for the secondary walls. The predominant component will determine the secondary wall thickness. For ^{60}Co teletherapy installation, the scatter predominates over leakage and the scatter determines the shielding thickness.

Since the Compton scatter probability decreases with increasing energy, leakage radiation becomes more prominent compared to scatter for >10 MV beams.

10.5.3 Other Radiations of Interest

Photons of energies >8 MeV can produce neutrons in photonuclear interactions with the various components in the linac head. So for >10 MV linac beams, neutrons also require shielding consideration. Electron beams of >10 MeV energies can also interact with the nuclei, producing neutrons, but the beam current is three orders of magnitude less in electron therapy mode and so no separate consideration is required for neutron shielding in electron beam therapy. Since the photon beams are more penetrating compared to the neutron component (photons TVL > neutrons TVL) adequate photon shielding of primary walls will automatically ensure adequate neutron shielding. Concrete has a high hydrogen content and hence high neutron absorption cross section. As in the case of primary barriers, a secondary barrier for scatter and leakage will also provide adequate protection for neutrons and the capture gammas. If concrete has been replaced by other materials (e.g., laminated barriers), then additional protection may be required. It is always desirable to verify that the calculated shielding thicknesses are adequate for protection against photoneutrons and capture gamma rays.

The treatment room door is the problem area in neutron shielding. These aspects will be considered in a later section.

Any nuclear interaction can leave the nucleus in a radioactive state. So, in the beam-off condition, there is a small dose the staff may receive from the neutron activation of the components of the linac head, if they are close to the linac head soon after a patient exposure. NCRP 151 lists the radionuclides (produced by gamma activation) and the resulting dose from routine exposure to them. However, there is no long-term buildup of activity and the radionuclides have short half lives so the exposure goes to background levels in a day or two. There is also a small probability of inducing radioactivity in the patient the leakage radiation of the equipment should be reduced so it is as low as reasonably achievable.

10.5.4 Designation of Areas around a Radiation Room

The areas surrounding the treatment room may be occupied by radiation workers (designated as controlled areas) or non-radiation workers (referred to as uncontrolled areas). Non-radiation workers include patients, relatives, visitors, and employees not related to the radiation work. The characteristics of a controlled area are defined in Chapter 5. The uncontrolled areas are accessible to the public and no entry restrictions are imposed. As explained elsewhere, some countries (e.g., UK) divide the areas into uncontrolled (<1 mSv/y), supervised (1–6 mSv/y), and controlled (>6 mSv/y) areas and derive design values according to these dose limits. The designation of areas must follow the country's regulations.

10.5.5 Design Parameters

10.5.5.1 Use Factor U

Use factor U is defined for each primary barrier as the fraction of the beam-on time the machine (i.e., the primary beam) faces a particular barrier during treatment. For an isocentric machine rotating in a plane, the machine can face any of the four barriers assuming equal usage U = 0.25 for each of the four incident surfaces. This was the traditional assumption made for shielding calculations for a long time; total use factor must be 1. However, if any direction is preferentially used, U increases for that barrier and correspondingly U can be less for other barriers, keeping the total use factor = 1. Sometimes, larger use factors are adopted to be more conservative in the calculations, in which case the total use factor can be greater than 1. The following are typical U factors used by many in external beam TDE: 1 for the floor, and 0.25 for the walls and the ceiling. For the secondary barriers, U is always 1 since scatter is always seen by the secondary walls irrespective of the direction of the primary beam.

If the use factor for a barrier is <1, the permitted dose in the shielded area can be correspondingly increased to (P/U) where P is the design value dose in the shielded area.

The U factor will change for special techniques like IMRT and total body irradiations (TBI), depending on the frequency of orientation of the linac head toward any particular wall. Typical use factors for IMRT have been defined in some recommendations (e.g., NCRP 49 [10]: gantry pointing to floor, U = 1; gantry pointing to ceiling, U = 0.25; gantry pointing to wall, U = 0.25). More realistic U factors can be defined by the user department based on their treatment experience, through a statistical study of the linac head orientations during treatment.

10.5.5.2 Occupancy Factor T

T is defined as the fraction of the time (i.e., fraction of an 8 h day or a 2000 h year) any area around the treatment room is occupied by a maximally exposed individual. If there is little occupancy, there is less constraint on the dose in the area. For instance, the ceiling need not be provided adequate shielding to reduce the dose to the design dose level usually set for uncontrolled areas, thus reducing the cost of construction. However, when higher radiation levels are allowed due to very low or no occupancy, the access to those areas must be controlled so that people do not use those areas for any activities, thus increasing the T value. If the area is occupied full time, T = 1. Since the areas around the treatment room could be a staircase, a rest room, or a visitor area (which are not occupied by the same people all the time), realistic occupancy factors must be assumed. Typical values are given in NCRP 151, for different types of occupancy (e.g., work areas like offices, laboratories, treatment planning areas, control rooms, living quarters and play areas, T = 1; patient examination room, T = 0.5; corridors, lounges and staff rest rooms, T = 0.2; toilets, waiting rooms, and car parks, T = 0.05; parking lots and pedestrian traffic areas, T = 0.025) which can be assumed for shielding calculations. If usage in an area is expected to be very heavy, more conservative values can be assumed for occupancy. The location of the radiation oncology department must be properly planned during the design stage of the hospital, so that there is less occupancy around the department. An easy way to remember occupancy factor is to assign 1 for full occupancy, 1/4 for partial occupancy, and 1/16 for occasional occupancy.

As stated earlier, if the occupancy is low, the dose in that area can be correspondingly increased. So, taking U and T into account, that area can be planned using a dose of (P/UT). T factors are typically used for uncontrolled areas since for the controlled areas occupied by the radiation workers, T is always assumed to be 1.

10.5.5.3 Primary Workload

Workload is the total monitor units (MUs) delivered by the TDE in a week. Since the linac beams are usually calibrated to deliver about 1 cGy per MU (i.e., MU/cGy ≈ 1), workload is generally defined as the cumulative dose per week, at 1 m from the source or the target, at the isocenter of the treatment equipment. Each patient treatment session delivers typi-cally 2–2.5 Gy to the tumor center at the isocenter of the beam delivery equipment. Correcting for the patient attenuation (i.e., dividing by tissue maximum ratio (TMR) or tissue-phantom ratio (TPR) values for typical tumor depth) the average equilibrium dose in free space at the isocenter can be estimated. For shielding calculation purposes, patient attenuation factor of 0.4 or 0.5 can be assumed, giving a typical dose of 4 Gy/patient. Four × maximum number of patients treated per week then gives the maximum workload in Gy/wk. Assuming 50 patients are treated per day, which depends on how busy the hos-pital is, the workload comes to 1000 Gy/wk. NCRP 151 recommends 30 patient treatments per day for a busy department, but in developing countries the number can be higher. A realistic estimate of the workload is essential for shielding calculations and is the respon-sibility of the medical physicist of the department. The workload must always be assumed on the higher side so that any future increase in the workload will not require replanning of the installation. It will also avoid the risk of undershielding.

For dual energy machines, NCRP 151 suggests a workload of 500 Gy/wk for the higher energy with the reminder workload attributed to lower energy. Alternatively, all the work-load can be assumed for higher energy and the shielding calculations can be done. This gives a conservative estimate of shielding thicknesses. In case of space limitation, a more realistic workload apportioning of the low- and high-energy beams must be done or the high-energy workload may be restricted to reduce the shielding thickness. The total work-load may be apportioned based on the usage of the two beam qualities. If w_X gives the fraction of patients treated with modality X, N_T is the total number of patients treated per week and D_f, the fractionated dose (or dose per fraction or per patient) corrected for patient attenuation, the total workload is given by:

$$W_{pri} = \sum_X w_X N_T D_f \text{ Gy/wk at isocenter}$$

Since the design goal for shielding purposes is usually expressed in mSv/wk, it is conven-ient to express both the workload and the design goal dose in the same units. This is not strictly correct since the dose equivalent (DE) and Sv apply to radiation doses at protection level and not doses at clinical level. It is, however, convenient for shielding calculations to express both the numerator and denominator in the same units. As the Q factor is 1 for pho-tons, there is no change in the numerical value of the workload when expressed in Gy/wk or Sv/wk, but care needs to be taken when talking of neutron dose. For instance, if the neutron dose of the linac is expressed in Gy/wk (say 0.1% of the X-ray dose at the isocenter) then the appropriate neutron Q factor must be used to determine the neutron DE. The neutron dose at the machine isocenter per unit X-ray dose can be expressed in Sv/Gy or Sv/Sv.

The workload determined for patient treatment is the clinical workload. However, the machine remains ON for other activities as well like the beam calibration, patient and machine QA activities, and during machine servicing and maintenance activities. This workload too must be considered for shielding calculation purposes. So, the total acceler-ator workload is the sum of clinical workload and the nonclinical workload.

10.5.5.4 Leakage Workload

Since the machine the leakage factor is L (= 10^{-3}), the leakage workload is given by:

$$W_L = (L \, W_{pri})$$

Example 10.3

A treatment room houses a dual energy (6 MV and 15 MV) linac and treats 50 patients per day, with 60% of the patient load for treatment with the 6 MV beam. The dose delivered at the isocenter per patient (corrected for patient attenuation) is 4 Gy. Calculate total clinical workload per week.

$$6 \text{ MV patients} = 0.6 \times 50 = 30$$

$$6 \text{ MV workload} = 4 \times 30 \times 5 = 600 \text{ Gy/wk} = 6 \times 10^5 \text{ mSv/wk}$$

$$10 \text{ MV workload} = 4 \times 20 \times 5 = 400 \text{ Gy/wk} = 4 \times 10^5 \text{ mSv/wk}$$

$$\text{Total clinical workload per week} = 10^6 \text{ mSv/wk}$$

Example 10.4

In the above example, assuming 10% of the 6 MV clinical workload as the physics and servicing workload of the linac, calculate the total linac workload per week.

$$\text{Physics and servicing workload} = 0.6 \times 10^5 \text{ mSv/wk.}$$

$$\text{So, the total linac workload per week} = 1.06 \times 10^6 \text{ mSv/wk}$$

Example 10.5

In a radiation oncology department, the patient load is 70% for IMRT and 30% for conventional treatments. The IMRT factor is 5. Calculate the total workload to be considered for secondary radiation shielding against leakage. The typical workload W_{pri} for primary barrier is 10^6 mSv/wk.

$$W_{sec} = 0.3 \ W_{pri} + 5 \times 0.7 \times W_{pri} = 3.8 \ W_{pri}$$

The workload to be considered for leakage calculations is 3.8×10^6 mSv/wk.

10.6 Concept of HVL and TVL

Half-value layer (HVL) and TVL values are the important parameters for shielding calculations. The HVL and TVL are the thicknesses of the attenuating shielding materials that decrease the incident beam intensities by 50% and 10%, respectively. The shielding thicknesses are usually derived in terms of the TVL and HVL thicknesses. In radiation protection, broad beam geometry values should be used, which would take into account the scatter in addition to the primary beam at the calculation point. So, the HVLs and TVLs are different for broad beam and narrow beam geometries. Also the broad beam TVLs are different for the primary and the scattered radiations because scatter reduces the energy of the scattered radiation.

The data for typical *primary beam qualities* used in radiation oncology for the most commonly used shielding materials are given in Table 10.2 [16].

For heavy concrete (density 3.8 g/cm^3), the TVLs can be scaled in inverse proportion to density. (This is due to the dominance of Compton interactions in this energy range,

TABLE 10.2

TVL Values for the Common Shielding Materials and
for Common Beam Qualities

Linac BQ	TVL and TVL$_e$ Values in cm (NCRP 151)					
	Lead		Steel		Concrete	
6 MV	5.7	5.7	10	10	37	33
10 MV	5.7	5.7	11	11	41	37
15 MV	5.7	5.7	11	11	44	41
Density (g/cm^3)						
	11.3		7.9		2.35	

Note: NCRP, National Council on Radiation Protection and
Measurements; TVL, tenth-value layer.

which is proportional to density.) The heavy concrete is 1.6 times denser compared to ordinary concrete. So, the heavy density concrete TVL is approximately the above TVL value divided by 1.6. The actual TVL value comes to less than this value so the scaled values are more conservative and add a margin of safety in the planned installation [15]. Since the beam hardens in the first TVL thickness (TVL$_1$), the subsequent TVLs become larger. So, NCRP 151 recommends the use of TVL$_1$ and equilibrium TVL, TVL$_e$, for the subsequent TVLs. For instance, if the shielding thickness is n TVL, the thickness to be used is TVL$_1$ + (n−1) TVL$_e$. Both the TVL values are given in Table 10.2.

The leakage TVLs are required for leakage shielding calculations. Leakage TVLs for lead, steel, and concrete, for the common beam qualities used in radiation oncology, 6, 10, and 15 MV, are given in Table 10.3 [16].

The TVL for other materials like earth and borated polyethylene can be found in NCRP 151, and also in the presentation by Melissa C. Martin [16].

The scatter TVLs are required for scatter shielding calculations. The scatter energy and hence the TVL depends on the scattering angle. The scatter TVL for 90° scatter are given in Table 10.4 for 6, 10, and 15 MV. For other angles, data can be found in NCRP 151.

TABLE 10.3

Leakage TVL Values

Linac BQ	Leakage TVL(TVL$_e$) (mm)		
	Lead	Steel	Concrete
6 MV	57	96	340 (290)
10 MV	57	96	350 (310)
15 MV	57	96	360 (330)
Density (g/cm^3)			
	11.3	7.9	2.35

Note: Lead & Steel TVL & TVL$_e$ are the same; TVL, tenth-value
layer.

TABLE 10.4

90° Scatter TVLs in Concrete

Linac BQ	TVL (cm)
6 MV	17
10 MV	18
15 MV	18

Note: TVL, tenth-value layer.

10.7 Shielding Materials and Shielding Effectiveness

The material most commonly used for shielding in radiation oncology is ordinary concrete. Concrete is most easily available, relatively inexpensive, and has high structural strength. It is used extensively in the construction industry. When the treatment room is constructed in a basement, earth can be used as natural shielding material for any of the walls. Using heavy concrete, the shielding thickness of the room walls can be reduced. This concrete contains the iron oxide magnetite and has a density of 3.8 g·cm^3. Another method of reducing barrier thickness, when there is a space constraint, is to use high Z materials like steel and lead in laminated barriers. For smaller thicknesses, lead sheets are used and for larger thicknesses interlocking lead bricks are convenient. Lead is denser than concrete but it is also more expensive, thus increasing the installation cost. Lead also does not have structural stability and is not self-supporting, so it must be suitably sandwiched.

TVL and the barrier thickness is approximately inversely proportional to the material density. This is because the Compton effect, the predominant mode of attenuation for megavoltage X-ray energies, depends on the material density and not on Z. The concrete density can vary from 2.25 to 2.45 depending on the raw material locally available [13]. The density locally measured for the concrete must be used in the calculations by adjusting the tabulated TVL values that refer to a standard density of 2.35 g/cm^2. A safety factor may be added to the permitted dose (i.e., the transmitted intensity) to allow for some variations in density when mixing concrete.

For >10 MV linacs, the interacting materials in the linac head eject photoneutrons. So, neutron shielding must also be given due consideration. Concrete contains hydrogen, so it is also an effective shielding material for neutrons. In the case of laminated barriers, though steel and lead are effective space-saving X-ray shielding materials, they are more transparent to neutrons and so the high Z materials must be on the inside of the shielding barrier so the neutrons produced can be moderated and absorbed by the concrete.

In the case of laminated barriers, special attention must be given to neutron shielding. A neutron shielding barrier essentially consists of efficient neutron slowing down material plus a thermal neutron absorber, plus a high Z shielding material (e.g., lead) to attenuate capture gammas produced in thermal neutron absorption. Hydrogenous materials (e.g., polyethylene) are efficient neutron slowing down materials but thermal neutron capture in hydrogen would produce high-energy gammas (in the MeV range) which are much more penetrating compared to the scatter radiation in the room, so lead or concrete is required to attenuate the capture gammas. If there are two adjacent linac rooms, lead or steel can be centered in the common wall. The larger the thickness of lead used, the

TABLE 10.5

Properties of Common Shielding Materials

Material	Density g/cm^3	Z	Relative Cost	Tensile Strength
Concrete	2.3	11	1	500
Heavy concrete	3.7–4.8	26	5.8	–
Low C steel	7.87	26	2.2	40,000
Pb	11.35	82	22.2	1900
Dry packed earth	1.5	–	Cheap	–

greater will be the neutron fluence and neutron capture gammas in concrete. For a ceiling, a base structure of concrete is necessary to support lead or steel. This reduces the concrete thickness on top making it less effective for neutron moderation. Here, steel may be preferred to lead since it produces fewer photoneutrons compared to concrete. This will increase the overall thickness of the ceiling due to larger TVL of steel compared to lead. Use of high-density concrete is another solution in this case to decrease the barrier thickness.

The properties of the shielding materials are given in Table 10.5 [26].

10.8 Permitted Dose and Design Value Dose

Since the annual dose limits (maximum permitted dose) for radiation workers and the public are different, the design values are also different for the areas occupied by them. The design values are actually the dose limits for these two categories of people, often optimized by an ALARA factor, for a realistic calculation of shielding thickness to reduce the cost of construction. Taking the average annual dose limits of the radiation worker and the public as 20 mSv and 1 mSv, the design value of the permitted annual dose at the point of interest (POI), beyond the shielding barriers, can be assumed to be one-tenth of these values (i.e., 2 mSv/y and 0.1 mSv/y, respectively). Making a conservative assumption of 50 weeks in an year, the weekly design dose values are 0.04 mSv/wk and 0.002 mSv/wk for the restricted access and public access areas, respectively. Canada used this procedure for setting the design values but not all countries use the same fraction of the annual dose for design values for room shielding. In the United States, a design value of 5 mSv/y is assumed for radiation worker areas. This figure can be derived from the maximum allowed annual occupational exposure of 50 mSv by taking one-tenth of this value as the ALARA design dose limit. It can also be interpreted in another way. The cumulative permitted dose limit according to the US regulations is "Age × 10 mSv" which implies an average annual dose of 10 mSv. So, the design annual dose value can be taken as a fraction of 10 mSv, say 5 mSv. This results in a design value of 0.1 mSv/wk. In the UK, the areas surrounding the radiation room are classified as controlled, supervised, and uncontrolled areas. The control area is defined as an area where the dose received is likely to be equal or greater than 3/10 of the annual dose limit of 20 mSv (i.e., 6 mSv/y). According to this definition, no area in a radiation oncology department occupied by staff can be classified as a controlled area. Only inside the treatment room can the doses be this high. British regulations also define a supervised area category where the dose is likely to exceed the public area dose limit (of 1 mSv/y) but unlikely to cross 6 mSv/y, the lower limit for defining a controlled area. According to this definition, all the areas occupied by the radiation workers, including the control room housing the treatment console that are generally referred to as controlled areas, qualify as supervised areas. Using the lower limit (1 mSv/y) as the design value for the supervised area, the weekly design dose limit for this area comes to be 0.02 mSv/wk.

Regarding the uncontrolled areas, the UK requires that the public exposure from any source must not exceed 0.3 mSv/y which makes $P = 6$ μSv/wk (or 0.006 mSv/wk) in the uncontrolled areas. NCRP uses an annual dose limit of 1 mSv (0.02 mSv/wk) and Sweden uses a more restrictive dose limit of 0.1 mSv (0.002 mSv/wk) for the uncontrolled areas. The usual practice in setting a design value is to scale down the annual dose limit or a fraction of the annual dose limit to weekly dose values by dividing by 50. In addition to weekly dose values, some countries have also issued hourly dose rates to comply with. This will be further discussed in the following sections.

The POI for shielding calculations is defined close to the barrier (1 foot from the barrier) and 1.5 feet above the floor of the ceiling. The point to remember in setting the design dose limit is that the design limit must be a fraction of the annual dose limit. It is not a good practice to assume the annual dose limit as the design value.

Once the design value for shielding is fixed, the shielding thickness directly depends on the workload and the shielding material. A conservative (but realistic) estimate of the workload is essential for adequately shielding the radiation installation without unnecessary cost escalation.

10.9 Calculation Principles

The principles of calculation for the primary, scatter, and the leakage radiation and the estimation of dose rates in the shielded areas are explained below.

10.9.1 Reduction Factor

Reduction factor (RF) is defined as the dose at the calculation point without and with shielding, respectively. The inverse of this factor is what is more commonly used, and is known as the transmission factor (TF or B). RF is a more convenient parameter to derive for manual calculations. The doses (i.e., the workload and the design values) are expressed per week.

$$RF = \frac{\text{Dose without shielding at the calculation point}}{\text{Planning dose at the same point}}$$

RF >1 and TF <1. Figure 10.9 shows the calculation point for a primary wall and a secondary wall, for which we will derive equations for the three types of radiations.

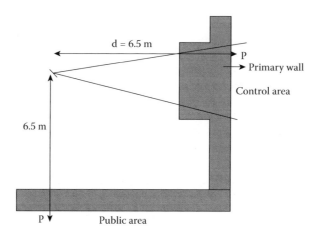

FIGURE 10.9
Shielding calculation for primary barrier.

10.9.2 Shielding for Primary Radiation

The most conservative estimate of the RF required for reducing the unshielded dose at the POI to the value P is:

$$RF_{pri} = \frac{(W_{pri}/d_{pri}^2)}{P}$$

d_{pri} is the distance between the focus point (not the isocenter) to the calculation point for the primary wall (30 cm from the wall on the outside) and for the primary beam.

$$d_{pri} = (\text{Isocenter to POI distance} + 1) \text{ m}$$

1 m is the focus to isocenter distance when measuring d_{pri} from the focus of the linac.

In order to avoid over shielding, the use factor of the barrier and the occupancy of the shielded location must be considered, the realistic RF is then given by:

$$RF_{pri} = \frac{(WT/d_{pri}^2)}{(P/UT)} = W_{pri}UT/P\, d_{pri^2}$$

So, the realistic unshielded weekly dose at the POI is not W, but W modified by U and T, and corrected for the inverse square law fall in dose. If the use factor or the occupancy factor is low, one has to shield only for a smaller value of the unshielded dose at the POI and hence a lower RF.

The above equation is the general equation used for the evaluation of the dose RF, for the primary barriers. If, for example, RF comes to 3×10^5, to reduce the unshielded weekly dose to the value of P, 5 TVL thickness shielding will be required for a reduction of 10^5. To further reduce the attenuation by a factor of 3, more than 1 HVL is required, and less than 2 HVL thickness. So, a conservative figure of 5 TVL + 1.5 HVL can be used for shielding purposes. The inverse of the RF is the transmission B or RF = (1/B or B^{-1}).

To get a more accurate value, the following simple method can be adopted. Each TVL reduces the radiation intensity by a factor of 10. If n TVLs are required for getting the required RF,

$$RF_{pri} = 10^n \text{ or } \log RF = n \quad \text{(in terms of B, n = } -\log B)$$

In the above example, no. of TVL required = $\log (3 \times 10^5) = 5$ TVL + log3 = 5 TVL + 0.47 TVL and 0.477 TVL is approximately 1.55 HVL, giving the same value. In terms of transmission factor, B = 0.333×10^{-5} and $-\log$ B = (0.477 + 5) TVL again giving the same value.

For example, the primary wall shielding thickness is 5 to 6 TVLs. The head leakage is of comparable beam quality to primary radiation but is less by a factor of 10^3. Similarly for scatter, the 90° scatter seen by the barrier is very low (scatter fraction $\approx 10^{-4}$). So, if the wall can shield against a workload W_{pri} it should be able to shield against the secondary workload. (i.e., $W_{pri} \times 10^{-3} + W_{pri} \times 10^{-4} << W_{pri}$ and can be ignored).

A typical thickness of primary barriers is about 5.5 to 6.5 TLVs or about 2 to 2.5 m of concrete.

10.9.3 Shielding for Leakage Radiation

The same procedure is used for the leakage calculations. The linac head shielding attenuates the leakage radiation by approximately a factor of 1000 or more. The maximum leakage (L) can be assumed to be 0.1% of the primary beam (output) at the isocenter. The workload is given by (W_L). The RF for the secondary barrier is then given by

$$RF_L = (W_L) UT/Pd_{sec}^2 = (LW_{pri}) T/P\, d_{sec}^2$$

where L = 10^{-3} and d_{sec} is the distance between the target and the POI for shielding. Leakage is assumed to be isotropic and is seen by all the walls; U = 1.

There is no field size for scatter or leakage since they are incident everywhere in the room. Because of the lower effective energy of leakage radiation, the leakage TVL is less than the TVL for the primary beam.

10.9.4 Shielding for Scatter Radiation

For scatter to occur, two components are important: (a) patient scatter and (b) primary scatter (from primary walls). The scatter fraction depends on the scattering angle. Scatter fraction and scatter energy decreases with the increasing scattering angle. Table 10.6 illustrates the variation of scatter fraction and scatter energy as a function of scatter angle. It can be seen that the scatter fraction is less than 2%, and is generally <0.1%. The scatter is significant for small angles of scatter that reach the primary wall just beyond the primary barrier.

Figure 10.10 shows how the reduction factor equation for patient scatter (ps) can be derived for the secondary walls.

$$RF_{ps} = \frac{(W/d_{sca^2})[a(\theta)(F/400)]T}{P\, d_{sec^2}} = \frac{W_{ps}a(\theta)(F/400)]T}{P\, d_{sec^2}}$$

d_{sca} gives the target to patient distance which is usually 1 m and hence ignored in the final equation. The scatter fraction is "a" (i.e., ratio of scattered radiation dose rate at 1 m from the patient to the primary beam dose rate on the patient). The scatter fraction a is a function of the angle of scatter and Figure 10.10 refers to 90° scatter. The tabulated scatter factors are for a field size of 20 cm × 20 cm. So, for a field area F, the scatter factor is "a (F/400),"

TABLE 10.6

Scatter Energy and Scatter Fraction as a Function of Scattering Angle

Beam Quality	Scatter Energy MeV (% Scatter Fraction)				
Angle	10°	20°	40°	70°	90°
6 MV	1.4 (1.004)	1.2 (0.67)	0.7 (0.22)	0.4 (0.07)	0.25 (0.04)
10 MV	2.0 (1.66)	1.3 (0.58)	0.7 (0.24)	0.4 (0.06)	0.25 (0.038)

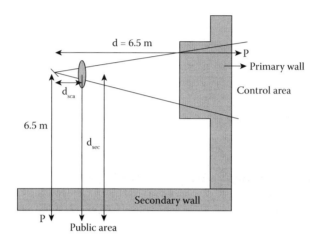

FIGURE 10.10
Concept of secondary wall shielding against patient scatter.

and the patient scattered dose at 1 m from the patient is given by W [a(90°) (F/400)] for 90 degree scatter. As a rough rule of thumb, the scattered dose, at 1 m from the patient, is approximately 0.1% to 0.2% of the weekly dose at the isocenter. By dividing the scattered dose by d_{sec^2}, the unshielded patient scattered dose rate at the POI is obtained. By multiplying by T, the planned dose rate at the POI for shielding purposes is found.

The Compton scattered photon energy is almost the primary energy for very small scattering angles and gradually decreases as the scattering angle increases. The TVL for the scatter radiation also decreases with increasing scattering angle.

10.9.5 Shielding for Secondary Radiation

The secondary barriers must shield for both the components of the secondary radiation—scatter and leakage. For linacs, head leakage takes precedence over scatter. After calculating the barrier thickness for leakage and scatter, as explained above, the necessary thickness to be used for the combined effect is estimated as per the following rule:

$$T_L - T_S < 1 \text{ TVL, T to be used} = \text{Larger thickness} + 1 \text{ HVL}$$

$$|T_L - T_S| \geq 1 \text{ TVL, T to be used} = \text{Larger thickness}$$

T_L and T_S refer to the calculated barrier thickness for leakage and scatter, respectively.

The same rule applies for determining the primary barrier thickness when dual energy (say 6 and 10 MV) linacs are used, if the barrier thickness for patient scatter and wall scatter differ, again the same formula is used to determine the barrier thickness for shielding against scatter.

The secondary TVLs are less than the corresponding primary TVLs due to the lower energies of the scatter and the leakage. Typical thickness for secondary barriers is around 3 to 3.5 TVL or about half the primary beam thickness.

10.9.6 Obliquity Factor

The Figure 10.10 refers to orthogonal scatter. For smaller angle scattering from the patient, the scatter radiation falls on the wall beyond the primary barrier, at an angle, as shown in Figure 10.11.

This increases the path length of the radiation compared to the actual thickness of the barrier. If θ is the slanting angle of incidence on the wall, t is the slanting thickness, and s is the actual thickness:

$$t \cos \theta = s$$

This relation is valid for all energies and for all angles up to 45° [22]. This reduces the actual barrier thickness required by the $\cos \theta$ factor (known as the obliquity factor), when the radiation is not incident normally on the barrier.

10.9.7 Instantaneous Dose Rate and Time-Averaged Dose Rate

The RF is computed for weekly workload and weekly doses permitted in the shielded region. For small values of W, U, or T, the RF will be smaller and hence the barrier thickness. This may increase the dose rate (in dose per hour) to a value which may be on the higher side. Hence many regulations have a dose in one hour (or IDR) constraint in addition to the weekly design dose value constraint.

The design goal P itself was not scaled down to hourly dose due to the following reason. Scaling the annual design dose to weekly dose (by dividing by 50) has some validity since the assumption that the weekly workload will be evenly distributed across the whole year is a reasonable one for shielding calculations. However, scaling down the annual or weekly (barrier transmitted) doses to hourly doses (say to compare with the installation survey results) will not be very accurate. A dose rate of say 20 μSv/h cannot be interpreted as

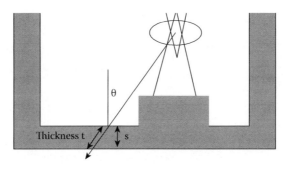

FIGURE10.11
Oblique incidence of radiation increases path length.

a dose of 20 μSv received in an hour since the beam is ON only for a fraction of 1 h. Also in the case of pulsed beams, the irradiation is not continuous but in pulses. So, in order to interpret the measured IDRs during survey, a time averaged dose (equivalent) rate (TADR) was introduced as a more reasonable dose rate to compare against survey results. These two quantities are defined below.

IDR is the measured value at 30 cm beyond the barrier, in the area being shielded. For accelerator measurements it is averaged over 30–60 s depending on the instrument response time, and the pulse cycle of the accelerator. The survey is usually carried out by directing the highest energy beam on to the primary wall for the maximum collimator field size to verify the shielding adequacy of these barriers. To maximize the radiation falling on the barrier, no phantom is placed at the isocenter for checking shielding adequacy of primary barriers and a scatter phantom is placed at the isocenter for measuring IDR for the secondary barriers. Thus all the primary and secondary barriers of the treatment facility must be checked for shielding adequacy.

Figure 10.12 shows the geometry for measuring IDR for the primary barriers.

The expected IDR for the plan is given by the transmitted dose rate at the POI, taking into account the use factor and a conservative occupancy factor of unity (implying maximal presence of someone at the POI).

$$IDR = DR_0 \ U \ B/(d+1)^2$$

where DR_0 is the machine output rate at the isocenter (1 m for a linac) and $[DR_0 \ U \ B/(d+1)^2]$ is the transmitted dose rate at the POI. B gives the barrier transmission factor which is the reverse of the RF (for a secondary barrier, U = 1).

NCRP 151 does not specify any constraint on IDR. Also, the US regulations do not specify any IDR constraint. The UK regulations, however, specify that the IDR must be less than 7.5 μSv/h for both controlled and uncontrolled areas. This implies that if the dose rate is higher than this value, the barrier thickness must be increased to bring the installation into compliance.

NCRP did not favor using IDRs based solely on the beam output rate, since a higher dose rate (e.g., flattening filter free [FFF] linacs) will lead to higher transmitted dose rate

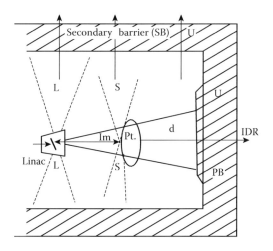

FIGURE 10.12
Instantaneous dose rate estimate for a primary barrier.

when not related to workload and the use factor of the barriers. So, a TADR was defined by NCRP for this purpose. The survey dose rate must be related to the TADR. $TADR_h$ (dose per hour) can be obtained by averaging over a day (8 h), week (40 h) or a year (2000 h) and can be represented as $TADR_d$, $TADR_{wk}$, and $TADR_{2000}$, respectively. $TADR_{2000}$ takes into account T factor (occupancy) as well.

$$TADR_h = IDR \times \text{fraction of time in an hour the machine is ON.}$$

$$= IDR \,\{(\text{machine on time per day})/8 \text{ h}\}\, \mu Sv/h$$

If DR_0 is the machine output dose rate at the isocenter, $W_d U/DR_0$ gives the machine-on time in a day. W_d is the daily workload.

$$TADR_h = IDR \,(\mu Sv/h)\; \times \frac{W_d U \; \mu Sv/d}{8 \; DR_0 \; \mu Sv/d}$$

For uncontrolled area, NCRP 151 specifies that $TADR_h$ must not exceed 20 µSv in 1 h. The US Nuclear Regulatory Commission also specifies that the dose in any unrestricted area shall be subjected to the same restriction. UK regulations specify $TADR_h$ (<0.5 µSv/h) and $TADR_{2000}$ (<0.15 µSv/h). Both the US and UK regulations do not impose any TADR constraint for controlled areas.

Similarly, TADR averaged over a week's time is given by

$$TADR_{wk} = IDR(\mu Sv/h)\; \times \frac{W_{wk} U \; \mu Sv/wk}{DR_0 \; \mu Sv/h}$$

$TADR_{wk} \times T$ should not exceed the design value P. This must be verified behind every primary barrier in the plan [25].

Since the daily workload can vary, $TADR_h$ can be derived from $TADR_{wk}$ using the following relation:

$$TADR_h = \frac{TADR_{wk}}{40} \times \frac{N_{max}}{N_{ave}}$$

where N_{max} and N_{ave} refer to the maximum and average number of patients treated in any 1 h. Many state regulations in the United States require that $(TADR)_h$ does not exceed 20 µSv in any 1 h.

The TADR constraints for shielded areas are:

1. $TADR_{wk} \times T$ should not exceed the design goal P
2. $TADR_h$ should not exceed 20 µSv in any 1 h

The constraints apply to all the shielded regions (i.e., all components of radiation—primary, scatter, and leakage). The room size chosen in the example may be slightly larger than a typical treatment room, but the calculation principles are identical.

A few examples are worked out below to illustrate the shielding calculation concepts.

Example 10.6

Compute the primary wall thickness in concrete, for a 6 MV linac beam using the data and the figure given below:

Focus to control point distance is 6.5 m; design value of P is 0.1 mSv/wk.

Workload is 10^6 mSv/wk; U = 0.25; T = 1. TVL = 35 cm (average of I TVL and TVL_e).

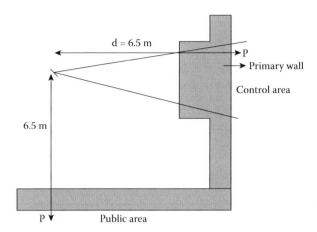

FIGURE 10.13
Barrier thickness computation for primary barrier.

$$RF = [(10^6 \times 0.25 \times 1)/(6.5)^2 \times 0.1] = 5.9 \times 10^4$$

So, 4 TVL (= 4 × 35 = 140 cm) would be required to reduce the radiation level by a factor of 10^4, and between 2 and 3 HVL (2.5 × 10.6 = 26.5 cm) for another reduction by a factor of 6. So the barrier thickness is roughly 160–170 cm (a more accurate figure is 0.77 TVL + 4 TVL = 0.77 × 35 + 4 × 35 = 167 cm) (Figure 10.13).

Example 10.7

In the example above, verify whether the shielding thickness complies with the IDR constraint, given in the UK regulations (IDR < 7.5 μSv/h). $DR_0 = 5$ Gy/min (300×10^6 μSv/h).

$$RF = \frac{DR_0 U}{P_{IDR} d^2} = \frac{300 \times 10^6 \times 0.25}{7.5 \times (6.5)^2} = 2.3 \times 10^5$$

To comply with the UK regulations, the required shielding thickness is 5.36 TVL. The barrier thickness in the above example, for the design goal of P, comes to only 4.77 TVL which will transmit more radiation. To comply with the IDR constraint, the barrier thickness must be increased by 0.59 TVL or by about 2 HVL. It must be noted from these two examples that the longer the period of averaging the workload (say per week instead of per day or per hour), the smaller is the required shielding thickness.

Example 10.8

In Example 10.6, assuming T = 1 behind the primary barrier, does the shielded area meet the criterion $TADR_{wk} \times T < P$. ($DR_0 = 5$ Gy/min)?

$$IDR = (DR_0\, U/d^2)\, B = (DR_0\, U/d^2)/RF$$

$$= \frac{300 \times 10^6 \times 0.25}{6.5^2 \times 5.9} \times 10^{-4} = 30 \text{ μSv/h}$$

$$TADR_{wk} = IDR \times \frac{W_{wk}U}{DR_0} = 30 \ \mu Sv/h \times \frac{10^3 \ Gy/wk \times 0.25}{300 \ Gy/h}$$

$$= 25 \ \mu Sv/h$$

$$TADR_{wk} \times T = 25 \ \mu Sv/h < P \ (100 \ \mu Sv/wk)$$

The shielded area meets the time averaged (weekly) dose rate criterion.

Example 10.9

In the above problem, compute the thickness of a secondary wall for head leakage using the data given below:

Focus to control point distance = 6.5 m; design value of P = 0.5 mSv/y.

The head leakage L is 0.1% (of the output); U = 1; T = 1.

The TVL for the leakage radiation is 28 cm in concrete.

Leakage dose = 0.001 × Workload = 0.001 × 10^6 mSv/wk

P = (0.5/50) = 0.01 mSv/wk (for protection purposes 1 y is taken to be 50 wk for ease of calculation and for a conservative estimate).

The RF is given by:

$$RF = [(WUT \ L)/d^2]/P = [10^6 \times 1 \times 1 \times 0.001/(6.5)^2]/0.01 = 2.4 \times 10^3$$

So the required secondary wall thickness for shielding against leakage is about 3 TVL + 1 HVL, or about 3 × 28 + 1 × 8.5 = 92.5 cm (taking log, 3.38 TVL or 94.6 cm).

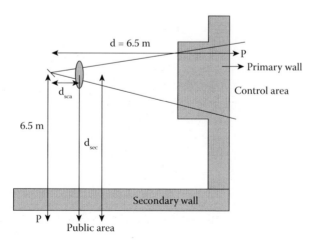

FIGURE 10.14
Shielding thickness for 90° patient scatter.

Example 10.10

Figure 10.13 is reproduced in Figure 10.14 for computing the shielding thickness for the secondary wall against 90° patient scatter. Calculate the secondary barrier thickness for the patient scatter using the following data for a 6 MV linac beam. What thickness must be provided for the combined effect of leakage plus scatter?

Patient to control point distance = 6.5 m; design value of P = 0.5 mSv/y. U = 1; T = 1; $a(90°)$ = 0.0006. (Assume a realistic field size of 20 cm × 20 cm). Workload is 10^6 mSv/wk.

TVL for 90 degree scatter = 28 cm in concrete

The field size factor is 20 × 20/400 = 1

$$RF_{ps} = \frac{W\, a(\theta)}{P\, d_{sec^2}} = \frac{10^6 \times 6 \times 10^{-4}}{0.5 \times 6.5^2} = 2.8 \times 10^1$$

Required wall thickness in concrete = 1 TVL + 1.5 HVL = 28 + 1.5 × 8.5 = 41 cm.

According to the rule of thumb $T_L - T_S$ = 92.5 − 41 = 51.5 cm which is >1 TVL. So, T to be used = larger thickness of 92.5 cm which will also take care of scatter.

It may be noted from the above examples that the leakage radiation dominates scatter for higher energy photon beams.

Example 10.11

Verify if the secondary wall (in Figure 10.14) complies with the IDR constraint criterion or not. (i.e., dose rate in the shielded area is <7.5 μSv/h). The same constraint applies to both controlled and uncontrolled areas).

$$RF = \frac{DR_0 U}{P_{IDR} d^2} = \frac{300 \times 10^6 \times 1}{7.5 \times (6.5)^2} = 9.5 \times 10^5$$

About 6 TVL or 6 × 28 = 168 cm. So, to comply with the IDR constraint the shielding thickness must be increased by 1.6 TVL (6 − 3.4 TVLs).

Example 10.12

For the same secondary barrier (Example 10.7), verify if the barrier meets the NCRP criterion of $TADR_{wk} \times T$ not exceeding the design goal P, and $TADR_h$ not exceeding 20 μSv in any 1 h. Assume the maximum and average numbers of patients seen in a day as 16 and 8, respectively. The dose delivered at the isocenter for a patient is 3 Gy. DR_0 = 5 Gy/min (300 Gy/h). Transmission factor B = 4.2 × 10^{-4} for leakage radiation and the attenuated dose rate at the point of interest = $DR_0\, U\, B/d^2$;

$$DR_0 \text{ is 5 Gy/min} = 300 \times 10^6 \text{ μSv/h}$$

$$IDR = [(300 \times 10^6 \times 0.25 \times 4.2 \times 10^{-4})/6.5^2] = 745.5 \text{ μSv/h}$$

$$TADR_{wk} = (TADR_{wk})_L + (TADR_{wk})_{sca}$$

TADR must be computed for both the components and summated.

$$(TADR_{wk})_L = IDR \times \frac{W_{wk} U}{DR_0} = 745.5 \text{ μSv/h} \times \frac{10^3 \text{ Gy/wk} \times 0.25}{300 \text{ Gy/h}}$$

$$= 621.3 \text{ μSv/wk (constraint: 100 μSv/wk)}$$

$$\text{TADR}_{wk} \times \text{T} (=1) = 621 \text{ mSv/wk}$$

$$\text{TADR}_h = \frac{\text{TADR}_{wk}}{40} \times \frac{N_{max}}{N_{ave}}$$

$$= \frac{621 \ \mu\text{Sv/wk}}{40} \times 2 = 5 \times 10^4 = 31 \ \mu\text{Sv/h} \text{ (constraint 20 } \mu\text{Sv in 1h)}$$

The barrier does not satisfy the criteria for IDR or TADR. In order to meet these constraints, the shielding thickness must be more than 3.4 TVL. The transmitted IDR criterion is much more restrictive compared to the TADR criteria.

10.10 Workload Enhancement in Advanced Treatment Techniques

If MU/cGy is greater than 1 for any of the treatment techniques or nonclinical activities, then this dose enhancement factor must be considered for computing the leakage workload for these activities. Techniques like TBI and IMRT involve delivering larger numbers of MUs for the same prescription dose, as in conventional treatment. In TBI, this occurs due to the increase in treatment distance. (1 MU $\approx$ 1 cGy calibration applies to the isocenter distance only. To deliver 1 cGy at larger distances, more MUs must be delivered, because of the lower dose rate at the extended distance.) In IMRT, the workload enhancement occurs due to the delivery of the same prescription dose (as in conventional treatment) through a large number of subfields (MUs) delivered over longer duration, instead of a single field as in conventional treatment. This increases the leakage workload many times. Thus, the assumption that the leakage workload is 10^{-3} times the primary workload is no longer valid, but needs to be multiplied by a leakage enhancement factor.

The percentage of IMRT treatments does not affect the primary or scatter workload since the same amount of dose is delivered at the isocenter as with conventional radiotherapy. To take this into account in the leakage workload for secondary barriers, a new factor called IMRT factor is defined as

$$\text{IMRT factor} = \frac{(\text{MU/PD})_{\text{IMRT}}}{(\text{MU/PD})_{\text{3DCRT}}}$$

where MU refers to the MU delivered, and PD the prescribed dose. The IMRT factor varies from 2 to 10 and an average IMRT factor, represented as C, of 5 can be assumed for IMRT treatments.

$$W_{L,\text{IMRT}} = W_{pri} \times C_{\text{IMRT}}$$

So the effective leakage workload is given by L $W_{pri} \times C_{\text{IMRT}}$. Increased leakage requires larger shielding and so the percentage of IMRT treatments will impact on the room shielding.

The IMRT factor applies only to the IMRT treatments carried out and not for 3D conformal radiation therapy (CRT) treatments. So the leakage workload is more complicated than the primary workload, as shown in the following equation.

$$W_L = 10^{-3} \sum_X w_X \ N_{wk} \ D_f (1 - w_{\text{IMRT}}) + 10^{-3} \sum_X w_X \ N_{wk} \ D_f \ w_{\text{IMRT}} \times C_{\text{IMRT}}$$

where w_{IMRT} refers to the fraction of treatments that are IMRT and $(1 - w_{IMRT})$, the remaining fraction involving conventional treatments; X refers to each modality in a dual modality machine N_{wk} gives the total number of patients treated per week.

For TBI, a prescription dose of say D_{TBI}, at the treatment distance d, corresponds to a dose of $D_{TBI} \times d^2$ at the isocenter. So, the TBI workload is given by $n \times D_{TBI} \times d^2$ where n is the number of patients treated per week.

The workload enhancement applies only to leakage workload and not for primary or scatter radiation workload, since the dose per week remains the same (i.e., $W_{IMRT} = W_{conv}$ for primary and scatter calculations). Similar workload enhancement factors apply for IMRT in other techniques such as cyberknife and SRS treatments. For more details on these aspects, the reader should consult the articles by Rodgers [9,24].

The workload for patient scatter gives the total weekly scattered dose at 1 m from the scatterer. The scatter fraction, $a(\theta)$, depends on the scattering angle and the amount of scatter is proportional to the incident field size. The tabulated $a(\theta)$ values refer to an incident field size of 20 cm × 20 cm, or 400 cm^2. The scatter fraction for a field area F is then given by $a(\theta) \times (F/400)$. For a maximum machine field size of F cm^2, the maximum weekly scatter workload is given by

$$W_{ps} = \sum_X w_X N_{wk} [D_f a(\theta) F/400]$$

where the notations have the same meanings as described earlier (d_{sca} = scaterer distance = 1m).

Example 10.13

A tomotherapy treatment involves delivery of a prescription dose 10.5 Gy at 3.5 m from the target of the linac. Two patients are treated per week. The conventional workload is 300 Gy/wk. What is the total workload that must be considered for the primary barrier used for TBI and conventional therapy? What workload will be considered for patient scatter to the secondary barriers?
TBI workload at 1 m distance is given by

$$W_{TBI} = 2 \times 10.5 \times (3.5)^2 \text{ Gy/wk} = 257.25 \text{ Gy/wk}$$

$$\text{Total primary workload, } W_{pri} = 300 + 257.25 = 557.25 \text{ Gy/wk}$$

The leakage workload for the secondary barriers also proportionately increases. However, patient scatter is only concerned with 300 Gy/wk and so scatter from patient to secondary barriers does not change.

10.11 Shielding Concepts for Tomotherapy

A tomotherapy system, developed by TomoTherapy Inc., USA, is shown in Figure 10.15. The tomotherapy equipment works on the principles of helical computerized tomography (CT) but the imaging X-ray unit has been replaced by a 6 MV linac, for the purposes of treatment delivery. Using a multileaf collimator (MLC), the equipment delivers the dose using the IMRT technique. So, this increases the treatment time compared to conventional treatment, for the same dose delivered to the tumor. So the enhanced leakage applies to

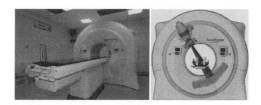

FIGURE 10.15
A tomotherapy treatment delivery system.

tomotherapy as well. A typical workload of 1000 Gy (or 10^5 MU) was assumed in conventional therapy.

For the TomoTherapy Hi-Art II® treatment system, a weekly workload of an order of magnitude higher than the conventional therapy workload (10^6 MU) can be assumed for a patient load of 30 patients per day and a treatment time of 5 min per patient, assuming a dose rate of 10 Gy/min at the isocenter.

The direction factor U, which is the maximum relative frequency to which the beam is pointed toward the place to protect, is set to 1. Indeed, the contributions of the different radiation components are evaluated when the gantry is continuously rotating, which is always the case during patient treatments. The occupancy factor T is set to 0 for unoccupied areas, 0.2 for partially occupied areas, and 1 for occupied areas. The design value doses are the same as in conventional therapy.

The standard shielding calculation methods discussed above do not apply for tomotherapy because of the special configuration of tomotherapy equipment. The tomotherapy system produces a slit beam of maximum field size 5 cm × 40 cm. It has a source to axis distance (SAD) of 85 cm and the field is calibrated to give 1 cGy/MU at the isocenter, at d_{max} = 15 mm. The machine dose rate is about 860 MU/min. The design of the tomotherapy system, from the shielding point of view, is as shown in Figure 10.16.

Assuming a patient load of 35 patients per day and a tumor dose of 3 Gy per fraction at the ioscenter, the typical primary workload comes to 35 × 3 × 5 or about 500 Gy/wk

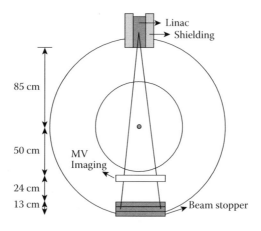

FIGURE 10.16
Inbuilt shielding in tomotherapy equipment.

or 5×10^4 MU/wk. Using the IMRT factor of 16, a conservative estimate of the leakage workload comes to 10^6 MU/wk.

In the tomotherapy system, the linac is shielded from the sides to reduce the leakage and a beam stopper (13 cm thick lead) attenuates the patient-transmitted primary radiation (by 4×10^{-3}) with virtually no patient-transmitted primary beam. Hence primary shielding is not the main issue in shielding tomotherapy rooms. The scatter is also confined to the slice region with some lateral spread. Therefore, head leakage is the most important component for shielding purposes for tomotherapy facilities. Since tomotherapy uses a 6 MV linac, there are no neutron shielding issues involved.

Figure 10.17 shows the method of mapping leakage radiation levels in the room [17].

Leakage measurements have been performed as a function of angle and as a function of distance from the isocenter (for each angle) and reported in the literature [17], for the treatment geometry (rotating gantry). The leakage measurements are performed keeping the MLCs completely closed. The measured leakage is specified as a percentage of the maximum open field dose at the isocenter (SAD) or referred to 1 m distance from the source (focal point). Maximum leakage is found in the gantry plane (90° and 270° directions) and minimum along the axial directions (0° and 180° directions).

Scatter measurements were performed for the maximum field size, with a large scatter phantom at the isocenter and as a function of angle and distance. The measurements yield scatter plus leakage and by deducting the leakage, the scatter contribution alone can be determined. The measurements show that scatter and primary radiation are negligible compared to leakage radiation [27]. The measured·radiation levels can be scaled to larger distances (>3 m) using the inverse square law.

The RF for shielding calculations are given by

$$RF_{pri} = \frac{(W_{pri}\, k_{att,BS})UT}{Pd^2}$$

where $k_{att,BS}$ refers to beam stopper attenuation factor.

$$RF_L = \frac{(W_L)\, T\, f_L}{Pd^2}$$

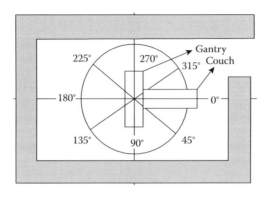

FIGURE 10.17
Leakage measurements as a function of room angle.

f_L refers to leakage fraction along the calculation direction, at 1 m distance from the isocenter. Head attenuation factor is not included for leakage here since the measured leakage fraction is used.

$$RF_{sca} = \frac{(W_{pri}\, f_{sca})\, T}{Pd^2}$$

f_{sca} refers to scatter fraction along the calculation direction, at 1 m distance from the isocenter.

Example 10.14

In the Figure 10.17, calculate the thickness of the secondary barrier (along 90° direction) required to shield the area for a design dose value of P = 0.1 mSv/wk (or 0.1 mGy/wk). The distance of the region from the isocenter is 4 m. The following data are given:

$W_{pri} = 5 \times 10^5$ mGy/wk; $W_L = 10^7$ mGy/wk U = 0.1; T = 0.5; $k_{att,BS} = 10^{-3}$;

$f_L(90°) = 1.36 \times 10^{-4}$; $f_{sca}(90°) = 8 \times 10^{-6}$; Assume TVL(concrete) = 34 cm for primary and 29 cm for leakage and scatter.

$$RF_{pri} = \frac{(W_{pri}\, k_{att,BS})UT}{Pd^2} = \frac{5 \times 10^5 \times 10^{-3} \times 0.1 \times 0.5}{0.1 \times 4^2} = 1.19 \text{ TVL}$$

where $k_{att,BS}$ refers to beam stopper attenuation factor.

$$RF_L = \frac{W_L\, T\, f_L}{Pd^2} = \frac{10^7 \times 0.5 \times 1.36 \times 10^{-4}}{0.1 \times 4^2} = 2.62 \text{ TVL}$$

f_L refers to leakage fraction along the calculation direction, at 1 m distance from the isocenter.

$$RF_{sca} = \frac{(W_{pri}\, f_{sca})\, T}{Pd^2} = \frac{5 \times 10^5 \times 8 \times 10^{-6} \times 0.5}{0.1 \times 4^2} = 1 \text{ TVL}$$

Along the 90° direction, leakage decides the leakage thickness giving a barrier thickness of 76 cm.

Example 10.15

For the same figure, calculate the thickness of the secondary barrier (along 180° direction) required to shield the area behind the barrier. Assume the same U, T, P, d, and TVL. Ignore primary contribution.

$$f_L(180°) = 7.9 \times 10^{-6}; f_{sca}(90°) = 1.76 \times 10^{-5}.$$

$$RF_L = \frac{W_L T f_L}{Pd^2} = \frac{10^7 \times 0.5 \times 7.9 \times 10^{-6}}{0.1 \times 4^2} = 1.4 \text{ TVL}$$

f_L refers to leakage fraction along the calculation direction, at 1 m distance from the isocenter.

$$RF_{sca} = \frac{(W_{pri}\, f_{sca})\, T}{Pd^2} = \frac{5 \times 10^5 \times 1.76 \times 10^{-5} \times 0.5}{0.1 \times 4^2} = 0.4 \text{ TVL}$$

In this case as well, only leakage decides the shielding thickness which is 1.4 TVL or about 41 cm thick concrete.

10.12 Shielding Concepts for Stereotactic Radiotherapy with Cyber Knife

A cyber knife (CK) is a 6 MV linac mounted on a robotic arm (see Figure 10.18) with more degrees of movement compared to other radiotherapy treatment equipment, offering unique advantages in stereotactic body radiotherapy treatments.

All walls and the floor are primary since it is equally likely to point in any direction with 80 to 100 beams being used in each treatment. The ceiling is a secondary barrier since the linac cannot point in the upward direction beyond a certain degree. Since the stereotactic fields are much smaller, the scatter is insignificant compared to primary and leakage radiations may be ignored in scatter calculations, but this needs to be verified for the particular layout. The leakage and primary become equally important in CK shielding calculations. The CK shows low leakage like the conventional linac system (<0.1% of the open beam isocenter output at 1 m distance).

The typical focus to target distance is 80 cm for cranial targets and can vary from 90–100 cm. The CK is calibrated to 1 cGy/MU at 80 cm distance, d_{max} depth for the maximum cone size (= 6 cm cone). The typical CK dose rate at 80 cm distance is 600 cGy/min with no field flattening filter. The treatment is based on few fractions (1–6) delivered in a single day, the maximum dose fraction being 12.5 Gy. Therefore, only one patient can be treated in a single day. The primary workload is, therefore, given by $5 \times 6 \times 12.5 = 3.75 \times 10^5$ mGy/wk (at 80 cm).

Example 10.16

The primary workload for CK is 3.75×10^5 mGy/wk, at 80 cm distance. What is the CK workload at 1 m distance?

$$W_{pri}(1 \text{ m}) = W_{pri} (0.8 \text{ m}) \times (0.8/1)^2 = 3.75 \times 10^5 \times 0.8^2 = 2.4 \times 10^5 \text{ mGy/wk.}$$

An IMRT factor of 15 is used to obtain the leakage workload (excluding the head attenuation factor). The recommended U factor for CK treatment is 0.05 (NCRP 151).

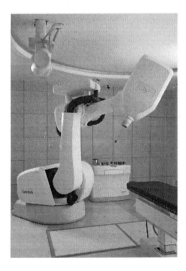

FIGURE 10.18
Cyber knife treatment delivery equipment.

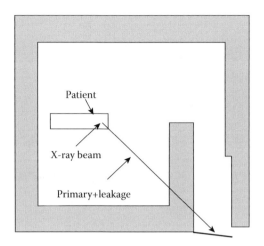

FIGURE 10.19
A CK layout showing primary and leakage transmission.

The same RF equations as used in conventional IMRT linac treatments are used here, not forgetting to include the head attenuation factor 10^{-3} in the calculation of RF for leakage. Both the primary and leakage transmission through the maze wall become important to assess the dose near the door, as shown in Figure 10.19.

For more details on shielding a CK facility, see the article by Rodgers [28].

10.13 Shielding Concepts for Gamma Knife

The GK system was developed for the single fraction treatment of small intracranial lesions using SRS procedures. The radiation is focused on the target using 192 miniature ^{60}Co sources arranged in rings on a hemispherical helmet, the center of the hemisphere being the treatment isocenter. By directing the beams through different angles, a large tumor dose can be delivered to a localized target with good sparing of normal tissues. Figure 10.20 shows a GK unit and a typical GK installation.

The total nominal activity in the unit is 6500 Ci (240 TBq). The head capacity is 6600 Ci. Because the patient's head is inserted into the unit there is no primary radiation coming out of the unit; the dose rates around the patient consist of leakage and scatter. When the machine is OFF (state 1), the dose mapping gives the leakage levels around the machine in the room (due to the continuous emission of the source). Except near the shielding door, the leakage levels at other locations are generally less than 10 μSv/h. There is negligible dose behind the machine due to the head shielding, and so not much protection is required for the maze wall. During the opening of the shielded door (state 2) or during treatment (state 3) the dose levels around the unit are different and indicate the leakage and scatter radiation intensities in these two states. The dose levels are maximum when the shielding door opens for placing the patient's head inside the unit. During patient treatment, the levels are intermediate between the above two states. The radiation level goes down as a function of time

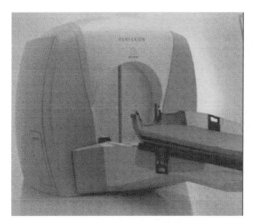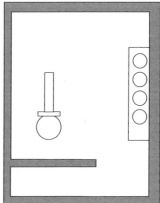

FIGURE 10.20
Gamma knife unit and an installation room.

due to the decay of the source with a half-life of 5.26 years; shielding must be provided for the maximum source activity. The dose rates around the unit can also be computed through empirical methods [29].

By mapping the dose levels, the dose pattern can be given in the form of isodose curves (curves of constant dose) or gradient lines which show the dose gradient along the lines. The data can be represented in the form of a grid, with each grid point giving the air kerma rate (AKR) value at that point (see Figure 10.21). This way of representation will help in finding the dose rate at any point by simple linear interpolation, and estimating the RF for the barriers. Alternatively, by normalizing the data to a standard distance (say 1 m from the isocenter), doses at other distances can be determined by applying inverse square law, which is of sufficient accuracy for shielding purposes. Because there is no primary radiation but only leakage and scatter, a typical barrier thickness for GK installations is less than 50 cm.

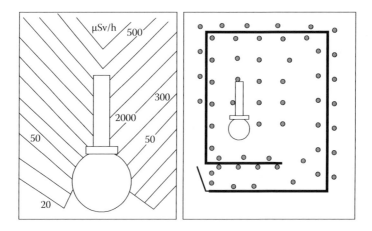

FIGURE 10.21
Dose mapping around a GK unit.

10.14 Workload Estimation for Advanced Techniques

Assuming a maximum daily patient load of 4, 20 patients will be treated in a week. The maximum treatment duration can be assumed to be 30 min (for the maximum activity of the sources). Typically, the patient position is adjusted four times during a treatment which implies the shielding door opens and closes four times, 1 min being the maximum duration of the opening or closing of the shielded door. This amounts to 8 minutes per patient (2.7 h/wk) when the machine is in state 2. The treatment duration is 10 h (20 × 0.5 h). The leakage duration is 40 working hours in a week. If dK_1/dt, dK_2/dt, dK_3/dt represent the AKRs at 1 m distance from the isocenter, during states 1, 2, and 3, the weekly dose at the point, K_{wk}, or the workload is given by [29]

$$W = 40\ dK_1/dt + 2.7\ dK_2/dt + 10\ dK_3/dt$$

Knowing the dose rates near the walls, the weekly doses near the barriers can be estimated. The design value on the other side of the barrier then gives the RF. The TVL for ^{60}Co is 26 cm of concrete.

Example 10.17

The dose rates were measured near one of the barriers of a GK room and the following values were obtained: Leakage = 0.25 μSv/h; with shielded door open, 129.6 μSv/h and during treatment 59 μSv/h. The shield open–close time and the treatment machine ON time are 2.7 hours and 10 hours, respectively. The other side of the barrier is a public area with a design value of 0.02 mSv/wk and occupancy of 1/5. Compute the barrier thickness required (assume ^{60}Co TVL as 26 cm concrete).

The weekly dose near the barrier = 0.25 × 40 + 129.6 × 2.7 + 59 × 10 = 950 μSv.

RF = WUT/P = 950 × 0.2/20 = 9.5 or 0.98 TVL or 26 cm of concrete (U = 1)

10.15 Shielding of a Simulator Room

A simulator simulates a linac in geometry to image the patient in the treatment geometry but uses diagnostic X-rays, and is capable of imaging in radiographic and fluoroscopic modes. A typical simulator facility is shown in Figure 10.22.

The shielding calculations are similar to linac shielding calculations. The following are the minor differences:

- Radiography and fluoroscopy involve an integral dose to the patient—dose rate (α current, mA) × time (min)—so the workload is expressed in the units of mA.min/wk at 1 m.
- Knowing the X-ray output in exposure (R) or air kerma (Gy), say at 1 m, for a set mA.min, the workload can be converted to Gy/wk as in linac shielding calculations.

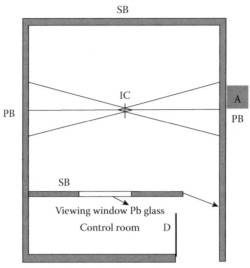

PB: primary barrier SB: secondary barrier
D: door (shielded) IC: isocenter

FIGURE 10.22
A typical simulator room.

- For isocentric units, the usage is about the same for all barriers so U = 1.
- The low-energy X-ray beam (simulator operates at 125 kVp) is attenuated by a factor of 10^3 by the patient so the primary radiation scattered by the wall is generally negligible and may be ignored. Only the patient scatter component needs to be considered for shielding purposes.

Assuming a maximum patient load of 50 patients per week and a maximum of 500 mAs per radiography:

$$\text{Total AK per week } \alpha \ 50 \times 500 \times (1/60) \text{ mA.min/wk} \approx 500 \text{ mA.min/wk}$$

Total dose delivered in fluoroscopy is proportional to the dose rate (mA) and the duration of fluoroscopy (min). Assuming the same patient load, maximum current of 5 mA and 1 min of fluoro time:

$$\text{Total AK per week } \alpha \ 50 \times 5 \times 1 \text{ mA.min/wk} = 250 \text{ mA.min/wk}$$

So a conservative total workload of 1000 mA.min/wk may be assumed for patient simulation. If X_0 is the exposure output of the unit in R/mA.min or K_0 is the AK output of the unit in mGy/mA.min:

$$\text{Weekly workload} = 1000 \ X_0 \text{ R/wk or } 1000 \ K_0 \text{ mGy/wk}$$

The simulator head attenuation factor is about 600 or L = 1/600. LW gives the leakage workload.

10.16 Reduction Factor for the Primary and Secondary Barriers

$$RF_{pri} = WUT/Pd^2 \ mA.min/Gy$$

$$RF_{pri} = LWUT/Pd^2 \ mA.min/Gy$$

For scatter radiation, $U = 1$; a gives the scatter fraction—scatter dose at 1 m from scatterer to incident dose—evaluated for a 20 cm × 20 cm field as a function of angle, at 1 m distance from the scatterer. The workload at the position of the scatterer is given by W/d_{sca}^2. If the scatterer is at 1 m distance, $d_{sca} = 1$. The scatter dose at 1 m from the scatterer, for a field size F, is given by $a \ (F/400) \ WT/(d_{sca}^2)$.

$$RF_{sca} = a \ (F/400) \ WT/ \ (Pd^2 \ d_{sca}^2) \ mA.min/Gy$$

The TVL for lead and concrete for 125 kV, heavily filtered beam under broad beam conditions, is given by 0.93 mm and 6.6 cm, respectively. The transmission factor is the inverse of RD. Transmission curves are available for the 125 kVp heavily filtered X-ray beam for concrete and lead, which give the required thickness from the computed transmission factor.

Example 10.18

Calculate the primary barrier thickness in concrete to shield the region A shown in Figure 10.22 above. The region is an office room occupied by the radiation oncology staff. Use the following data for the calculation.

The isocenter to the point of interest distance = 4 m;

P = 0.4 mGy/wk; workload W = 2000 mGy/wk at 1 m

Use factor = 0.25; Occupancy factor = 1; TVL = 6.6 cm concrete

$$RF = WUT \ / \ Pd^2 = 2000 \times 0.25 \times 1/ \ \{0.4 \times (3+1)^2\}$$

$$= 1.8 \ TVL \approx 12 \ cm \ concrete$$

Example 10.19

Calculate the secondary wall thickness for leakage shielding in the above example, for shielding the region B. The region B is a waiting room. Use the following data for the calculation.

The isocenter to the point of interest distance = 3 m;

P = 0.02 mGy/wk; workload W = 2000 mGy/wk at 1 m

Occupancy factor = 1/16; Leakage attenuation factor for the head = 1000. TVL = 6.6 cm of concrete

$$RF = LWT/Pd^2 = 10^{-3} \times 2000 \times (1/16)/\{0.02 \times 3^2\} = 0.16 \ TVL = 1.1 \ cm \ concrete$$

Example 10.20

Calculate the secondary wall thickness for scatter radiation shielding, in the above example, for shielding the region B. Use the following data for the calculation.

The isocenter to the point of interest distance = 3 m;

P = 0.02 mGy/wk; workload W = 2000 mGy/wk at 1 m

Occupancy factor = 1/16; TVL = 6.6 cm of concrete

RF = LWT/Pd^2 = 10^{-3} × 2000 × (1/16)/{0.02 × 3^2} = 0.16 TVL = 1.1 cm concrete

The door must have a shielding equivalent to the secondary barrier—the door must not be in the primary beam—which is about 1–2 mm lead lining to the wooden door.

A simulator room is enclosed by gypsum wallboards or concrete walls. Additional shielding can always be provided by lead lining. The simulator must be so installed that the primary beam does not face any low–shielded areas (e.g., windows, doors, protective screens).

10.17 Shielding of a CT Simulator Room

The CT simulator is a dedicated planning CT with virtual simulation for tumor localization. It operates at 125 kVp and 250 mAs. The primary beam is internally shielded and U = 1 for scatter and leakage. A typical CT simulator room is shown in Figure 10.23 with a grid.

Knowing the mAs, the number of slices per scan, and the number of patients per week, the workload can be expressed in terms of slices per week or mA.min per week.

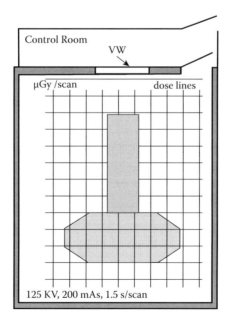

FIGURE 10.23
CT simulator room layout.

For shielding calculations the isodose plots around the scanner need to be known, which is usually provided by the manufacturer in terms of µGy/scan or in other units. Using these data, the dose at the grid points can be determined. The weekly dose (in air kerma = DE) at any point in the room, on the walls of the room, in the control room, or the public areas can then be computed. This is the unattenuated dose at the POI. The RF is given by

$$RF = (DE/wk \text{ at POI}) \times T/P$$

10.18 Technical Aspects of Treatment Delivery Room Design

The above explained concepts are effectively used for planning a room layout for treatment delivery. A typical treatment room design, as per the above shielding concepts, is shown in Figure 10.24.

The only problem with this kind of design is in providing adequate shielding thickness for a door. A laminated door, with adequate shielding for scatter, will be extremely heavy and must be motorized. If the door becomes jammed due to an electrical fault or other reasons, there must be an alternative escape door for getting the patient out. Though direct door facilities have been designed, they are not very practical and are rarely used in radiotherapy.

An alternative solution is to provide a maze so that the door does not see direct scatter but only photons that are multiply scattered, as shown in Figure 10.25. This is the normal design of the treatment room with the gantry plane of rotation parallel to the maze wall, and the maze region seeing only the secondary radiation. The maze area and the door do not see any primary radiation.

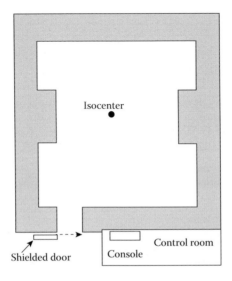

FIGURE 10.24
Linac facility with direct shielded door.

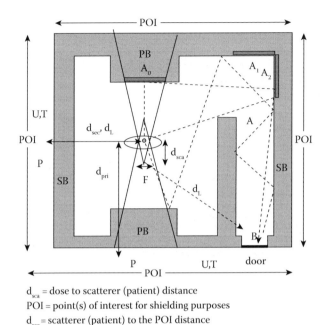

d_{sca} = dose to scatterer (patient) distance
POI = point(s) of interest for shielding purposes
d_{sec} = scatterer (patient) to the POI distance

FIGURE 10.25
Room scattering in a treatment facility.

The figure also shows the photon scatterings by the dotted lines. As can be seen from the figure, the door does not see direct scatter from the patient or the primary walls. The main components to consider for estimating the dose at the door position are the radiations reaching the maze region. These are:

1. Primary wall scatter into the maze region and reflected toward the door, H_{pri}
2. Patient scatter that reaches the maze area and is reflected toward the door, H_{ps}
3. Leakage scatter that reaches the maze region and is reflected toward the door, H_{LS}
4. Head leakage that penetrates the maze wall and reaches the door, H_{tr}

The dashed lines in Figure 10.25 represent all the above components. The figure also shows rays multiply scattered before reaching the door. Each scatter reduces the radiation intensity by a factor of roughly 10^3 and so the multiply scattered photon intensity at the door will be negligible. Therefore, the DE at the door needs to be computed due to the above four components only. The amount of scatter dose depends on the incident intensity, and the total surface area scattering or reflecting the radiation. The incident intensity varies as the inverse square of the distance from the source, and the amount of reflected radiation depends on the area of the scatterer.

10.18.1 Design of Primary Walls

The primary barrier width is determined from the maximum field size of the linac collimator that is projected on the wall. This can be easily determined from the geometry.

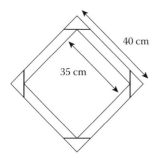

FIGURE 10.26
Maximum field size projected on the primary barrier.

The maximum width of the field size at the isocenter is the diagonal of the 40 cm × 40 cm field. This is given by $\sqrt{2}$ × 40 cm or 56 cm. Since the primary collimator in the linac is circular, it cuts the field at the corners reducing the maximum width. For this purpose, 35 cm × 35 cm is used as the maximum field size (see Figure 10.26) for estimating the maximum field size projected on the primary wall.

A margin of 30 cm is added on either side of the projected field on the wall, as a margin of safety, as shown in Figure 10.26. The small angle scatter will have nearly the same energy as the primary photon and hence will require larger thickness compared to a large angle scatter.

The extra bulge of the primary wall can be provided on the inside of the wall or outside. If the extra thickness is provided outside the room, then the field size must be calculated at the distal end of the barrier. Similarly if a uniform thickness of laminated barrier with lead or steel is used due to limitations of space, the field size is considered at the distal end of this material as shown in Figures 10.27 through 10.29.

For the ceiling, it is convenient to provide the extra thickness on the outside.

If a laser alignment device is fixed in the primary barrier, there will be a reduction in the shielding thickness by about an HVL. The supporting plate thickness of the device must therefore be enhanced with steel or lead to make up for this loss of one HVL [21,22].

For secondary barriers, no field size is defined since all parts of the barrier see scatter and leakage. Also, U = 1 since secondary walls see scatter and leakage whenever the machine is operating, irrespective of its orientation.

Example 10.21

In Figure 10.29 given above, using the following data, calculate the width on the wall that must be considered for providing a primary barrier. Maximum field size at the isocenter is 40 cm × 40 cm field. Target to the wall distance is 4 m. Target to isocenter distance is 1 m.

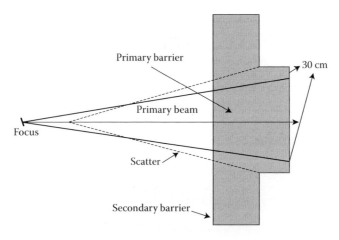

FIGURE 10.27
Projected field defined at the distal end of the barrier.

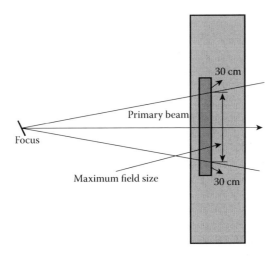

FIGURE 10.28
Projected field defined at the distal end of the laminated barrier.

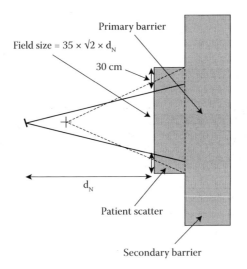

FIGURE 10.29
Projected field defined at the proximal end of the laminated barrier.

$$\text{Maximum diagonal field at isocenter} = \sqrt{2} \times 40 = 56.6 \text{ cm}$$

$$\text{From similar triangles, projected field dimension on the wall}/56.6 = (3+1)/1$$

$$\text{Projected field dimension on the wall} = 56.6 \times 4 = 226.4 \text{ cm}$$

$$\text{Adding a margin of 30 cm on either side,}$$

$$\text{Primary barrier total width of } 226.6 + 30 + 30 = 286.4 \text{ cm.}$$

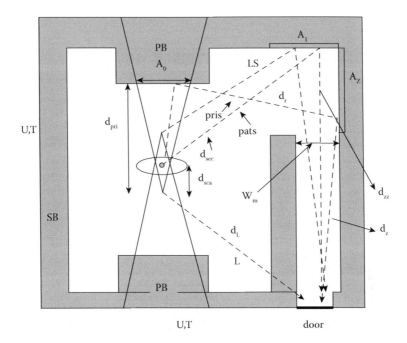

FIGURE 10.30
Scatter dose at the maze entry door from scatter.

10.18.2 Secondary Radiation Down the Maze and the Dose Near the Maze Entry Door

Door shielding is an important aspect of treatment room design and is more complex compared to barrier shielding. Since the radiotherapy technicians work close to the door, adequate shielding at the door is important to protect staff against both photon and neutron exposures. The photon dose at the door has four major components:

1. Scatter of the primary beam from the primary barrier to the maze outer wall (A_2) and a second scatter from the maze wall to the door (H_{pris})
2. Scatter of the head leakage reaching the maze wall (A_1) and scattering toward the door (H_{LS})
3. Patient scatter reaching the maze wall (A_1) and scattering toward the door (H_{pats})
4. Direct transmission of leakage radiation through the maze wall to the door (H_{Ltr})

Each dose component is calculated for the primary beam striking a primary barrier as shown in Figure 10.30.

10.18.3 X-Ray Dose at the Maze Entry

The total X-ray scatter dose at the door of the treatment room depends mainly on the primary wall scatter, patient scatter, and leakage scatter.

$$H_X = H_{pris} + H_{pats} + H_{LS} + H_{Ltr}$$

The method of calculating the dose components at the door is as follows.
For the primary wall:

$$\text{Incident intensity on the primary barrier} = (WU)/(d_{pri})^2$$

$$\text{Total incident radiation} = (WU)/(d_{pri})^2 \, A_0$$

$$\text{Scatter radiation from the primary wall} = [(WU)/(d_{pri})^2 \, A_0] \, \alpha_0$$

$$\text{Incident intensity on the maze wall} = [(WU)/(d_{pri})^2 \, A_0 \, \alpha_0]/(d_r)^2 = YY$$

$$\text{Radiation scattered by the maze wall (toward the door)} = YY \, A_z \, \alpha_z$$

$$\text{Radiation intensity at the door} = (YY \, A_z \, \alpha_z)/(d_{zz})^2$$

$$\text{Primary wall scatter reaching the door} = H_{pris} = \frac{WU\alpha_0 A_0 \alpha_z A_z}{(d_{pri}d_r d_z)^2}$$

The primary beam scatters twice before reaching the door. So, the inverse square law of variation of radiation intensity will appear twice (i.e., d_r and d_z). The inverse square law applies for a point source but here it is applied for the barrier-scattered radiation which simplifies calculations and gives a good estimate of the dose components at the door. A primary transmission factor of 0.25 may be used in the above equation (for 6–10 MV X-ray beams) since the beam is attenuated by the patient before striking the primary wall. The leakage scatter and patient scatter dose components are also evaluated in a similar fashion. The leakage involves single scattering at the maze wall but the patient scatter involves two scatters, one at the patient and the second at the maze wall. These two dose components are given by

$$H_{pats} = \frac{a(\theta)WU(F/400)\alpha_a A_1}{(d_{sca}d_{sec}d_{zz})^2}$$

(The source to patient distance is generally 1 m, $d_{sca} = 1$).

and

$$H_{LS} = \frac{LW_L U \, \alpha_1 A_1}{(d_{sec}d_{zz})^2}$$

L is the head leakage factor (10^{-3}). Since the linac head rotates in a circle, the source of leakage can be assumed to be the isocenter point. Because of the position of the inner maze wall only the area A_1 of the maze wall can provide single scatter for the leakage to reach the door.

Scattering reaching the door depends on the length of the maze wall, the maze width, scatter distances, and the linac beam quality. Multiple scatterings of radiation before reaching the door can be ignored since the intensity reduces by a factor of 3 in every scattering, and the scatter energy also becomes very low. The concrete wall reflection coefficient α is a function of incident scatter radiation energy and the incident angle. The patient scatter energy can be as low as 0.5 MeV, and the wall-scattered energies will be much lower. By using α for 0.5 MeV, a conservative estimate of the dose at the door can be obtained.

Direct transmission of the leakage radiation through the maze wall and reaching the door is another component of dose at the door.

$$\text{The leakage transmission component, } H_{Ltr} = \frac{LW_L U\, B}{(dL)^2}$$

B is the barrier transmission factor. Even scattered radiation can be transmitted through the maze wall but because of the much lower energies of large angle scatterings the scatter transmission contribution is very small and can be ignored.

The area A_0 is given by the projection of the 40 cm × 40 cm field size from the isocenter to the wall. $(d_{pri})^2$ gives the magnification factor for the isocenter field area. For IMRT, a maximum field size of 15 cm × 15 cm can be used. A_1 gives the area that provides single scattering route for the leakage to reach the door. A1 is given by the width (that is slightly larger than the maze cross section) x wall height.

Since the beam can strike any of the four barriers (right and left walls, floor, and ceiling) the total dose (for all the four barriers) is given by

$$H_{T,X} = 2.64 \times H_X$$

Figure 10.30 illustrates the various components of the X-radiation contributing to the dose near the door. The experimental figure of 2.64 (NCRP 151) is more realistic than a factor of 4 for the contribution from the four barriers, since they all contribute differently to the dose at the door and do not make equal contributions. If the gantry angles are not uniformly distributed and any beam orientation is preferentially used, then the 2.64 may not be valid [18].

10.18.4 Neutron and Capture Gamma Dose at Maze Entry

10.18.4.1 Production of Gamma Rays

For >10 MV beams, photoneutrons are produced which are moderated by the linac head to a minor extent, but predominantly by the room walls, floor, and ceiling. The interaction of the thermalized neutrons with concrete gives rise to γs in the reaction $^1H(n, \gamma)^2H$. In the case of a laminated barrier with Pb, photonuclear interactions with Pb also produce neutrons and a gamma component from the metastable excited state of the Pb atom, as shown by the following equation.

$$^{206}Pb(X,n)^{205m}Pb \rightarrow\ ^{205}Pb + \gamma$$

So, for the high-energy linacs, neutrons and capture gammas contribute two additional radiation components which need to be considered in shielding calculations. The capture gamma ray spectrum in concrete extends to several MeV (average 3.6 MeV).

As mentioned earlier in the chapter, the primary and secondary barriers that shield for primary, secondary, and leakage photons also effectively shield for the neutrons and the capture gammas, if the barriers are made of only concrete and no special consideration is necessary. (For laminated barriers, high Z materials are relatively transparent to neutrons andso additional shielding may be necessary to shield against neutrons and capture gamma rays.)

10.18.4.2 Neutron Fluences in the Linac Room

The linac neutron spectrum resembles the fission neutron spectrum as shown in Figure 10.31. The target, the primary collimator, and the flattening filter are the main components producing photoneutrons in the linac head. These materials are relatively transparent to neutrons though a degree of moderation of neutrons occurs in the linac head. The angular emission from the head is approximately isotropic so the inverse square law for the neutron yield (or emanation from the head) can be applied to determine the

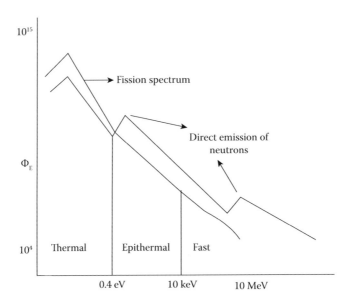

FIGURE 10.31
Comparison of linac neutron spectrum and fission spectrum.

direct neutron fluence at any distance d from the target of the linac. The yield is approximately 2×10^{12} n/Gy of X-ray dose at the isocenter. The neutron yield per treatment dose fraction (2 Gy) can therefore be assumed to be 10^{12}–10^{13} neutrons.

The neutrons are further scattered by the room and maze walls, floor, and the ceiling leading to further energy degradation. The thermal to fast neutron component is a function of location in the treatment room. The thermal to fast neutron ratio is approximately 1% near the linac head, about 20% near the walls, about 30% in the maze area, and nearly 70% in the control room. The mean energy of the spectrum is <300 keV in the treatment room, <100 keV in the maze area, and <5 keV in the control room.

So, the neutron fluence at any location can be represented by three components—the direct neutrons, the room scatter neutrons, and the thermal neutrons.

$$\Phi_{n,t}(d) = \Phi_{n,d}(d) + \Phi_{n,sc}(d) + \Phi_{n,th}(d)$$

R.C. McCall [31] has undertaken extensive studies on this topic and their results are summarized in this section, for a concrete vault. The fluence of direct neutrons, room scattered neutrons, and the thermal neutron fluences are given by

$$\Phi_{n,d}(d) = Q_n/4\pi d^2$$

$$\Phi_{n,sc} = 1.26\,Q_n/S$$

$$\Phi_{n,th} = 5.4\,Q_n/S$$

S gives the average size of the room excluding the maze area. Q_n is the neutron yield of the linac head expressed in number of neutrons emitted from the head per unit X-ray dose at the isocenter. Lead is relatively transparent to neutrons, so for a head shielded by lead, neutron head attenuation factor $k_{natt} = 1$. For tungsten shielding $k_{natt} = 0.85$. For Q_n values for the medical accelerators in the range 6–25 MV, the Q_n factors are available in the literature [19,30,31].

10.18.4.3 Neutron Dose Determination from Fluence

For monoenergetic neutrons, kerma factors (kerma per unit neutron fluence) are available in the literature. The neutron tissue kerma is given by:

$$K_t(d) = \Phi_n(d)\ E\ \mu_{en.tr}(E) \text{ or } [K_t(d)/\Phi_n(d)] = E\ \mu_{en.tr}(E)$$

The kerma factors are energy dependent. The above equation can also be used for the average energy of a spectrum. A method of estimating the neutron average energy is described in NCRP 79 [12]. For the linac spectrum, the average energy of the direct neutrons exiting the linac head is less than 1 MeV, and the average energy of the scatter spectrum is about 0.24 MeV. Using the NCRP 79 method, the average energy of the neutron spectrum in the treatment room (excluding the thermal neutron fluence) comes to about 0.34 MeV [13].

The TVL of 0.34 MeV neutrons in concrete is about 21 cm, about half the TVL of the primary beam. For concrete walls, X-ray shielding automatically ensures neutron shielding. This also applies for the leakage shielding since the leakage is about 1 mGy/Gy for photons, and about 1–2 mSv/photon Gy for neutrons. So as long as the thickness is greater than a TVL, shielding for X-ray leakage automatically ensures shielding for neutron leakage as well. (This however is not true for laminated barriers, because the high Z materials are relatively transparent to neutrons compared to photons and special attention must be given to neutron shielding as well. A moderated material may be required for slowing down the neutrons coming out of high Z materials like lead or steel, followed by lead to attenuate the capture gammas.)

The index of biological harm (Q factor) is much higher for neutrons compared to X-rays. For X-rays, Q being unity, the dose and dose equivalent are numerically the same and so it did not matter whether the dose was expressed in Gy or in Sv. For neutrons, the dose must be expressed in Sv and not in Gy since the Q factor is about 10 and not unity. So, the neutron leakage must be expressed in units of Sv per Gy of X-ray dose at the isocenter.

10.18.4.4 Neutron Dose at Maze Entry

The neutron dose (actually DE) and the capture gamma dose near the door depend on the dimensions of the linac room, the distance of the inner maze points from the isocenter, the scattering room surface area, maze entrance cross section, the maze cross section, and the length of the maze. The parameters like the beam quality, the field size, and beam orientation also influence the neutron dose, though to a lesser extent.

Figure 10.32 explains the neutron and capture gamma dose estimation at the door of a linac facility.

The total neutron fluence at point A in the maze area can be estimated from the following equation:

$$\Phi_{n,t}(A) = \Phi_{n,d}(A) + \Phi_{n,sc}(A) + \Phi_{n,th}(A)$$

$$= Q_n/4\pi d^2 + 5.4\ Q_n/2\pi S + 1.26\ Q_n/2\pi S$$

The fraction 2π in the last two terms account for only a fraction of the room surface contributing to neutron fluence at A. The neutron fluence in the maze area and near the door can be reduced by increasing the distances d_1 and d_2, and reducing the maze entrance cross section. The neutron leakage entering the maze area is at the maximum when the linac head is closest to the maze entry area, minimum when the head is in the opposite end, and intermediate when the head is pointed toward the floor. So, the head pointing to

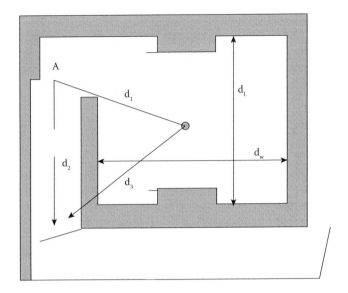

FIGURE 10.32
Computation of neutron and capture gamma dose at the door.

the floor is an appropriate position for shielding calculations against leakage. The neutron dose increases with decreasing field size, and maximum neutron doses are obtained when the collimators are completely closed. For a conservative estimate of neutron doses, the survey must be made with the collimators completely closed, and for realistic estimates, the survey may be performed in the clinical conditions of use.

The neutron dose along the maze is attenuated by a typical TVL_n of 5–7.5 m for many linac rooms. Wu and McGinley [32] found that, for a room construction as in Figure 10.32, the neutron dose decreases along the maze with a TVL given by

$$TVL_n = 2.06 \times \sqrt{S_1}$$

where S_1 is the cross sectional area of the maze in m^2. They also found that the neutron DE along the maze can be given by the following equation:

$$H_n = 2.4 \times 10^{-15} \times \phi_A \times \sqrt{A_r/S_1} \times \left[1.64 \times 10^{-\left(\frac{d_2}{1.9}\right)} + 10^{-\left(\frac{d_2}{T_n}\right)}\right]$$

where H_n is the neutron DE at the maze entrance, in Sv per X-ray Gy at isocenter, ϕ_A, the neutron fluence at point A in the figure. T_n is the neutron TVL, A_r the cross section of the maze entry in the treatment room, and d_2 the distance along the maze from the point A.

There is another equation arrived at by Kersey [33] for estimating the neutron dose near the maze entry point which is given below.

$$H_n = H_1 \times 10^{-3} \times (A_r/S_1) \times (1/d_1)^2 \times 10^{-d_2/5}$$

where H_n is the neutron DE at the maze entrance and H_1 is the neutron DE at 1 m from the X-ray target, expressed in mSv per X-ray·Gy at the isocenter. Values of H_1 are tabulated in

Table 10 of the IAEA Safety Report Series 47 [13]. A_r and S_1 are cross sectional areas, in m^2, of the inner maze entrance and the maze, respectively.

The Kersey method gives a more conservative estimate of dose compared to the McGinley equation, but both equations are being used for neutron dose estimates.

The weekly dose due to neutrons is given by $W \times H_n$.

10.18.4.5 Capture Gamma Dose at Maze Entry

The gamma dose (per Gy of X-ray dose at the isocenter) at the door is given by [31,33]

$$H_{n,\gamma} = 5.7 \times 10^{-16} \times \Phi_{n,t}(A) \times 10^{-d_2}/6.2$$

$$= W \times D_\gamma \; Gy/wk$$

10.18.5 Total DE at Maze Entry

The total dose at the door position, for >10 MV linacs is given by

$$H_{T,X+n} = H_{T,X} + H_n + H_{n,\gamma}$$

For linacs >10 MV, neutrons and capture gammas dominate the dose near the door. If the maze wall is kept thin and the outer wall thicker, then the direct leakage transmission will dominate the X-ray dose at the door and the scatter contributions will become insignificant.

The RF for estimating the door shielding can be obtained from ($H_{T,X+n}/P$). For more details regarding the evaluation of doses near the door, the reader should study the excellent presentation of Melissa C. Martin [30]. The following examples illustrate some of the concepts described above [13].

Example 10.22

Figure 10.33 refers to an 18 MV linac facility. Calculate the X-ray dose at the door, for the linac beam, due to

1. Patient scatter
2. Primary wall scatter
3. Leakage transmission

(shown by the lines drawn in red, green, and brown, respectively). The following data are given:

a. For patient scatter, $W = 600$ Gy/wk; $U = 0.25$; A_1 width = 2.8 m; wall height = 4.2 m;

$$a(45°) = 8.64 \times 10^{-4} \; 45° \text{ scatter}; \; \alpha_1(0°) = 2.03 \times 10^{-2} \text{ for 0.5 MeV}$$

Isocenter to wall distance $d_1 = 8.4$ m; wall to door $d_2 = 8.54$ m

Field size incident on patient = 40 cm × 40 cm

b. For primary wall scatter, $W = 600$ Gy/wk; $U = 0.25$; $A_0 = 2.8 \; m^2$ (projected maximum field area); wall height = 4.2 m;

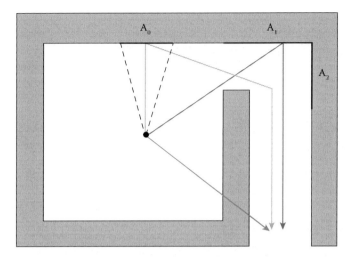

FIGURE 10.33
Computation of X-ray dose near the maze entry door.

$$\alpha_0 = 1.62 \times 10^{-3} \text{ for 18 MV; } \alpha_z(0°) = 7.54 \times 10^{-3} \text{ for 0.5 MeV;}$$

$$A_z = \text{maze entry cross section} \times \text{incident wall height} = 10.2 \text{ m}^2$$

$$\text{Isocenter to primary wall distance } d_{pri} = 4.2 \text{ m;}$$

$$\text{wall reflection point to door } d_r = 5.9 \text{ m}$$

$$\text{reflection point to door } d_z = 6.8^2$$

c. Calculate the head leakage primary wall scatter reaching the door for the same patient scatter path given in example (a) above (d_1 and d_2, U, W are the same). The reflection coefficient must be chosen for the incident angle and the reflection angle, for the beam quality of the radiation. Use the following additional data for the calculation:

$$\alpha_1 = 4.11 \times 10^{-3} \text{ (45° incidence, 0° reflection angle, 18 MV); } L = 10^{-3}$$

d. Also calculate the X-ray dose at the door due to direct head leakage transmission. Assuming the isocenter to be the origin of leakage, isocenter to door distance, d is 7.1 m. Barrier transmission factor, $B = 2.31 \times 10^{-4}$. What is the total X-ray dose at the door due to scatter and leakage transmission (ignore patient attenuation). Based on the calculation, state whether the door requires any X-ray shielding.

Answer

a. $$\text{Area } A_1 = 2.8 \text{ m} \times 4.2 \text{ m} = 11.8 \text{ m}^2$$

$$D_{dr,ps} = \frac{aWU\alpha_1 A_1}{(d_1)^2 (d_2)^2} = \frac{8.64 \times 10^{-4} \times 600 \times 10^6 \times 0.25 \times 2.03 \times 10^{-2} \times 11.8}{(8.4)^2 \times (8.54)^2}$$

$$= 60 \text{ μGy/wk}$$

b. Primary wall scatter reaching the door $= D_{dr,wls} = \dfrac{WU\alpha_0 A_0 \alpha_z A_z}{(d_{pri}d_r d_z)^2}$

$$= \frac{600 \times 10^6 \times 0.25 \times 1.62 \times 10^{-3} \times 2.8 \times 7.54 \times 10^{-3} \times 10.2}{(4.2)^2 \times (5.9)^2 \times (6.8)^2}$$

$= 1.8 \ \mu Gy/wk$

c. $D_{dr,LS} = \dfrac{LWU\alpha_1 A_1}{(d_1 d_2)^2} = \dfrac{10^{-3} \times 600 \times 10^6 \times 0.25 \times 4.11 \times 10^{-3} \times 11.8}{(8.4)^2 \times (8.54)^2}$

$= 1.4 \ \mu Gy/wk$

d. Dose due to leakage transmission

$$= D_{dr,LT} = \frac{LWUB}{d^2} = \frac{10^{-3} \times 600 \times 10^6 \times 0.25 \times 2.31 \times 10^{-4}}{(7.1)^2}$$

$= 0.6 \ \mu Gy/wk$

Total X-ray dose at the door $= 2.64 \times (60 + 1.8 + 1.4 \times 0.6) = 63.8 \ \mu Gy/wk$.

The design value for the control room is 100 μGy/wk. So, no extra shielding is necessary and an ordinary wooden door can be installed. In fact, no door is actually necessary since the dose at the door is less than the design value, but for physical and interlocking safety purposes a door needs to be provided.

Example 10.23

A room design for installing an 18 MV linac is shown in Figures 10.34 and 10.35 (not to scale).

a. Determine the total surface area available for neutron scattering from the isocenter position view.

The following data are given:

Ceiling barrier thickness = 1.1 m; room height = 4.2 m; room length = 9 m; width = 7.8 m

b. Calculate the weekly neutron dose equivalent at the maze entry point, using the equations of Kersey, and Wu and McGinley.

The following data are given:

The yield of the neutron head is 1.22×10^{12} neutrons per Gy of photon dose at isocenter. Workload = 600 Gy/wk; The total scattering surface area S = 236 m^2;

Isocenter to maze center point A, d_1 = 6.4 m; maze point A to maze entry d_2 = 8.5 m

Maze entry width in the linac room = 2.43 m

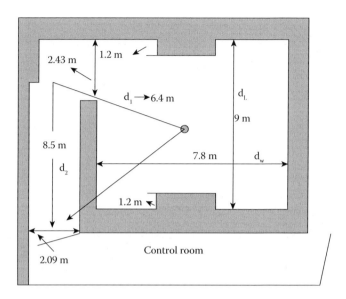

FIGURE 10.34
Horizontal cross section for an 18 MV linac layout.

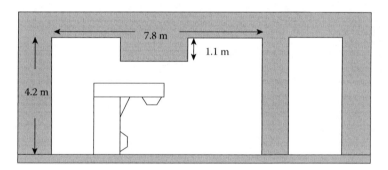

FIGURE 10.35
Vertical cross section for an 18 MV linac layout.

Maze width = 2.09 m; room height = 4.2 m

Neutron DE at 1 m from the target H_1 = 3.81 mSv/photon Gy at isocenter

c. In the above example, calculate the total capture gamma dose at the maze entry point.

d. In the above example, calculate the shielding to be provided for the maze entry door which opens into the control room, for the neutron dose at the door. Use the following data:

Neutron TVL in borated polyethylene (BPE) = 4.5 cm

The design value for the control room = 0.1 mSv/wk

e. In the above example, calculate the shielding to be provided for the maze entry door which opens into the control room, for the capture gamma dose at the door. Use the following data:

TVL in lead for the photons in the maze area = 6 mm

Total X-ray dose (excluding capture gamma dose) = 63.8×10^{-6} Gy/wk.

Answers

a. The room surface area A is made up of floor and ceiling, front and back walls, and the two lateral walls.

The average room height (h) = 4.2 m − 0.55 m = 3.65 m

The average room width = 9 − (0.6 + 0.6) = 7.8 m

Room length = 7.8 m

The total surface area S = 2 × (7.8 × 3.65 + 7.8 × 3.65 + 7.8 × 7.8) = 236 m^2

b. Step 1: Determine $\Phi_{n,t}(A)$

$$\Phi_{n,t}(A) = \Phi_{n,d}(A) + \Phi_{n,sc}(A) + \Phi_{n,th}(A)$$

$$= Q_n/4\pi d^2 + 5.4\, Q_n/2\pi S + 1.26\, Q_n/2\pi S$$

$$= [1.22 \times 10^{12}/4\pi\, (6.4)^2] + [5.4 \times 1.22 \times 10^{12}/2\pi \times 236] + [1.26 \times 1.22 \times 10^{12}/2\pi \times 236]$$

$$= 10^9\, (2.37 + 4.45 + 1.037) = 7.85 \times 10^9\, \text{n/cm}^2$$

Step 2: Determine the neutron dose near the door using the Kersey, and Wu and McGinley equations.

Area of cross section of maze entry A_r = 2.43 × 4.2 = 10.2 m^2

Maze area of cross section S_1 = 2.09 × 4.2 = 8.78 m^2

Kersey equation:

$$H_n = [H_1 \times 10^{-3}/(d_1)^2] \times (A_r/S_1) \times 10^{-d_2/5}$$

$$= [3.81 \times 10^{-3}/(6.4)^2] \times (10.2/8.78) \times 10^{-8.5/5}$$

$$= 10^{-6} \times (7.76 \times 1.18 \times 0.2) = 1.8 \times 10^{-6}\, \text{Sv/Gy}$$

Weekly neutron DE at the maze entry = $600 \times 1.8 \times 10^{-6}$ = 1.1×10^{-3} Sv/wk.

Neutron DE at maze entry using Wu and McGinley equation:

This requires neutron $TVL_n = 2.06 \times (S_1)^{0.5} = 2.06 \times (8.78)^{0.5} = 6.1$ m.

$$H_n = 2.4 \times 10^{-5} \times \Phi_{n,t}(A) \times (A_r/S_1)^{0.5} \times [1.64 \times 10^{-d_2/1.9} + 10^{-d_2/TVL_n}]$$

$$= 2.4 \times 10^{-15} \times 7.85 \times 10^9 \times 1.079 \times [3.36 \times 10^{-3} + 4.04]$$

$$= 0.82 \times 10^{-6} \text{ Sv/Gy}$$

Weekly neutron DE at the maze entry $= 600 \times 0.82 \times 10^{-6} = 1.1 \times 10^{-3}$ Sv/wk.

$$= 4.9 \times 10^{-4} \text{ Sv/wk}$$

The Kersey method overestimates the neutron dose values.

c. Capture gamma dose

$$H_{n,\gamma} = 5.7 \times 10^{-16} \times \Phi_{n,d}(A) \times 10^{-d_2/6.2}$$

$$= 5.7 \times 10^{-16} \times 7.85 \times 10^9 \times 10^{-8.5/6.2}$$

$$= 1.9 \times 10^7 \text{ Gy/Gy at isocenter (or Sv/Gy at isocenter)}$$

Weekly gamma dose at maze entry $= 600 \times 1.9 \times 10^{-7} = 1.14 \times 10^{-4}$ Sv/wk.

d. Neutron shielding for the door

$$RF = 4.9 \times 10^{-4} \text{ Sv/wk} / 1 \times 10^{-4} \text{ mSv/wk} = 0.7 \text{ TVL}$$

Required BPE thickness $= 0.7 \times 4.5 = 3.2$ cm of BPE

e. X and γ dose shielding for the door

Total X and γ dose $= 63.8 \times 10^{-6} + 114 \times 10^{-6} = 1.8 \times 10^{-4}$ Sv/wk

$$RF = 1.8 \times 10^{-4} \text{ Sv/wk} / 1 \times 10^{-4} \text{ mSv/wk} = 1.8 = 0.26 \text{ TVL}$$

Required lead thickness $= 0.26 \times 6 = 1.6$ mm of lead

10.19 Design of the Door

For low-energy linacs, the door that requires X-ray shielding is typically a lead barrier with steel encasement for rigidity. For higher energy linacs, the door that requires neutron and capture gamma ray shielding is typically made of borated polyethylene (BPE, boron incorporated about 5% by weight) sandwiched between lead sheets to stop neutrons and capture gammas. Cadmium is also a good thermal neutron absorber like boron, and can be used. Borated polyethylene is commercially available in this form for door shielding purposes. A typical design of the door is as shown in Figure 10.36.

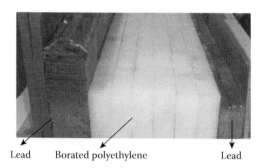

Lead   Borated polyethylene   Lead

FIGURE 10.36
Design of linac room door shielding.

The rationale of the door design, for neutron shielding, is as follows:

1. Lead on the inside shields against gammas and also moderates faster neutrons to some extent.
2. Hydrogenous material (here polyethylene) slows down faster neutrons to thermal energies.
3. Incorporated boron absorbs the thermal neutrons resulting in (n,γ) reactions.
4. The outer lead lining attenuates the capture gammas.

Figure 10.37 shows the maze area of a 15 MV linac installation with the door on the side of the maze wall (with a 90° turn to the maze). The length of the maze is about 11 m. This design (long maze and 90 degree turn for the door) considerably attenuates the neutrons reaching the door since one extra scatter will be required for the radiation to reach the door. A survey conducted near the door of this installation showed negligible neutron or X-ray dose, requiring just an ordinary wooden door for the maze entry.

Another way of reducing the neutron and gamma dose near the door, to avoid a shielded door, is to provide a short length to the maze after the 90 degree turn. A third alternative is to reinforce the barriers in the maze area with BPE so more neutrons are thermalized before reaching the door.

FIGURE 10.37
Maze design with wall on the side. (Picture Courtesy Fortis Memorial Research Institute, Gurgaon, India.)

10.20 Door Interlocks and Emergency OFF Provisions

The treatment room door must be interlocked to the treatment unit so that the machine will go off and the treatment terminated if the room door is opened by anyone during exposure. This is one of the most important safety checks that must be verified during the equipment QA testing. It must also be ensured that the treatment can be resumed only after resting the unit at the console and not by just closing the door again. It is the responsibility of the technician to ensure there is no one in the room before closing the door and initiating the treatment. A reset switch may be provided near the exit of the treatment room at a position that gives a clear view of the interior of the treatment room (as is the practice in some countries) so that the technician can activate this switch before leaving the room. There are in fact two switches—one inside, and one outside the room. Both switches must be activated, within a certain time period, for initiating treatment. Without activation of the two switches, the radiation cannot be turned on [14].

Emergency off switches must be available on the machine and also inside the treatment room so the treatment can be immediately terminated and the machine shut off if required for any emergency purpose. An emergency may arise while staff are inside or outside the treatment room, so emergency switches are provided at several places (e.g., on either side of the treatment couch, on the wall of the bunker, at the maze entry point, on the machine console in the control room). The emergency may arise if someone is left behind in the treatment room and the treatment has been initiated, or the patient has a medical condition like a cardiac arrest, or there is some fire hazard.

The interlocks, the emergency off systems and the patient communication systems must be regularly tested for proper functioning as per the frequency recommended by the protocols or by the regulatory bodies.

10.21 Patient Safety

The TDE has been designed to incorporate many safety features to ensure patient safety. The following features prevent inappropriate exposure of the patient during treatment:

- Defense in depth: This is to prevent any overexposure (e.g., dual timer or monitoring systems to terminate a treatment exposure so that if one fails, the other will do the job).
- Emergency stop systems: These enable the technician to stop treatment in the event of any impending collision of the linac with the patient, or if the patient needs immediate attention during treatment.
- Interlocks: These prevent treatment from being delivered when the parameters set on the machine are at variance with the planning parameters.
- Record and verify system: This can ensure, among other things, that the treatment settings are faithfully reproduced during each treatment session.

10.22 Public Safety

Public safety is ensured by following means:

- Shielding of the radiation installation
- Locating the radiation oncology wing in a corner of a hospital, in low-occupancy surroundings
- Providing safety interlocks to prevent any inadvertent entry of any member of the public into the treatment room during treatment
- Warning signs and sign boards indicating the waiting areas

10.23 Staff Safety

Staff safety is ensured by the following means:

- Safe design of the TDE
- Safe design of the treatment room facility
- Personal monitoring

10.24 Linac Equipment Leakage Survey

The leakage is expressed as a percentage Gy (or ADE in Sv) per Gy (= Sv) delivered at d_{max} at isocenter. The leakage regulations require that the X-ray leakage of the linac must be measured at 1 m distance from the target and also from the electron path in the waveguide. Figure 10.38 illustrates the concept showing the measured leakage rates at 1 m from the target (at four points) and along the waveguide at distances marked with a gap of 0.25 m

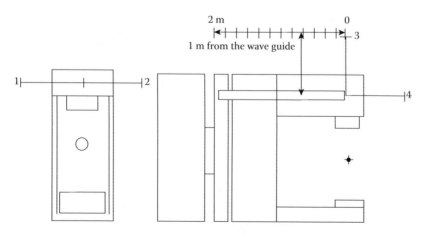

FIGURE 10.38
Leakage survey of a linac.

TABLE 10.7

X-ray Leakage at 1 m from the Target (10 MV Beam)

Points Marked from the Target Position			
1	2	3	4
0.0046	0.0059	0.0022	0.0051

TABLE 10.8

X-ray Leakage along the Waveguide at 1 m from the Guide

Distance at the Points Marked 0–2 m								
0	0.25	0.5	0.75	1	1.25	1.5	1.75	2
0.02	0.04	0.03	0.05	0.03	0.06	0.04	0.02	0.04

between the points. The measured values are expressed in units of percentage Gy per Gy delivered at the isocenter (Tables 10.7 and 10.8).

Example 10.24

The X-ray leakage at 1 m from the target for a 10 MV linac beam was measured as 0.02% Gy/Gy. The X-ray beam output for a 10 cm × 10 field at the isocenter = 400 cGy/min. What is the X-ray leakage in Gy/min at the measurement point?

$$0.02\% \text{ of } 400 \text{ cGy/min} = (0.02\ /100) \times 400 = 0.08 \text{ cGy/min}$$

10.25 Installation Survey

Compliance with the design limits can be ensured by surveying around the treatment room facility and also on top of the ceiling and below the floor (if there is any occupancy below). Once the linac becomes operational, the first measurement to be performed is the installation survey.

The survey must be carried out in the integrate mode of the survey meter by monitoring the radiation levels for a finite time period and then deducing the weekly or annual doses. When the radiation levels are very low, the survey meter may take some time to respond depending on the RC time constant of the circuit. So, by moving quickly across the wall compared to the response time of the instrument, the survey meter may read less than the actual, or zero. This must be kept in mind during survey. The integrate mode of operation is useful for measuring the dose rates near the door inside the treatment room.

Figure 10.39 shows a typical treatment facility and the beam facing a primary barrier for conducting a survey in the occupied area outside the barrier. For surveying behind the primary barriers, the machine must be operated for the maximum clinical dose rate and maximum field size conditions, and the beams must strike the primary barriers with no phantom in the beam path (as shown in Figure 10.39).

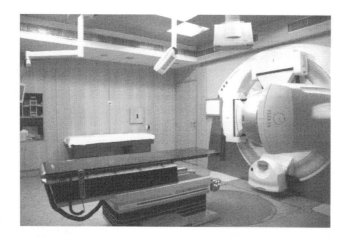

FIGURE 10.39
Beam facing one of the primary barriers for survey (Picture Courtesy Fortis Memorial Research Institute, Gurgaon, India.)

For surveying the secondary barriers, the radiation survey must be carried out for the worst case scenario or for the maximum scatter dose conditions. For this purpose, a phantom representing the patient (minimum 30 cm × 30 cm × 30 cm size phantom) must be centered at the isocenter with the machine operating at maximum dose rate conditions (highest energy and field size), before surveying behind secondary barriers. The control room is an important area to survey since the technicians are present here all the time during patient treatment (Figure 10.40).

Figures 10.41 and 10.42 show the radiation level monitoring points (indicated by a star) around this installation. The survey must be performed at points 30 cm away from the

FIGURE 10.40
Scatter geometry and secondary barrier survey. (Picture Courtesy Fortis Memorial Research Institute, Gurgaon, India.)

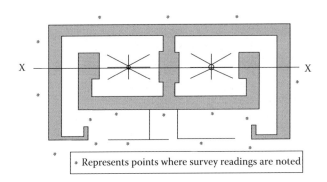

FIGURE 10.41
Radiation survey around the linac facility.

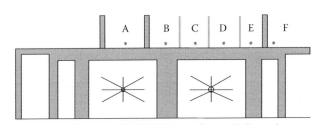

A: PET-CT machine room B: PET-CT control room C: Doctor's room
D: TMT stress room E: Gamma console room F: Gamma machine room

FIGURE 10.42
Longitudinal (X–X) view of the above facility. (The figure is not to scale but is given only to illustrate the survey concept.)

barriers; no one stands leaning on the wall all the time. So, this is the closest occupancy position beyond the barriers and design calculations are carried out for this distance from the barriers.

Figure 10.42 also shows the rooms occupied on top of the treatment room and the locations where radiation monitoring was carried out.

The important locations for the survey are (a) near the door (b) at the control console, and (c) at the conduit where the dosimeter cables are taken out of the treatment room. The cable conduit must always be at an angle so that the photons will be multiply scattered before reaching the control room. Even if the radiation level at the conduit is higher if this point is inaccessible (e.g., a control console at this position will increase the staff distance from the conduit which is now inaccessible) then it may be allowed if the radiation level at the closest accessible point from the conduit is in compliance.

Shielding calculations can be verified only by radiation survey. If there is any shielding inadequacy, the survey will reveal this and the shielding must then be reinforced for compliance. Often, some walls may carry conduits for cables or there may be AC ducts which will reduce the shielding. Any such deficiency will be revealed during radiation survey. If the radiation levels are slightly higher on the outside which could be a public area (like a playground) it will be easier to enclose this area in such way that it becomes inaccessible, rather than reinforcing the walls.

Suitable survey meters must be used for surveying for the photon and neutron fields. Two important characteristics to be considered in the choice of survey meter are the energy dependence, and the pulsed nature of the linac beam. In the case of photon survey, the pulsed nature of the beam is much more important than the energy dependence. Geiger–Muller (GM) types of survey meters must be avoided and only ion chamber-based survey meters must be used with pulsed beams, as explained in the chapter on radiation detectors. Most therapy linacs operate with a pulse repletion frequency (PRF) that varies between 100 and 400, that have pulse widths ranging from 1 to 10 μs (AAPM 1986). So, the actual machine ON time ranges between 100×1 μs to 400×10 μs per second. This is known as the duty factor of the linac. The small duty factors imply high IDRs (average dose rate/duty factor) which put a severe limitation on the operation of the survey meter. Pulse counting survey meters with long dead times (e.g., GM survey meters) tend to become saturated and may read zero or the PRF. Also, scintillation survey meters may not be a suitable survey instrument due to photomultiplier limitations for high instantaneous currents. Ion chamber-based survey meters must therefore be used for linac survey. Other types of survey meters can be used for the survey of telecobalt installations or brachytherapy installations where the radiation is not pulsed. There are several ion chamber-based survey meters (e.g., RTI, Ludlum, or Fluke survey meters) that can be used for photon survey. The survey meter must have sufficiently high chamber voltage to prevent recombination at such high IDRs.

If the linac energy is >15 MV, then a neutron survey is also required to be carried out. For 10 MV beams, the neutron doses may be negligible. Neutron survey must be carried out for both primary and secondary barriers, especially if the barriers are laminated since steel or lead are not very effective in neutron shielding. Neutron rem meters can be used for monitoring neutron levels outside the treatment room. The active detectors cannot be used inside the treatment rooms since the photon fluences are three to four orders of magnitude larger compared to neutron fluences, which will interfere with neutron monitoring due to a pileup of photon pulses.

Survey records must be maintained by the department for the purposes of regulation. The survey report must contain details regarding the instrument make, model calibration date, the linac (model, installed room, etc.), date of survey, survey conditions (linac parameters like beam energy, field size, and dose rate), and person doing the survey. A sketch of the room layout showing the surveyed locations and the survey values at the locations must be a part of the survey report. The radiation survey report is very important for the regulatory agency to certify the installation, from radiation safety point of view, for commissioning treatment.

The survey must be carried out periodically to ensure continued radiation safety. Many national regulations require a radiation protection survey to be carried out on an annual basis. This is particularly important in the case of laminated barriers where the lead or steel shield could have moved, affecting the shielding integrity.

Example 10.25

Calculate the duty factor (DF) of a therapy linac operating at a PRF of 100 and a pulse width of 1 μs. What will be the peak intensity of the beam compared to the average intensity?

$$DF = 100 \text{ pps} \times 1 \text{ μs} = 1 \times 10^{-4} = 0.0001$$

$$\text{Peak intensity} = \text{Average intensity}/DF$$

Peak intensity is 10^4 times the average intensity.

A survey behind a secondary barrier of a linac installation gave a reading of 3 mR/h. The surveyed area is uncontrolled with a dose limitation of 1 mSv/y. Does the area comply with the design goal? The following data are given:

The beam-on time = 5 h/wk; Assume 50 weeks of treatment per year.

$$3 \text{ mR/h} \approx 30 \text{ } \mu\text{Sv/h}$$

Annual dose = $30 \times 5 \times 50 = 7500 \text{ } \mu\text{Sv/y} = 7.5 \text{ mSv/y}$

The dose is higher and barrier requires additional shielding.

10.26 Patient Observation and Communication

The control room is always situated adjacent to the treatment room so that the technician can keep an eye on the entrance door for any unauthorized entry. The control room also has a TV monitor for visual observation of the patient, and an audio control for audio communication with the patient. Some countries require a mirror system for visual observation of the patient in addition to the TV monitor system, so that the patient can be observed even if the monitor system fails. If the monitor system fails, patient treatment must be terminated and cannot be resumed until the TV monitoring system is repaired. This provision will avoid accidental exposure to the staff and also aid patient monitoring.

10.27 Warning Signs

Posting warning signs is a legal requirement as mentioned in Chapter 5. An illuminated light outside the treatment room is also a regulatory requirement. Usually, there will be two illuminations: yellow when the treatment machine is activated, and red when the treatment beam is ON. Clear signs in patient and visitor areas, and patient change areas will avoid unauthorized entry into treatment areas.

- There should be a visible signal at the entrance of the maze, control area, and in the treatment room, when radiation is being produced.
- There should be an audible signal in the treatment room just prior to radiation being produced.

10.28 Ventilation

For high-energy linacs (>15 MV), there is a likelihood of ozone buildup and radioactive ^{15}O and ^{13}N gases in the treatment room as explained in Chapter 2. About 8–10 air changes per hour will prevent this buildup in the room.

References

1. *IAEA Basic Safety Standards*, BSS 115, IAEA, Vienna, Austria.
2. *IEC Part 2-1:Particular Requirements for the Basic Safety and Essential Performance of Electron Accelerators in the Range 1 MeV to 50 MeV*, Rep. IEC 601-2-1, IEC, Geneva, 2009.
3. *IEC Part 2-11:International Electro Technical Commission, Medical Electrical Equipment, Particular Requirements for the Safety of Gamma Beam Therapy Equipment*, Rep. IEC 601-2-11, IEC, Geneva, 1997.
4. *IEC Part 2-17:Particular Requirements for the Basic Safety and Essential Performance of Automatically-Controlled Brachytherapy Afterloading Equipment*, Rep. IEC 601-2-17, IEC, Geneva, 2013.
5. *IEC Part 2-8: Particular Requirements for Basic Safety and Essential Performance of Therapeutic X-ray Equipment Operating in the Range 10 kV to 1 MV*, Rep. IEC 601-2-8, IEC, Geneva, 2010.
6. *IEC Part 2-29: Particular Requirements for the Basic Safety and Essential Performance of Radiotherapy Simulators*, Rep. IEC 601-2-29, IEC, Geneva, 2008.
7. IAEA TECDOC 1040, *Design and Implementation of a Radiotherapy Programme: Clinical, Medical Physics, Radiation Protection and Safety Aspects*, IAEA, Vienna, Austria, 1998.
8. European Commission, *Radiation Protection 162, Criteria of Acceptability of Medical Radiological Equipment used in Diagnostic Radiology*, Nuclear Medicine and radiotherapy, 2012.
9. NCRP151, *Structural Shielding Design and Evaluation for Megavoltage X- and Gamma-Ray Radiotherapy Facilities*, NCRP, Bethesda, MD, 2005.
10. NCRP 49, *Structural Shielding Design and Evaluation for Medical Use of X-Rays and Gamma-Rays up to 10 MeV*, NCRP, Bethesda, MD, 1976.
11. NCRP 51, *Radiation Protection Guidelines for 0.1–100 MeV Particle Accelerator Facilities*, NCRP, Bethesda, MD, 1977.
12. NCRP 79, *Neutron Contamination from Medical Electron Accelerators*, NCRP, Bethesda, MD, 1984.
13. IAEA Safety Report Series 47, *Radiation Protection in the Design of Radiotherapy Facilities*, IAEA, Vienna, Austria, 2006.
14. IPEM Report 75, *The Design of Radiotherapy Treatment Room Facilities*, IPEM, York, UK, 1998.
15. P.W. Horton, *An Introduction to Installation Shielding in Radiotherapy*, Parts I&II, Royal Surrey County Hospital Guildford, UK. https://indico.cern.ch/event/277160/contributions/1621681/attachments/504535/696722/Installation_Shielding_in_Radiotherapy_Part_1_bw.pdf (accessed on 1 November 2016).
16. M.C. Martin, *Shielding Design Methods for Linear Accelerators*, AAPM Summer School, 2007. https://indico.cern.ch/event/277160/contributions/1621681/attachments/504535/696722/Installation_Shielding_in_Radiotherapy_Part_1_bw.pdf (accessed on 1 November 2016).
17. M.C. Martin, *Tomotherapy Vault Design*, 2007 AAPM Summer School, USA, https://www.aapm.org/meetings/07SS/documents/MartinShielding_Tomotherapy.pdf (accessed on 1 November 2016).
18. R. Kudchadker, *Radiation Vault Design and Shielding*, M D Anderson Center, Houston, TX.
19. P.H. Mcginley, *Shielding Techniques for Radiation Oncology Facilities*, Medical Physics Publishing, Madison, Wisconsin, 1998.
20. P.J. Biggs, Primary Beam Widths in Ceiling Shielding for Megavoltage Linear Accelerators, *Radiation Protection Management*, 19(4), 2002.
21. P.J. Biggs, *Linear Accelerator Shielding: Thirty Years Beyond NCRP 49*, Researchgate, 2015. https://www.researchgate.net/publication/265452237_Linear_Accelerator_Shielding_Thirty_Years_Beyond_NCRP_49 (accessed on 1 November 2016).
22. P.J. Biggs, *Basic Principles of Radiation Therapy Shielding Design*, AAPM Physics Review, 2014. http://www.aapm.org/meetings/2014am/reviewcourses/documents/t08biggsslidesshielding_2014.pdf (accessed on 1 November 2016).
23. R.K. Wu, *Overview & Basis of Design for NCRP Report 151*, AAPM Meeting, 2007. https://www.aapm.org/meetings/07SS/documents/revNCRP151AAPM.pdf (accessed on 1 November 2016).
24. J.E. Rodgers, Radiation therapy vault shielding calculation methods when IMRT and TBI procedures contribute, *Journal of Applied Clinical Medical Physics*, 2(3):157–64, 2001.

25. J.E. Rodgers, *Shielding II:Practical Examples including IMRT, TBI, SRS*, AAPM Meeting, 2007. http://www.aapm.org/meetings/amos2/pdf/29-7857-79725-296.pdf (accessed on 1 November 2016).

26. McGill University Lecture, *Radiation Oncology Treatment Room Design: Linear Accelerator Bunkers*, Lecture presentation, McGill University, Montreal, Canada. http://www.bic.mni.mcgill.ca/~llchia/HP_lectures/shielding_1.pdf (accessed on 1 November 2016).

27. J. Balog, et al., Helical tomotherapy radiation leakage and shielding considerations, *Medical Physics*, 32(3), 710–719, 2005.

28. J.E. Rodgers, *Shielding Cyber Knife Shielding Design*, AAPM Meeting, 2007. https://aapm.org/meetings/07SS/documents/RodgersCyberKnifeShieldingpostmeeting.pdf (accessed on 1 November 2016).

29. P.N. Mcdermott, Radiation shielding for gamma stereotactic radiosurgery units, *Journal of Applied Clinical Medical Physics*, 8(3), 147–157, 2007.

30. M.C. Martin, *Maze Calculations/Door Design*, AAPM Summer School, 2007. https://www.aapm.org/meetings/07ss/documents/MartinMazes.pdf.

31. R.C. McCall, *Neutron Yield of Medical Electron Accelerators*, Rep. SLACPUB-4480, Stanford Linear Accelerator Center, Stanford, CA, 1987.

32. R.K. Wu and P.H. McGinley, Neutron and capture gamma along the mazes oflinear accelerator vaults, *Journal of Applied Clinical Medical Physics*, 4(2), 162–171, 2003.

33. R.W. Kersey, Estimation of neutron and gamma radiation doses in the entrance maze of SL 75–20 linear accelerator treatment rooms, *Medicamundi*, 24, 151–155, 1979.

Review Questions

1. Which international body sets standards to ensure electrical, mechanical and environmental safety of radiotherapy equipment?

2. The leakage dose of X-rays and neutrons of a high-energy linac is 0.1% Gy/Gy. Will they cause same amount of harm to the exposed individual?

3. What three parameters control the radiation hazards? Which parameter mainly controls the radiation exposure in external beam therapy?

4. Should we consider neutron shielding issue for a 6 MV linac treatment facility? Explain your answer.

5. What is the approximate leakage factor for a linac head?

6. What are the design values for a controlled room and an uncontrolled area, around a linac facility, according to NCRP 151 recommendations?

7. What are primary and secondary barriers?

8. For the shielding of a high-energy (>10 MV) linac beam facility, what radiations must be considered?

9. Define workload (W), use factor (U), and occupancy factors (T). What U factor would you will assign for the secondary barriers?

10. How do you calculate the workload for conventional and IMRT treatments?

11. Explain how shielding calculations are done for 90 degree patient scatter falling on a secondary wall that protects a control room occupied by the therapy technician.

12. After determining the barrier thickness for shielding against both scatter and leakage, how do you determine the thickness for the combined effect?

13. Write down the reduction factor (RF) equation for the primary barrier and define the terms in the equation.

14. Define TVL. The RF for a secondary barrier was determined as 2.7×10^3. How many TVLs will be required for the barrier?

15. Why is the machine leakage much larger for IMRT treatments, compared to conventional treatments, for the same prescription dose to the tumor?

16. What is the most common material used for shielding linac rooms? What are the advantages of laminated barriers that use steel or lead?

17. Define instantaneous dose rate and time averaged dose rate, and how are they elated to a prospective survey of a linac installation.

18. What are the disadvantages of directly shielded doors of a linac facility? How do you overcome these disadvantages?

19. Explain how the door is designed for a linac facility to shield against neutrons, capture gamma rays, and scatter X-rays.

11

Radiation Protection in Brachytherapy

11.1 Introduction

Brachytherapy means treatment at short distance. In brachytherapy gamma radioactive sources are

1. directly imbedded in the tumor (e.g. interstitial sites like head and neck, breast etc.)
2. kept close to the surface of the tumors using moulds or plaques (to treat skin lesions or the mucosal surface) or
3. placed into accessible body cavities (e.g. cervix, vagina etc.) or into the lumens (e.g. bronchus, esophagus, bile duct, blood vessels etc.) to treat these organs or tissues.

In this kind of treatment, the tumor receives maximum dose and the dose falls off outside the target volume, but is generally used for radical treatment of smaller localized tumors or as a local boost to EBT. The disadvantage of brachytherapy is that it can be only used for small localized tumors, with moderate gamma energies preventing too much radiation from 'spilling' out of the target volume. In radiotherapy, about 15–20% of the patients are treated by brachytherapy.

11.2 Treatment Dose Rates

There are three different dose rates provided by brachytherapy sources at the tumor position which are classified as high-dose rate (HDR), medium dose rate (MDR), and low-dose rate (LDR). The International Commission on Radiation Units and Measurements (ICRU) 38 [1] defines these dose rates as:

- LDR 0.4–2 Gy/h
- MDR 2–12 Gy/h
- HDR >12 Gy/h

The HDR treatment is delivered at a dose rate much greater than 12 Gy/h (HDR ^{192}Ir is delivered at >400 Gy/h) and a treatment fraction is delivered in minutes. The MDR range requires about 1–5 h to deliver a fraction. This range is seldom used in brachytherapy since treatment time is much longer compared to HDR, and the biological effects are also less favorable. The LDR range corresponds to the classical dose rates that have been used in brachytherapy for nearly a century since the start of the practice, with treatment times in the range of 24–144 h. In addition to the above three categories, there is also very low dose rate (VLDR) brachytherapy practiced with ^{125}I or ^{103}Pd sources, in permanent implant therapy. The dose rates in VLDR are in the range 0.01–0.3 Gy/h. Since HDR delivers a very high dose rate, the treatment is fractionated as in EBT to allow the normal tissues to recover. Still the classical dose rate (in radium therapy) is what the radiation oncologists are comfortable with since the biological response is very well known from the previous experience. So a pulsed dose rate, about one-tenth of the HDR dose rate treatment was introduced. In this treatment, the tumor is treated with short HDR pulses, lasting 10–30 minutes each, at hourly intervals, over a day or two. This allows time for normal tissue recovery during the course of treatment.

11.3 Development of Afterloading Techniques

When brachytherapy practice started 100 years ago, there was no choice of dose rates since Ra and Rn were the only two sources available for the treatment of brachytherapy, with very low dose rate output. The radium photon energies were also highly penetrating and so the source was not ideally suited for brachytherapy. The sources also delivered high doses to the physicians practicing brachytherapy because of direct handling of the sources. In addition to these disadvantages, there was also an additional risk of radon escape from the ruptured radium needles or radon tubes, giving rise to internal exposure problems.

The disadvantages of direct loading brachytherapy can be summarized as

- Unnecessary radiation dose to staff
- Hasty application to minimize staff dose
- No provision for correction when the geometry of placement was different from the intended geometry affecting treatment

With the advent of nuclear reactors in the 1950s, there was an opportunity to choose more suitable radioisotopes for brachytherapy in place of Ra or Rn. The first Ra substitute that became available for brachytherapy was ^{137}Cs, followed by ^{60}Co. Rn and Ra sources were replaced by ^{137}Cs tubes and ^{60}Co needles. However, the staff doses were still high, compared to EBT, since the sources were being handled directly. The sources were also too bulky to attempt afterloading techniques. (For instance, ^{137}Cs recovered from spent fuel elements was of low specific activity making the source bulky for the required activity.)

The production of artificial radioactive isotopes by neutron irradiation of stable elements in the nuclear reactor gave rise to several radioisotopes with more desirable properties (suitable half-life, suitable photon energies, high specific activity, etc.) for use in brachytherapy. The availability of high specific activity sources led to source miniaturization (e.g., modern ^{192}Ir or ^{60}Co sources have activities in curies but with a size of just a few mm diameter and a fraction of 1 cm in length), a prerequisite for practicing afterloading techniques. Miniaturization of the sources also resulted in smaller diameter applicators which were more comfortable to the patients during implantation or insertion. Soon

LDR sources and HDR sources were introduced in brachytherapy for manual afterloading (MAL) and remote afterloading (RAL) brachytherapy.

In the MAL technique, the applicator placement and source insertion became two separate activities. This gave rise to the following advantages:

- Elimination of staff radiation doses during applicator placement
- Better accuracy in applicator placement
- Verification of applicator placement
- Scope for correcting any errors in applicator placement

So MAL not only reduced staff doses but also increased treatment accuracy compared to the direct loading technique. The nursing staff, however, received some dose due to the presence of source in the patient. The physician dose also could not be completely eliminated since some dose would be received during source insertion and source removal. Protective shielding barriers were, however, made use of to reduce the staff doses considerably.

The development of RAL devices was a great step toward the reduction of staff exposures compared to direct loading or MAL brachytherapy. It also greatly improved the accuracy of patient treatment since the source loading applicators can be placed in the patient with no urgency, and the position can be adjusted through imaging means before loading the sources into the applicators. HDR brachytherapy also reduced the treatment time making it possible to treat patients as outpatients like in external beam therapy. Today HDR brachytherapy has mostly replaced LDR brachytherapy in treatments involving removable implants. The most common HDR RAL sources used in brachytherapy are ^{192}Ir and ^{60}Co. More than 90% of the HDR units are ^{192}Ir HDR units, though the principles of operation of the two units are very similar. Figure 11.1 illustrates the three loadings mentioned.

The principle of treatment of a patient using RAL unit is illustrated in Figure 11.2.

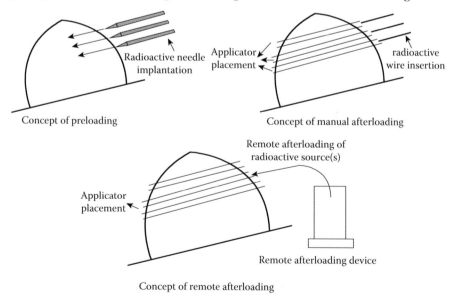

FIGURE 11.1
Breast implant illustrating the three implantation techniques.

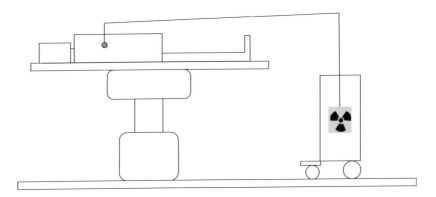

FIGURE 11.2
Principle of RAL brachytherapy.

Figures 11.3 and 11.4 illustrate ^{192}Ir HDR treatment of gynecological cancers and ^{125}I implant treatment of prostate cancer.

Two of the commercially marketed HDR units used extensively in HDR brachytherapy are shown in Figure 11.5. The ^{192}Ir HDR unit makes use of a single source of ^{192}Ir which can simulate a line source (e.g., ^{192}Ir wire) through programmed movement of the source along the source length it is trying to simulate. The source moves in a finite number of steps (called step lengths) to cover the full source length. In the case of planar implant or multiplanar implants, an applicator for each line source is placed in the patient, the number of applicators being equal to the number of source lines to be imbedded (to cover the target volume for irradiation). The source is then programmed to move through each applicator in steps to simulate each source line.

Each applicator is connected to one channel of the HDR unit through a transfer tube, as shown Figure 11.6. The HDR unit has several channels as shown in the figure and an indexer system directs the source to any channel (or applicator) as per the treatment program.

There are different types of applicators available to treat the different sites.

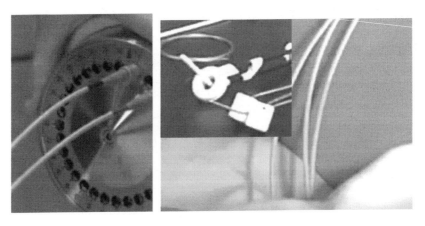

FIGURE 11.3
Treatment of gynecological cancers (applicator shown in the inset).

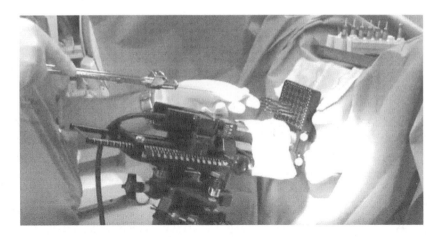

FIGURE 11.4
Treatment of prostrate cancer.

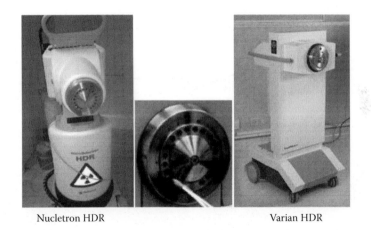

Nucletron HDR   Varian HDR

FIGURE 11.5
Remote afterloading ^{192}Ir HDR units. (Pictures courtesy Nucletron, Varian.)

The HDR ^{192}Ir source is a very tiny source (<1 mm diameter × 3.5–10 mm length, depending on the model) which is welded to a flexible drive cable. The source remains in a shielding container inside the HDR unit. According to the Nuclear Regulatory Commission (NRC) regulations, the leakage with the machine in the OFF position cannot exceed 1 mR/h at 10 cm distance from any accessible surface of the HDR unit.

Before the introduction of HDR RAL brachytherapy, the LDR radium therapy was first replaced by an LDR RAL brachytherapy device (called Selectron) with radium substitute sources (of ^{137}Cs). The device contained 48 pellets of 2.5 mm diameter each, 20 active, and 28 inactive sources. Each line source was replaced by a source train of ^{137}Cs spherical sources and inactive sources. The activity of each source was around 30 mCi. The device was manufactured by Nucletron of the Netherlands. The unit is shown in Figure 11.7; it has been largely replaced now by the ^{192}Ir HDR unit because of the versatility of the latter, so this device will not be considered in this chapter.

FIGURE 11.6
Each channel of the HDR connects to an applicator.

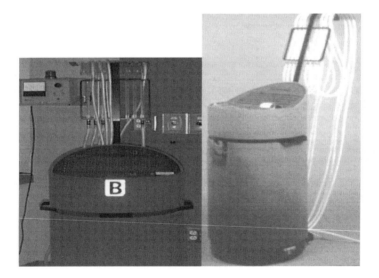

FIGURE 11.7
^{137}Cs LDR RAL unit.

11.4 Temporary and Permanent Implants

There are two modes of brachytherapy implants in brachytherapy—temporary and permanent implants. Permanent implant is mostly used for the treatment of prostate cancer using VLDR brachytherapy sources ^{192}I or ^{103}Pd. Because of the very low photon energies, the patient absorbs much of the energy and the dose rates around the patient are relatively very low. The patient is usually discharged and sent home when the radiation levels around them is considered acceptable so that the public or family members around the patient would not exceed the dose limits recommended for them by the regulatory authorities (the patient discharge criteria will be discussed in another section). Since the sources decay to negligible levels, there is no need to surgically remove the sources from the patient.

Temporary implants involve placing the sources in the treatment site only for a specific duration so the tumor receives the prescribed dose. The temporary implant can be an LDR treatment (e.g., treatment of breast cancer using ^{192}Ir wires or gynecological cancers using intracavitary applicators) or an HDR treatment—most of the sites are treated this way. The temporary sources need to be replaced periodically depending on their half-lives. The permanent implant sources also must be procured regularly depending on the number of patients treated.

Most of the Ra and Rn sources have been withdrawn from brachytherapy practice. Presently small ^{103}Pd or ^{125}I seeds are being used as permanent implants to treat prostate cancer and ^{192}Ir sources (^{192}Ir wires or HDR ^{192}Ir) and ^{60}Co HDR sources are commonly used as temporary implants.

11.5 Source Characteristics

The brachytherapy sources are available in various forms (e.g., tubes, seeds, pellets, wires). All the sources are doubly encapsulated to filter out particulate radiations and also to prevent the escape of radioactive material into the ambient which will pose health risk to the staff. One of the safety checks to be carried out periodically in brachytherapy is to ensure that the sources used are not leaking (i.e., the source integrity is maintained).

The characteristics of the most commonly used brachytherapy sources are given in Table 11.1.

TABLE 11.1

Properties of the Most Commonly Used Brachytherapy Sources

Source	Eff. Energy	T1/2	HVL Pb mm	TVL Concrete cm	AKRC* µGy/h.m²/MBq	Source Form	Application
^{60}Co	1.25 MeV	5.26 years	12	22	0.309	Spherical	HDR
^{137}Cs	0.66 MeV	30.2 years	6.5	17.5	0.079	Tubes, needles	LDR interstitial
^{192}Ir	0.38 MeV	73.8 days	6	14.7	0.116	Seeds, wire, ribbon	LDR/HDR
^{125}I	28 keV	59.6 days	0.025	–	0.034	Seeds	Permanent implant
^{103}Pd	20 keV	17 days	0.02	–	0.035	Seeds	Permanent implant

Note: AKRC, air kerma rate constant; HDR, high-dose rate; HVL, half-value layer; LDR, low-dose rate; TVL, tenth-value layer.

11.6 Source Output Specification

The source output (in air kerma rate [AKR] or exposure rate [ER]) at distance d from a point source of activity A, in a scatter-free room, is given by

$$K_a(d) = A \; \Gamma_{Ka}/d^2 \; (Gy/h); \; X(d) = A \; \Gamma_x/d^2 \; (R/h)$$

where Γ_{Ka} and Γ_x are known as AKR and ER constants. If A is expressed in mCi, AKR and ER in Gy/h and R/h and d in cm, the units of the rate constants are Gy cm^2/mCi·h and R cm^2/mCi·h. Specifying d in m, the source output at 1 m is given by

$$K_a(1 \; m) = A \; \Gamma_{Ka} \; Gy/h; \; X(1 \; m) = A \; \Gamma_x \; R/h$$

This quantity is referred to as the reference air kerma rate (RAKR) or air kerma strength (AKS, S_k) of the source. RAKR/A gives the RAKR per unit activity. If the latter quantity is given, as in the International Atomic Energy Agency (IAEA) Safety Series No. 47, it must be multiplied by the source activity to get the source output at 1 m distance. RAKR or S_k corresponds to the dose rate D_0 used for specifying the linac dose rate at 1 m distance, in an earlier chapter. There is some confusion in the use of the units for RAKR and AKS but they are numerically the same. RAKR is actually measured at 1 m distance from the source and so it gives the AKR at 1 m distance in units of μGy/h at 1 m. On the other hand, the definition of AKS gives the opportunity of measuring the AKR at any convenient distance d from the source. When referring to the measured value to one 1 m distance to define AKS, as K(d) d^2, the units are μGy m^2/h but numerically RAKR and AKS have the same value since they give the AKR at 1 m distance from the source. RAKR is used in Europe while the United States and few other countries use AKS.

For radiation protection purposes, the brachytherapy sources can be treated as point sources and the above equation can be used to determine the source output (neglecting the room scatter correction to the computed source output).

Sometimes, the source output is specified in units of RHM, which refers to the output at 1 m in R/h per unit activity. For instance, the exposure rate constant (ERC) of the ^{192}Ir source is 4.7×10^{-1} R/Ci h at 1 m. So, the RHM of the source is 0.47.

AKR or ER can be converted to tissue dose rate using appropriate conversion factors. For shielding calculations, however, air kerma or exposure can be used directly instead of converting to tissue dose, as will be illustrated in the next section.

11.7 Radiation Protection in LDR Brachytherapy

Many of the LDR techniques, especially those involving MAL, have become obsolete. The only procedures used in LDR brachytherapy are the ^{192}Ir wires for interstitial therapy (e.g., breast implant), a temporary implant procedure, and the use of ^{125}I seeds (or ^{103}Pd seeds) in prostate cancer treatment, a permanent implant procedure. Permanent implant therapy is preferable for sites that are difficult to access (e.g., prostate) though removable implants may give higher accuracy in planning and treatment. Some basic safety aspects relating to these procedures will be described next.

11.7.1 Removable Implant LDR Brachytherapy

In this LDR MAL technique, the applicator is fixed in the patient on an operation table; the accuracy of applicator placement is confirmed by imaging procedures and if not according to plan, the applicator positions are readjusted. The implant dosimetry is then done using the treatment planning system and the treatment duration computed. The sources are then prepared in the source preparation room, their AKSs determined using a well-type ionization chamber, then they are loaded into a shielded container and transported to the treatment room for afterloading.

Figure 11.8 illustrates the secure MAL procedure with minimum risk of the wire leaving the treatment position.

The applicator used in the patient is a rigid needle or flexible tubing. The advantage of a rigid applicator is that there is no need for a dummy wire for localization. With a rigid template, the needles can be placed according to plan and no localization may be necessary. Rigid templates are often used in breast implants (with a template on either side) or in the perineum (with a single template). The disadvantage of flexible tubing is that the implants are rarely straight.

A template-based breast implant is shown in Figure 11.9.

After the completion of treatment, the radiation oncologist removes the sources (generally along with the applicator) and places them in the shielded container. The sources are

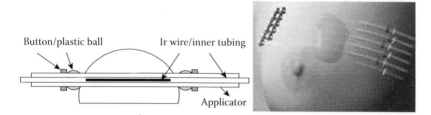

FIGURE 11.8
MAL procedure in LDR brachytherapy.

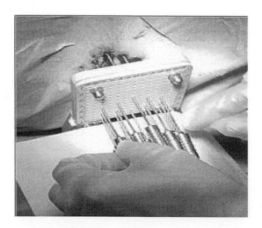

FIGURE 11.9
Template for source implant (Picture courtesy IAEA website).

then removed from the applicator in the source preparation room, under the L-bench. If the tubing is cut to remove the sources from the patient, the patient must be surveyed to ensure that no part of the source is left behind by accident while cutting. (The survey meter must have been checked beforehand for proper functioning to be sure of the validity of the patient survey reading.)

11.7.2 Permanent Implant in LDR Brachytherapy

In the case of a permanent implantation, ^{125}I seeds are generally used. The seeds are procured as loose seeds for individual implantation, or in strands with 1 cm spacing between the seeds. The strands are usually cut to the required length in the source preparation room and preloaded into the seeding needle before implantation. The sources are directly loaded by the radiation oncologist using a template for directing the needles according to plan (as shown in Figure 11.4). The seeds are placed at the desired positions using a rectal ultrasound probe (Figure 11.10).

FIGURE 11.10
^{125}I source handling.

Use of electronic personal dosimeters (EPDs) for monitoring staff doses during brachytherapy procedures is useful for optimizing staff exposures. It is mandatory to wear personal monitoring badges during these procedures.

11.7.3 Source Log

Details of all the sources (source type, identification, activity on a date, measurement data, etc.) available in the department must have been entered into a log book. The source usage data (date of use for a patient, patient room number, treatment duration, date and time of the source leaving or returning to the source room, etc.) also must be entered into the book. The log must be maintained for a minimum period of 3 years, or per as per the country's regulatory requirements. Today, there are not many discrete sources used in brachytherapy. Apart from the HDR source which is confined to the machine, only ^{125}I sources, ^{103}Pd sources, and ^{192}Ir wires are used in brachytherapy which need to be properly accounted for. Fortunately, they are of low activity and can be easily stored in a small lead safe.

11.7.4 Source Preparation

The source preparation room is used for visual inspection of sources, source storage, source preparation for implantation, and source assay before transporting it to the treatment room for placing in the patient. In the case of ^{192}Ir wires, they need to be cut from the coil to the required length, properly identified and placed in an inner tubing and closed at the ends with a seal before they are placed in the shielded container, ready for afterloading. Commercial loading devices are also available for this purpose. Often dummy wires are used for localization purposes before loading the actual wires for treatment.

An L-bench, used for source handling, and a lead pot for storing the sources are shown in Figure 11.11.

The L-bench is made of lead with a lead glass top to observe source preparation and source assay using a re-entrant type of chamber. The chamber must have a calibration factor

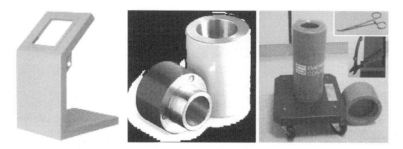

FIGURE 11.11
L-bench and lead pot, and emergency container for storing sources.

for each model of the source for determining the source activity. Generally only a random sample (say 10%) of the seeds is assayed to establish their activity. All the source handling must be carried out behind an L-bench to reduce the dose to the staff member. The sources must always be in the lead pots (for transport) or in a lead safe and not on the workbench except during source preparation. The storage container must have sufficient shielding thickness to restrict radiation levels around the container surface (at 10 cm distance) to <1 µSv/h. The container must carry a radiation symbol to indicate the presence of radioactivity.

The tools used for source preparation (long-handled forceps, scissors, etc.) must not be used for any other work since they may have some radioactive contamination. Contamination checking of the workbenches must be routinely carried out so that no small pieces of active wire remain on the table which may get lost causing a radiation risk, however small the risk may be.

The room must have a zone monitor to indicate the radiation levels in the room at all times, and a contamination monitor to detect any contamination on the surface of the workbench used for source preparation, or for surveying the room garbage before it is sent out. A high sensitivity survey meter (Geiger–Muller (GM) or preferably NaI scintillation based) must be available for the purposes of survey.

11.7.5 Source Transportation

The sources are carried to the application room from the source handling room using a (lead) transport container. They come on a cart and can be easily wheeled to the treatment room. Before leaving the source room the container must be surveyed to ensure that it is adequately shielded. The radiation levels must be <1 mSv/h at 1 m from the surface of the container and <2 mSv/h on the surface of the container, respectively. The source handling room and the treatment room must be close to each other, preferably in an exclusive brachytherapy suite so that the source need not be taken through public areas.

The source must be under the control of the authorized person at all times from the time of leaving the room until it returns to the room, and it cannot be left unattended at any time. The transport container must be locked securely or latched, so that the source will not fall out of the container during transport if it topples, for any reason. The container must have a radiation symbol on the outside of the container. The container usually stays in the patient room during treatment and can serve as an emergency container for storing the source in case the source comes out of the patient for any reason. The required tools must be available for transferring the source to the storage container. After the treatment, the source is returned to the source room in the same container.

11.7.6 Source Leak Tests

Before the sources are put to clinical use it is important to check, at regular intervals, that the sources are not leaking and that the distribution of activity in the source is uniform, as expected.

- All the brachytherapy sources procured for clinical use must come with a leak test certificate (along with the date of testing) to ensure that the newly purchased sources are not leaking.
- The common method of leak testing is to wipe the source capsule with a tissue moistened with alcohol and look for any radioactivity contamination, using a GM detector or a scintillation detector that is capable of detecting 5 nCi (or about 200 Bq) which is taken as the criterion for source leakage testing.
- Any source showing a wipe test detected activity >5 nCi must be isolated and should not be used.
- For higher activity sources, a wipe test can be carried out inside the source container or inside a source carrying applicators and any detected activity >5 nCi points to a leaking source.
- Similarly, if the source forms a closed circuit such as the source in HDR equipment, the source leakage can be indirectly assessed by checking the applicators when the sources are replaced.

The frequency of check recommended is at 6-month intervals. When the sources are replaced within that period, no intermediate checking is necessary. In the case of sources used in RAL units, the sources are not accessible for carrying out any direct checks.

11.7.7 Activity Distribution in the Source

- The activity distribution can be checked by autoradiography method.
- In the case of RAL units, the applicator can be placed on a film and the source is programmed to stay on the film at a predetermined position to take the autoradiograph (this method can also be used for testing the dwell position accuracy and the step length in the case of RAL units).

11.7.8 Prevention of Radioactive Contamination

Any airborne radioactivity can be inhaled or ingested into the body. So, it is important to check brachytherapy sources periodically for any breach of source integrity (e.g., the welding of the encapsulation is not correct) so that in case of any loss of integrity, escape of radioactive material can be detected before serious contamination of the facility, equipment, or personnel occurs. The facility must maintain the records of sealed source leak tests. Contamination may also occur during source preparation (e.g., cutting of ^{192}Ir wires for a patient implant may lead to contaminated tools). So these tools should not be taken out of the source preparation bench for any other purpose. Following the cutting of ^{192}Ir wires, very small pieces of the wire, too small to the naked eye, may be left behind. A contamination monitor or a scintillation-based detector should be used to detect any radiation on the preparation table and safely dispose of these pieces.

11.7.9 Source Replacement

^{137}Cs, ^{60}Co, and ^{192}Ir sources require periodic replacement. Short half-life sources like the HDR ^{192}Ir sources need to be replaced every 3 months. This requires standard procedures for source procurement and transfer, and also for the disposal of the used sources. Usually, there is an agreement between the user and the supplier so that the supplier takes back the used sources during replacement. There are also national waste management facilities in many countries, which may accept all the used sources or at least the ones purchased within the country.

11.7.10 Security of Radioactive Sources

The majority of serious accidents and incidents with the use of sealed radioactive sources, or source containing devices, in medicine are as a result of loss or theft. Security of radioactive sources is a major concern today and regulatory agencies are more inclined to recommend application of alternative technologies instead of radioisotope technologies in society. For example, there is some concern regarding the continued use of ^{60}Co telecobalt machines in cancer treatment, when linac technologies are available for the same purpose. All efforts must be made to ensure the security of all the brachytherapy sources so they are not lost or stolen. In this connection, the following points must be taken into consideration:

- Radioactive sources must be under the control of an authorized staff member at all times.
- The authorized person will be chiefly responsible for ordering, receiving, storage, handling, usage, and disposal.
- Access to brachytherapy sources will be generally limited to the radiation oncologists, medical physicists, and the radiation safety officer (RSO).
- A source inventory log must be maintained in the source (storage) room where details of outgoing sources (source identity, number of sources, date and time of leaving the room, etc.) and their destination (patient's name and room identification) will be entered and tallied against the sources returned to the storage room.
- The source room must always remain locked and only authorized personnel can have access to the room keys.
- The source room must be posted as a restricted area with a radiation symbol.
- The transported sources must be carried in a shielded transport container securely latched, so that no source can come out of the container if the container is dropped or turned over.

11.8 Principles of Radiation Hazards Control

The objective of radiation protection is the same as in EBT namely protection of the patients, staff, and the public. The staff involved in brachytherapy practice are the radiation oncologists, physicists, nursing staff, and the technicians. The principles of radiation hazards control time/distance/shielding and as low as reasonably achievable (ALARA) play a more significant role in brachytherapy compared to external beam therapy.

Since brachytherapy sources are often handled by staff for calibration or for implantation, the dose to the extremities (mainly hands) also become important, in addition to the whole body dose limitation.

11.8.1 Time

The staff dose is proportional to the duration of radiation exposure, so the MAL of sources must be performed in the minimum possible time. This requires ready availability of source handling tools and training. By practicing the loading procedures using dummy sources, the loading time can be optimized.

11.8.2 Distance

The longer the source handling tool, the less is the staff dose due to the approximate inverse square law of the fall in the dose with distance. At the same time, the source handling tool (e.g., forceps or tongs) must be of a convenient length. The source should never be touched with bare hands since the contact dose is extremely high due to near zero distance. When many sources have to be loaded, the sources that are not being handled must be in a source container and not outside it even if this makes the loading more convenient.

11.8.3 Shielding

Movable shielding materials are mainly used in LDR brachytherapy to reduce doses to the radiation oncologist during the loading of the ^{192}Ir wires or other isotopes (e.g., ^{137}Cs tubes). The shield is also used during nursing procedures to reduce staff doses. A movable lead shield is shown in Figure 11.12.

FIGURE 11.12
Movable lead barriers placed between staff and the patient.

The shield is typically 40 inches × 24 inches in size, though other sizes are also available. The thickness is about 1 inch which is roughly the tenth-value layer (TVL) in lead for [137]Cs. It will cut down the dose by 90% and higher for other isotopes like ^{192}Ir which are less penetrating. L-benches are used on the source preparation table as explained before, and lead pots are used for storing or transporting discrete brachytherapy sources. In the case of RAL units, the sources can be remotely retracted before attending to the patient, and hence there is no radiation safety problem with RAL units.

11.8.4 Patient Discharge Criterion (for Permanent Implant Patients)

Before discharging the patient, the implant activity must be less than the recommended discharge activity or the radiation level around the patient less than the recommended dose rate. These two parameters are determined to ensure that the dose received by the public or the family members coming in contact with the patient, does not exceed the recommended dose limits recommended for the two categories of people. The general regulatory requirement is that the dose to the public should not exceed 1 mSv (the annual dose limit). Since the public can be exposed to more than one source of radiation (e.g., someone living close to a nuclear installation, and also a cancer center, may receive a small amount of dose from each of these sources), some regulations require that the public dose limit should not exceed 0.3 mSv from a single source. So some countries may impose 0.3 mSv as the maximum dose that can be received by the public from the patient. If the family members are treated as caregivers, then a special limit of 5 mSv is imposed on them since they receive special benefit by attending their relative and may exceed the public dose limit. But in most cases, the dose received by those around the patient is very low.

The method of setting this limit is as follows. Figure 11.13 illustrates the concept.

The dose rate around the patient is proportional to the implant activity, so the limits set in terms of these quantities are interrelated. The total implant activity at the time of implant, A_0, is known. Since the activity decreases with time in an exponential manner with a typical half-life, the dose rate around the patient also follows the same law. The implant activity or the dose rate after a certain time t, following the implant is given by

$$A(t) = A_0 (1 - e^{-0.693t/T_{1/2}})$$

$$DR(t) = DR_0 (1 - e^{-0.693t/T_{1/2}})$$

where DR is the dose rate at a specified distance from the implant.

The dose rate can also be computed for the implant imagined to be in free space, but the actual dose rate would be less because of the patient attenuation of the radiation which depends on the photon energies. Since the photon energies of ^{125}I and [103]Pd are very low (28 keV and 21 keV, respectively), much of the radiation will be attenuated by the patient (say by a factor of 30 or more). So, the dose rate around the patient is usually measured using a survey meter.

The effective activity of the implant activity, τ, is given by $1.44 \times T_{1/2}$, where $T_{1/2}$ is the physical

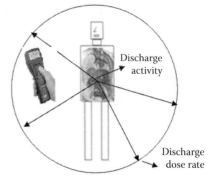

FIGURE 11.13
Dose rate measurement at 1 m distance from the patient body.

half-life of the source. The total disintegrations or the total dose at a specified distance, for a time interval "t" is given by

$$N(t) = A_0 \times 1.44 \ T_{1/2} \ (1 - e^{-0.693t/T_{1/2}})$$

$$D(t) = DR_0 \times 1.44 \ T_{1/2} \ (1 - e^{-0.693t/T_{1/2}})$$

The total number of disintegrations, N, or the total dose at a specified distance, D, for the complete decay of the activity can be obtained by integrating the above equations for infinite time.

$$\text{Total exposure at specified distance } D = (DR_0 \times 24) \times 1.44 \ T_{1/2}$$
$$= D_0 \times 34.6 \times T_{1/2} \ (\mu Gy)$$

This is the dose anyone coming close to the patient will receive, if they stay at the specified distance all the time.

In the above equation, since $T_{1/2}$ is in days, DR_0 has been converted to $\mu Gy/day$ by multiplying by 24 hours/day.

$$DR_0 = \frac{A_0 \Gamma_{ka}}{d^2} (\mu Gy/h) \ \times 1(\mu Sv/\mu Gy)$$

Assuming a resonable occupany factor T for the person concerned, the total dose that will be received by him is given by

$$D = \frac{A_0 \Gamma_{ka}}{d^2} \times 34.6 \times T_{1/2} \ T$$

A typical conservative value of the occupancy factor can be taken to be 0.25. A_0 is the activity at the time of release if the patient is released soon after the implant. If the activity is considered higher than the discharge limit recommended by the national bodies, then the patient will be released when the discharge limit activity is reached. The activity at the time of release will be recorded by the staff before releasing the patient. The typical activity of the implant is in the range of 10–50 mCi; if the cumulative dose at 1 m from the implant comes to less than 1 mSv, the patient can be discharged. Similarly, if the measured dose equivalent (DE) rate at 1 m distance from the patient is <10 μSv/h (<1 mR/h), the patient can be discharged (NRC regulation). Some national bodies have a higher patient release DE rate of 25 μSv/h or less. The patient, however, will be discharged only after the staff survey the patient and ensure that the patient meets the regulatory release requirements.

If the dose to the family member is likely to exceed the regulatory limits, which is very unlikely, the patient is given specific instructions regarding contact with the family members.

Example 11.1

The total activity of a ^{125}I implant was 1.48 GBq. What will be the dose rate at 1 m from a point source of the same activity as the implant, assuming no attenuation in the implant or in the patient? The air kerma rate constant of the source is 1.27 μGy m^2/mCi·h (1 Gy = 1 Sv).

The dose equivalent rate (DER) at 1 m distance from a point source of 40 mCi = 1.27 × 40 = 50.8 μSv/h.

The measured DER at 1 m from the patient is usually much lower in the range of 5–10 μSv/h.

11.8.5 General Instructions for the Discharged Patients

- ^{125}I is a low-energy brachytherapy source and is significantly attenuated by the patient's body. So, a patient released by the hospital poses no radiation risk to the public or the family members, including children and pregnant women. So no radiation precautions are generally needed for the patient.
- However, in keeping with the ALARA principle, the patients are advised not to allow children on their laps or prolonged close contact of pregnant women, for the first 2 weeks after the implant.
- Anyone can be allowed for any length of time at a distance of 1 m from the patient.

11.8.6 Considerations for Postmortem/Funeral of Permanent Implant Patients

In the case of death of a patient carrying permanent implant of ^{125}I seeds, the action to be taken depends on the magnitude of the radiation levels around the patient. Cremation can be allowed if 1 year has passed following the implantation (3 months for ^{103}Pd) which will have negligible effects on the crematorium staff or the public (ICRP 98 [2]). If it is less than the period stated, then the sources may have to be removed before cremating the patient. In the case of ^{125}I seed implant, if the implant activity is low (i.e., the ambient DE at 1 m from the body is <25 μSv/h), it is unlikely that the morticians or embalmers would be exposed to significant doses during these processes [3]

Example 11.2

The DE rate constant of ^{125}I point source is 0.037 μSv·m^2/h MBq. The ash collected from the crematorium of an ^{125}I implant patient, who died soon after the implant, contained three seeds. The seed activity during implant was 0.4 mCi each. Assuming the person collecting the ashes is holding it at a distance of 1 foot from the body, for 2 h, how much dose he will receive? Is it an acceptable dose for the person?

$$1 \text{ mCi} = 37 \text{ MBq}$$

$$0.4 \text{ mCi} = 0.4 \times 37 = 14.8 \text{ MBq}$$

DE at 1 m from 0.4 mCi ^{125}I seed source = 0.037 × 14.8 = 0.55 μSv/h.

Actually the DE will be less than this due to self-attenuation in the seed source. DE at 1 foot from the body will be more by a factor of $(100^2/30^2) \approx 10$, that is, 5.5 μSv/h. If the ashes contain three seeds the dose received in 2 h is given by

$$3 \times 5.5 \times 2 = 33 \text{ μSv}$$

The permitted dose for the public is 1000 μSv annually. So 33 μSv is acceptable dose for the person. Though the patient implant might have contained 50–60 seeds or more, one can expect only a couple of seeds to show up in the collected ashes.

Example 11.3

The ^{125}I implant of a patient contained 70 seeds of 15 MBq each. The patient dies immediately after the implant. If 1 MBq is the exempt activity for cremation, how long one should wait before cremating the patient? (Half-life of ^{125}I is 60 days).

$$\text{Total activity of the implant} = 70 \times 15 = 1050 \text{ MBq.}$$

To become <1 MBq, the activity must go down by 1050 times. Since each half-life will reduce the activity by half, one requires 10 half-lives to reduce the activity by 1024. This will take 10 × 60 or 600 days or about 20 months. However, this is the time required for the cumulative activity of all the seeds in the implant to become 1 MBq. Since the ashes collected will contain only a couple of seeds a collection of, say, 2 or 3 seeds should be calculated. For an individual seed, in six half-lives, the activity will become 15/64 = 0.23 MBq. So, for a concentration of three seeds, the total activity would be 3 × 0.23 = 0.7 MBq (<1 MBq). So, the body can be cremated after six half-lives or after 1 year from the date of implant. So if the patient dies within a year, the seeds must be removed before cremating the body, as per the exempt criterion given here.

11.8.7 Nursing of Brachytherapy Patients

In RAL HDR brachytherapy, radiation exposure to staff members and nursing staff is virtually eliminated. The treatment can always be interrupted briefly for nursing care and then the treatment resumed. In the case of LDR MAL treatments, the dose to staff and nursing personnel can be reduced by increasing distance from the patient, reducing time spent in the radiation field, and by making use of bedside shields. Availability of audiovisual contacts with the patient will prevent the nurse from entering the patient room unnecessarily which would increase their dose.

LDR brachytherapy patients may have temporary (say ^{192}Ir) or permanent (^{125}I or ^{103}Pd) implants. Temporary LDR implants stay in place for a predetermined treatment duration (that can last few days) before being removed. These patients must be placed in special rooms and are to be on bed rest. The nurse should be aware of the following rules and nursing instructions while attending on the patient:

- The nurse must wear the thermoluminescence dosimeter badge (and a ring dosimeter to assess the dose to the hands).
- The patient is a source of radiation exposure.
- The patient room door must have a sign stating *Caution: Radiation Area*.
- The patient chart also should have a form indicating *Caution: Radioactive material* for both the temporary and permanent implant carrying patients.
- The nursing instructions must be affixed to both the room door and the patient chart.
- The sources and catheters should never be touched with bare hands.
- The patient must not get out of bed and must be advised accordingly. If this happens, the nurse should tell the RSO.
- In case the patient needs to be moved out of the room (say for X-raying or for any other purpose), the wheelchair or the stretcher used to transport the patient must prominently carry the radiation warning sign.
- Portable bedside shields help in reducing the dose received from the patient to staff and visitors.
- Sometimes the bed shields may be employed to reduce the dose rates near the door or in some regions outside the room. These shield positions should not be disturbed for any reason.
- The nurse should stand behind the mobile shield whenever possible (lead aprons are not useful for ^{192}Ir and higher energy sources).

- No one must be allowed inside the room without any reason.
- If any source falls on the floor it must be picked using a long forceps and placed immediately in the shielded container, before informing the RSO.
- Written procedures for handling anticipated emergency situations must be available in the nursing station.
- Patient must be surveyed after the sources are removed and before being taken out of the room.
- Room linen and dressings must be surveyed for any sources before sending them out of the room.
- Cleaning staff must not clean the patient room or must be accompanied by the nurse who will ensure that only essential cleaning is done (not linen and other items).
- Pregnant visitors and visitors under 18 years should not be allowed to visit the patient.
- Adult visitors must stay a couple of meters away from the patient for a specified period (say a couple of hours or as recommended by the institution). Often the "stay times" are posted on the entry doors, estimated from the monitored dose rate at the visitor position, and the time needed to reach the permitted dose for the visitor (usually 0.1 mSv/wk or 0.02 mSv/wk corresponding to an annual dose of 5 mSv or the public dose limit of 1 mSv, respectively, for a 50-week year).
- Prior to the release of the patient, staff will account for all sources implanted in the patient by tallying them with the sources removed from the patient, then surveying the patient and also the room to ensure no source is left behind, and to check for any lost sources.
- Removal of the radiation sign following the release of the patient indicates that there are no entry restrictions for the room any more.
- If the patient dies during treatment, they must remain in the room until the temporary implants are removed and the patient released by the physician concerned, and the RSO.

In the case of permanent implants (e.g., ^{103}Pd or ^{125}I seeds) used in the treatment of prostate cancer, the implants are routinely performed on an outpatient basis. The nurse must be aware of the following points regarding these implants, if the patient is hospitalized for nursing care or for other reasons such as HDRs of implant.

- The patient must be admitted in a separate room meant for brachytherapy patients.
- The gamma energies are very low and the body attenuates most of the radiation. So, the patient is not a significant source of radiation exposure.
- The seeds are extremely small (about 5 mm in length) and can be found in urine or also in linens, dressings, or trash.
- If a seed is found it must be picked up using tweezers, placed in a container, and the RSO must be advised.
- Lead aprons can be used for ^{125}I and ^{103}Pd source energies as a personal protective barrier to reduce the staff or nurse's dose (but they are totally ineffective for ^{192}Ir or higher gamma energy sources, and should not be used for these isotopes).

11.9 Safety Features of Remote Afterloading Devices

The RAL technique takes three steps involved in MAL technique irrelevant, thus virtually eliminating the staff occupational exposures. These are

1. Source preparation
2. Source placement
3. Source removal

In addition, the dose received by the oncology nurse is also virtually eliminated since the treatment can be interrupted for nursing procedures for LDR RAL therapy. With the discontinuation of RAL LDR in favor of RAL HDR treatment that lasts just a few minutes, there is no nursing care required for the patient.

There are three well-known ^{192}Ir HDR models in the market and one ^{60}HDR model. All the RAL devices have the following common features to ensure radiation safety:

1. A well-shielded primary storage safe with leakage dose rates as low as <1 μSv/h
2. A drive mechanism to move the source to the treatment position and retract it back to the safe at the end of treatment
3. An inbuilt radiation detector to ensure that the source indeed returned to the safe along with the drive cable to which the source is welded, on termination of treatment session
4. A check cable to check that the path from the safe to the treatment position is a closed path and not open or having no obstruction (e.g., transfer tube is not connected to the indexer end or to the applicator end or a kink in the path)
5. Automatic retrieval of the source back to the safe in case of power failure or other emergencies
6. An emergency shut off in case the automatic retrieval fails
7. A manual retraction mechanism if the above two mechanisms fail

Additional safety features in the HDR device (that must be confirmed during quality audit testing) are:

- If the transfer tube is not connected to the programmed channel, the source must not come out (this is tested by programming a channel and then initiating treatment without connecting the transfer tube).
- If the transfer tube is connected to a channel other than the one programmed, the source must not come out. (This is tested by programming a channel and then connecting the transfer tube to a different channel and initiating treatment.)
- The source must go to the programmed position and move in steps as per the programmed step length (this can be verified by placing the applicator on a film and programming the console to move the source to the desired position, and other positions, with predetermined step lengths, as shown in Figure 11.14).
- If there is any kink in the transfer tube, the source must retract to the RAL equipment (this is tested by deliberately creating a kink in the transfer tube and initiating treatment).

FIGURE 11.14
Source position and step length verification.

11.10 Shielding Calculations for Remote Afterloading Brachytherapy Room

A typical LDR/HDR RAL treatment room layout is shown in Figure 11.15. The shielding calculations are similar to the EBT treatment room calculations as discussed in Chapter 10. The shielding thickness of the barriers depends on the activity and energy of the sources used, the treatment duration, and the number of patients treated per week. These parameters must be maximized to arrive at a conservative estimate of the barrier thicknesses. Since source implantation or catheter placement is an invasive surgical procedure, the brachytherapy treatment room must be preferably adjacent to the operating theater and also the X-ray and CT facilities.

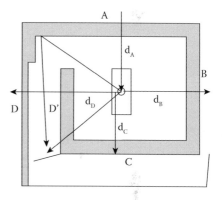

FIGURE 11.15
A typical RAL brachytherapy facility.

The shielding concept is illustrated here taking a typical example of ^{192}Ir HDR facility. The typical maze design has been used, as in the case of EBT. If a maze is not incorporated, the door will require additional lead shielding to reduce the radiation levels outside the treatment room. Unlike the EBT rooms, all the barriers are primary barriers in a brachytherapy facility as the source emission from the patient is isotropic. The position of the source (in the patient) is assumed to be at the center of the room for shielding calculations and so the same geometry must be used for the treatments. If the RAL unit is moved around in the room, it may affect the shielding thickness of the barriers and can make the walls unnecessarily thick. If ^{60}Co HDR is used for HDR brachytherapy, the shielding calculations must be performed for the ^{60}Co HDR source activity and energy. If the HDR unit is used in a linac treatment room then no calculations may be necessary, since the barriers are sufficiently thick for HDR brachytherapy.

The number of brachytherapy patients treated per week is always limited because of the smaller fraction of the radiotherapy treatments referred to brachytherapy and the availability of a surgical room for the preparation of the patient for brachytherapy.

Since low occupancy or low workload can increase the dose rate in the regions outside the treatment room, further constraints are imposed on the dose rates in these areas, as in the case of EBT. For an uncontrolled area, the UK regulations impose the conditions that the instantaneous dose rate (IDR) must be <7.5 μSv/h. The time-averaged dose rate in a day TADR$_8$ must be less than 0.5 μSv/h.

11.10.1 Workload Estimate

Using the source specification concept used earlier, the workload W is given by

$$W = [RAKR \text{ (or AKS)} \times t \times n] \text{ Gy/wk at 1 m from the source}$$

where
  t = treatment duration (in hours)
  n = no. of patients treated per week
  So, tn gives the total treatment hours per week.
  If the source output is in reference ER or exposure strength, the ERC can be used, and the air kerma to exposure conversion factor which is 0.87 cGy/R, or simply assume 1 cGy $\approx$ 1 R, which is a good approximation for the purposes of radiation protection.
  The patient will attenuate the radiation and so an attenuation factor can be incorporated into the workload. Not taking it into account adds a margin of safety to the barrier thickness computed.

Example 11.4

For ^{192}Ir RAL HDR treatment, the typical treatment time is about 5 minutes and in a day a maximum of four patients may be treated (or 20 patients per week). For a 5 day, 8 h/day working week, calculate the weekly workload.

Total treatment time/wk = 20 × 5/60 h/wk = 1.66 hours.

For conservative calculations, 3 h/wk can be assumed.

Take the activity as 10 Ci and Γ_x for ^{192}Ir as 4.69 × 10^{-4} R m^2/mCi·h.

Weekly workload = 3 h × 10 × 10^3 × 4.69 × 10^{-4} = 14 R/wk at 1 m

A conservative value of 20 R/wk at 1 m can be assumed for the workload (i.e., 2 × 10^4 mR/wk) of an ^{192}Ir HDR unit. In terms of AK, the workload can be taken as 20 cGy/wk.

11.10.2 Evaluation of Reduction Factors

The same formula used in Chapter 10 applies to brachytherapy facility as well:

$$RF = \frac{WUT}{Pd^2} = \frac{WT}{Pd^2}$$

  The patient being the source of radiation due to source implant, all the barriers receive radiation at the same time and must be considered as primary barriers of use factor U = 1. Therefore, only the occupancy needs to be considered for the areas around the facility. For controlled areas, T = 1; for an uncontrolled area, the occupancy depends on the nature of the area.
  The scatter calculations for determining the dose rates at the maze entrance to the treatment room are similar to the linac room maze calculations. The two components contributing to dose rate at the entry point are the primary incident on the maze wall through the maze opening and reflected toward the maze entry point, and the primary transmitted through the maze wall and reaching the door.

The shielding calculations will be illustrated below in a few examples for the typical facility shown in Figure 11.15. The tenth-value thickness (TVT) values have been taken from NCRP 40 and NCRP 49 [4,5]. The TVT values and the air kerma rate constant (AKRC) values may slightly vary depending on the sources of the data, but the main objective of this chapter is to illustrate the shielding calculation methodology.

11.11 HDR ^{192}Ir Shielding Calculations

Example 11.5

Calculate the RF for the barrier in the B region, in Figure 11.15 for a ^{192}Ir HDR unit. The following data are given.

$$U = 1$$

$$\text{Occupancy } T = 1$$

$$d_B = 2.57 \text{ m; TVL of } ^{192} \text{ Ir} = 13.5 \text{ cm concrete}$$

$$P = 0.02 \text{ mSv/wk (2 mR/wk) for the B region, which is not controlled.}$$

$$RF = WUT/ Pd^2 = 2 \times 10^4 / 2 \times (2.57)^2 = 1.51 \times 10^3$$

$$\text{Log}_{10}(RF) = 3.18$$

$$\text{Thickness required for wall B} = 3.18 \times 13.5 = 42.93 \text{ or } 43 \text{ cm}$$

Example 11.6

Calculate the RF for the barrier in the C region (shown in Figure 11.15) which houses the control console of the HDR unit. The following data are given.

$$U = 1; \text{Occupancy } T = 1; d_C = 2.675 \text{ m; } P = 0.1 \text{ mSv/wk (10 mR/wk)}$$

$$RF = WUT/ Pd^2 = 2 \times 10^4 / 10 \times (2.675)^2 = 279.5$$

$$\text{Log}_{10}(RF) = 2.45$$

$$\text{Thickness required for wall B} = 2.45 \times 13.5 = 42.93 \text{ or } 33 \text{ cm}$$

Example 11.7

Calculate the RF for the barrier in the A region (shown in Figure 11.15). The following data are given. (The design value for the public area has been changed in this example).

$$U = 1; \text{Occupancy } T = 1; d_A = 2.82 \text{ m; } P = 6 \text{ μSv/wk (0.6 mR/wk)}$$

$$RF = WUT/ Pd^2 = 2 \times 10^4 / 0.6 \times (2.82)^2 = 4.19 \times 10^3$$

$$\text{Log}_{10}(\text{RF}) = 3.6$$

Thickness required for wall B = 3.6 × 13.5 = 48.6 or 49 cm

Since the design value (as used in the UK and other European countries) for the public area is much smaller compared to the previous examples, the shielding thickness has increased.

Example 11.8

Calculate the RF for the barrier in the D region (shown in Figure 11.15). The following data are given.

$$U = 1$$

$$\text{Occupancy T} = 1/4$$

$$d_D = 5.45 \text{ m}$$

$$P = 0.02 \text{ mSv/wk (2 mR/wk)}$$

$$\text{RF} = \text{WUT}/\text{Pd}^2 = 2 \times 10^4 / 4 \times 2 \times (5.45)^2 = 84.16$$

$$\text{Log}_{10}(\text{RF}) = 1.93$$

Thickness required for wall B = 31.93 × 13.5 = 26.05 or 26 cm

The shielding thickness has decreased here due to the greater distance between the source and the shielded region. This thickness will be shared with the maze wall.

Example 11.9

The above HDR room is to be used for gynecological treatments. The ^{60}Co multisource HDR unit contains 20 ^{60}Co sources each of activity 18.5 GBq. The RAKR per unit activity of the source is 0.308 μGy/MBq·h at 1 m. A maximum of 15 sources are used in the intra-uterine tube with ovoids, in a single treatment. The duration of treatment is 6 minutes. A maximum of 30 patients are treated per week. Is the thickness of wall B sufficient for the ^{60}Co HDR unit used here? The following data are given.

U = 1; Occupancy T = 1; d_B = 2.57 m; P = 0.02 mSv/wk (2 mR/wk) for the B region, which is not controlled; ^{60}Co TVT in concrete = 20.6 cm

Total activity in the patient = 15 × 18.5 = 277.5 GBq or 277.5 × 10^3 MBq

Workload = 0.308 × 277.5 × 10^3 × 0.1 × 30 = 2.56 × 10^5 μGy/wk at 1 m

$$\text{RF} = \text{WUT}/\text{Pd}^2 = 2.56 \times 10^5 / 2 \times (2.57)^2 = 1.51 \times 10^3$$

$$\text{Log}_{10}(\text{RF}) = 4.29$$

Thickness required for wall B = 4.29 × 20.6 = 88 cm

The wall thickness is inadequate for this treatment. This is because of the greater total activity of ^{60}Co, the deeper penetration of the higher energy photons and higher output of the source.

Example 11.10

In the above example, is the shielding thickness computed for ^{60}Co enough to satisfy the dose constraints for IDR and TADR$_8$ for the region B (uncontrolled area), 7.5 µSv/h and 0.5 µSv/h, respectively, for the worst case scenario (i.e., using all 20 sources and maximum workload of five patients per day)?

$$RF = DR_0 / P_{IDR} \, d^2$$

$$DR_0 = 0.308 \times 20 \times 18.5 \times 10^3 = 113.96 \times 10^3 \, \mu Gy/h$$

$$RF = 113.96 \times 10^3 / [7.5 \times (2.57)^2] = 2.2932 \times 10^3$$

$$Log_{10}(RF) = 3.36$$

$$\text{Thickness required for wall B} = 3.36 \times 20.6 = 69 \text{ cm}$$

So no additional thickness is required to satisfy the IDR constraint. TADR$_8$ is given by

$$TADR_8 = IDR \times \text{use factor for the wall} \times \text{fraction of time machine was ON in 8 h day.}$$

$$TADR_8 = 7.5 \, \mu Sv/h \times (0.5 / 8) = 7.12 \, \mu Gy/h = 0.47 \, \mu Sv/h$$

This is less than the constraint imposed (0.5 µSv/h) and so no additional shielding or designation of the area (into a controlled area) is necessary.
In the above example, the occupancy factor was 1. A lower occupancy, as often occurs for the public areas, would have increased the allowed dose in this area and thus reduced the RF and hence the wall thickness. In such a case, a greater thickness may be necessary to meet the dose rate constraints.

11.12 LDR ^{137}Cs Shielding Calculations

Example 11.11

^{137}Cs is generally used for LDR treatments. In LDR treatment the patient treatment time is 18 h or more, but a radiation worker works only for 8 h. Assuming a ^{137}Cs source activity of 0.5 Ci and an RHM of 0.33, the workload is given by

$$W = 40 \times 0.5 \times 0.33 = 6.6 \text{ R/wk at 1 m}$$

If two patients are treated in the room per week,

$$W = 2 \times 6.6 = 13.2 \text{ R/wk}$$

Assuming that the room layout shown in Figure 11.15 is used for ^{137}Cs LDR treatment of a single patient, the method of calculation is illustrated below for region C.

Region C (control room)

$$W = 6.6 \text{ R/wk}; U = 1; \text{Occupancy } T = 1; d_D = 2.675 \text{ m}$$

$$P = 0.1 \text{ mSv/wk (40 mR/wk)}; \text{TVT for } ^{137}\text{Cs is 16.3 cm.}$$

$$RF = WUT/ Pd^2 = 6.6 \times 10^3 / 40 \times (2.675)^2 = 23.06$$

$$Log_{10}(RF) = 1.93$$

$$\text{Thickness required for wall B} = 1.36 \times 16.3 = 22.2 \text{ cm}$$

The wall already has a thickness of 43 cm and so no additional shielding will be required for an LDR brachytherapy patient. Since a maximum of only three patients can be treated per week, due to the much longer treatment duration, the shielding thickness is still adequate for treating three LDR patients in a week.

11.13 HDR ^{60}Co Shielding Calculations

Example 11.12

Assuming that a single source ^{60}Co HDR unit is installed in the ^{192}Ir HDR room, check, say for the same region B, whether the shielding thickness provided is adequate. Use the following data:

The maximum activity of ^{60}Co in the HDR unit is 2 Ci. $\Gamma_{x,Co}$ is 1.32 R/Ci·h at 1 m. Maximum treatment time of 4 h/wk; U = 1; T = 1; d_B = 2.57 m; P = 2 mR/wk. TVT for ^{60}C0 is 20.6 cm.

$$W = 4 \times 2 \times 1.32 = 10.5 \text{ R/wk at 1 m}$$

$$RF = WUT / Pd^2 = 10.5 \times 10^3 \times 1 \times 1 / 2 \times (2.57)^2 = 7.95 \times 10^2 = 2.9 \text{ TVL}$$

Required wall thickness = 2.9 × 20.6 = 59.8 cm. An additional 16 cm thick concrete must be added to the wall B, if a single source ^{60}Co HDR unit is installed in the same room.

11.14 HDR ^{60}Co Maze Scatter Calculations

Example 11.13

A ^{60}Co HDR unit with a source activity of 2 Ci is installed in a brachytherapy facility. The layout is shown in Figure 11.16.

The AKRC of the source is given by 0.380 μGy·m^2/MBq·h. The source activity is 74 GBq. The maze entry width in the treatment room is 1.5 m. Maze width is 1.5 m. The wall height in the maze area is 2.2 m. Source to wall distance d_1 is 4.5 m. Maze length is 3.75 m and the next turn in the maze is 1 m. The reflection coefficients for ^{60}Co for concrete are 1.02 × 10^{-2} for 45° incidence and 0° reflection (first scatter), and 4.06 × 10^{-3} at normal incidence and 75° reflection (second scatter) that reaches the door. Calculate the scatter dose at the maze entry point.

$$\text{Dose rate at the maze wall} = 0.308 \times 74 \times 10^3/(4.5)^2$$

$$\text{Reflected radiation dose rate at the second wall} = [0.308 \times 74 \times 10^3/(4.5)^2]$$

$$\times [1.02 \times 10^{-2} \times (1.5 \times 2.2)/(3.75)^2]$$

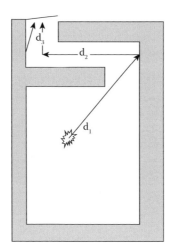

FIGURE 11.16
A ^{60}Co HDR layout.

Dose rate at the door entry point = $[0.308 \times 74 \times 10^3/(4.5)^2]$

$$\times \, [1.02 \times 10^{-2} \times (1.5 \times 2.2)/(3.75)^2]$$

$$\times \, [4.06 \times 10^{-3} \times (1.5 \times 2.2)/(1)^2]$$

$$= 0.04 \; \mu Gy/h$$

11.15 Safety Features of the RAL Treatment Room

HDR room postings and entry restrictions are similar to the linac rooms. Designation of controlled areas with signs, posting of the treatment room door with a warning light for radiation, and communications are as for EBT rooms. Other common features are an electrical interlock system to terminate treatment (and retract the source to the parking position of the unit) on opening the treatment room door, and viewing and intercom systems for patient observation, as discussed in Chapter 10.

HDR treatments are often carried out in the linac room itself, instead of in a separate brachytherapy treatment facility. In this case there is an additional safety feature to ensure that only one of the treatment delivery equipment can be operated at a time. This is achieved by providing a three-way switch which has positions indicating OFF, linac, and HDR, as shown in Figure 11.17. When the key is in the HDR position, the linac is disabled, and vice versa.

There is one patient risk that does not exist in linac therapy—the possibility of the source getting lodged in the patient and not returning to the RAL unit. To address this, there is an additional safety feature in the brachytherapy rooms (that did not exist in linac rooms) namely the installation of wall-mounted area monitors.

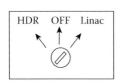

FIGURE 11.17
Three-way switch for a linac-HDR room.

The area monitor must be visible to the technician before entering the treatment room, after the termination of treatment. This monitor will indicate higher radiation levels in case the source has not gone back to the parking position but is stuck in transit for some reason. This is an emergency situation.

A well-documented procedure must be available in the department for handling such situations and the document must be available in the control room, not elsewhere in the department. All the therapy technicians must be trained in emergency handling and be familiar with the procedures to handle such situations. All the necessary tools and equipment (e.g., emergency source container, GM based, or other type of survey meter, forceps, surgical clamps, suture removal kit) must be readily available to handle any kind of emergency. The probability of such situations may be low, but is not zero; several emergency situations have occurred in HDR brachytherapy.

Some of the important radiation safety checks are:

- The camera and monitor must be in working order.
- The intercom must be working.
- The survey meter must be working (which can be checked by observing if the survey meter responds when taken near a check source, or close to the RAL equipment, in the OFF condition).
- The area monitor must be working (which can be tested with a check source). It must indicate higher radiation levels or the "source out" condition by turning the monitor light ON during the initiation of treatment.
- The door open interlock must be tested by initiating a dummy treatment plan and deliberately opening the treatment room door to check whether the treatment is terminated or not.
- By pressing the emergency off switches, the initiated dummy treatment run must be terminated.

These tests are listed in Table 11.2. All the tests must be passed before starting patient treatment.

The safety devices required for fulfilling the above radiation safety requirements are shown in Figure 11.18.

TABLE 11.2

Radiation Safety Daily Checks for RAL Brachytherapy Room

Safety Checks	Status		
	Pass	Fail	Comments
Camera/monitor function			
Intercom function			
Survey meter function (with check source)			
Area monitor function (with check source)			
Area monitor function (during treatment)			
Door open interlock function			
Wall emergency off switch function			
Console emergency off switch function			

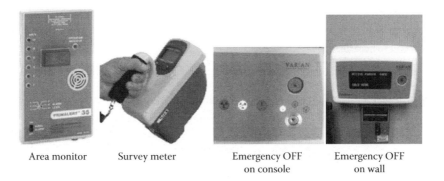

Area monitor   Survey meter   Emergency OFF   Emergency OFF
                                on console   on wall

FIGURE 11.18
Radiation safety devices required for RAL brachytherapy room.

11.16 Emergency Situations and Handling of Emergency Situations

The HDR source automatically returns to the safe following the completion of a treatment session. When this does not happen (and the machine sounds an alarm) it indicates an emergency situation. The emergency could arise as a result of (1) the source getting stuck in the applicator or transfer tube, (2) electrical or computer failure, and (3) source disconnected from the drive cable. The handling of the situation very much depends on the position of the source following the emergency. In HDR brachytherapy the risk probability is low, but the consequences are severe complications to death.

There are three separate indications to alert the presence of radiation in the treatment room, or that the source is not in the shielded location of the HDR device. The glowing warning light on the treatment room door is an indication that the treatment is ON. This should prevent any member of the public, visitor, or patient from entering the room. If a person does enter the room, the door interlock will terminate treatment retracting the source to the parking position. The other two indications are to alert the staff that, following the termination of the treatment session, the source has not returned to its usual position. The radiation monitor on the treatment room wall and the other on the treatment unit, will both show higher levels of radiation in the room. The incorporation of a radiation monitor on the treatment unit has an interesting story. In one of the HDR radiation accidents, the source became disconnected from the drive cable and remained in the patient, so only the drive cable was retracted into the device on termination of treatment. The wall-mounted monitor indicated the presence of radiation in the room but this indication was ignored by the staff since the monitor was prone to false alarms. The patient later died of over exposure. Following this accident, a GM detector was incorporated into the treatment unit so it would indicate radiation presence if only the drive cable retracted into the unit without the source. Another redundant check in this case is to survey the patient for the presence of any radiation.

Before starting HDR brachytherapy treatments in a hospital, the department must be well equipped to handle such situations. In this connection, the following points are important to note:

- An emergency plan must be ready to handle emergencies.
- The execution of the plan must be well documented and must be displayed in the HDR control room.

- All necessary emergency tools (emergency shielding container, long forceps, cutter, suture removal kit, survey meter, etc.) must be available in the control room.
- All the technicians/physicists must have undergone training in handling emergency situations.
- The emergency procedures must be practiced regularly (many of the accidents in brachytherapy resulting in unintended radiation exposures were due to lack of training and procedures) [6,7].

11.17 Handling an Emergency When the Source Is Stuck

Details of how to handle the above emergency situation (i.e., source not returning to the parking position on termination of treatment), mainly based on the recommendations of the manufacturer (Nucletron Operations BV), are briefly given below.

11.17.1 Emergency OFF

When the source is not retracting to the parking position on termination of the treatment session, the technician must try to press the emergency OFF switch on the control panel to retract the source using an independent circuit. If this too fails, the technician (trained in HDR emergency handling operations) must first inform the RSO of the department then enter the treatment room to rectify the situation.

11.17.2 Manual Source Retraction

Before entering the treatment room, the technician should wear an EPD and enter the room with a survey meter in hand to note the radiation levels. The first objective on entering the treatment room, is to make sure that the source is retrieved from the patient safely, by turning the emergency crank (see Figure 11.19) on the treatment unit manually, which will retract the cable and return the source to the storage safe of the HDR unit.

11.17.3 Disconnecting the Applicator from the Machine

If manual retraction fails, the patient must be surveyed to make sure that the source has moved into the transfer tube and is not in the patient. If so, the applicator can be disconnected from the transfer tube, the drive cable can be cut and by pulling it out of the transfer tube, it can be placed in the shielded container. This will reduce the dose received by the patient and staff compared to surgical removal of the applicator. Following this, the patient and the HDR unit must be surveyed to reconfirm that the source has returned to the unit and is not left in the patient, before taking the patient out of the room. According to some, a much safer procedure would be to remove the applicator from the patient without disconnecting the applicator from the transfer tube, so the source is not exposed. On the other hand, if the patient survey reveals that the source is still in the patient, the surgical removal of the applicator may take some time during which the patient would be receiving radiation exposure.

The patient is then taken out of the room and the treatment room is locked, restricting access, and the RSO is informed who will decide further action.

FIGURE 11.19
Manual crank shown on the HDR unit for source retraction.

NOTE: If the source is jammed for a short period of time (say 5–6 minutes) and not able to retract no clinical consequences are expected for the patient. On the other hand, if the source is in the patient he can receive a large dose during the same period of time with significant clinical complications.

For more details on the safety aspects of the ^{192}Ir HDR unit, the reader is advised to consult the AAPM Report No. 41 [8].

Example 11.14

A 10 Ci ^{192}Ir source stays in the patient for 6 minutes without retracting on termination of treatment. What is the dose received by the patient to the tissues at 1 cm from the source? What is the order of dose received by the technician standing at 1 m distance from the patient? Assume the following:

$$\text{AKRC of the source} = 0.116 \ \mu\text{Gy m}^2/\text{h MBq}$$

$$1 \text{ Gy (air kerma)} = 1 \text{ Sv (DE)}$$

$$10 \text{ Ci} = 10 \times 37 \times 10^9 \text{ Bq} = 37 \times 10^4 \text{ MBq}$$

AKRC refers to the AKR at 1 m from the source (as is clear from the units). At 1 cm distance the AKR will be 10^4 times greater.

So, AKR at 1 cm distance = $0.116 \times 10^4 \times 37 \times 10^4 = 0.116 \times 37 \times 10^2$ Gy/h (or Sv/h)

AK at 1 cm distance in 6 minutes = $0.116 \times 37 \times 10^2 \ (\times 6/60) \approx 43$ Sv

which is greater than the injury threshold dose

The dose at 1 m distance is 10^4 times less (since 1 m = 100 cm). So the approximate dose received by the technician for the same duration is 43×10^3 mSv/10^4 or 4.3 mSv which is much less than the regulatory annual dose limit prescribed for the technician.

11.18 Radiation Surveys

The general concepts regarding radiation room survey discussed for linac rooms also apply here. In a brachytherapy room, unlike a linac room, all walls, floor, and ceiling are continuously irradiated by primary radiation since the patient is an isotropic radiation source. The shielding adequacy of a brachytherapy room must be tested by keeping all the source(s) at the approximate patient position in the room. The sources are not kept in a phantom to simulate the treatment condition, to get a conservative estimate of the dose rates in the occupied areas.

The survey report must contain the following information:

1. Survey meter details (type, model, calibration, etc.)
2. Details of shielding parameters (W, U, T, etc., used)
3. Radiation unit, maximum output at 1 m, room location, room layout details
4. Survey results and conclusions

If the survey shows that additional shielding may be required for any of the barriers, the available solutions are:

- Providing additional shielding
- Restricting access to the area (if possible)
- Designating an uncontrolled area as controlled area and preventing access to public

11.18.1 HDR Equipment Leakage Survey

The dose rate due to leakage radiation around the HDR unit must not exceed the following limits:

- 100 μSv/h (10 mR/h), at a distance of 5 cm from the surface of the container
- 10 μSv/h (1 mR/h), at a distance of 1 m from the surface of the container

Figure 11.20 illustrates how the HDR equipment must be surveyed for leakage, at the two specified distances from the surface of the HDR unit.

11.18.2 Source Storage Container Survey

Brachytherapy source storage and transport containers must meet the following limits for leakage radiation level around the container (Table 11.3):

The container survey report must contain all the relevant information like the type of container (for storing temporary or permanent sources) source identification (^{125}I, ^{192}Ir wires, etc.), the activity (in mCi) stored at the time of storage, measured survey values.

11.18.3 Patient Survey

Immediately following the source implant in the patient, the medical physicist or the trained technician must make a radiation survey of the patient and the surrounding areas of use to ensure that the sources are in the patient and no source is left behind. An ion chamber–based survey meter such as Victoreen 450 P may be used for this purpose.

The radiation levels must be monitored at 1 m from the patient bed, at the recommended visitor position, behind the bed shield and near the door and the adjacent uncontrolled areas which will help to determine the stay times for the visitors and the nurse in the patient room, and the doses received in uncontrolled areas.

Immediately following the removal of sources from the patient, the medical physicist or the trained technician must again make a radiation survey of the patient to ensure that all sources have been removed.

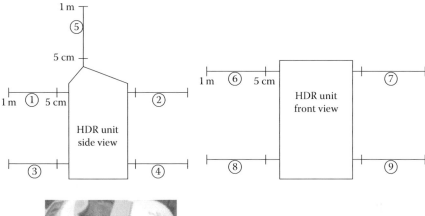

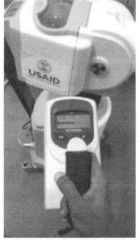

Position	Radiation level (μ Sv/h)	
	at 5 cm	at 1 m
Position 1		
Position 2		
Position 3		
Position 4		
Position 5		
Position 6		
Position 7		
Position 8		
Position 9		

FIGURE 11.20
Leakage survey of RAL equipment.

TABLE 11.3

Leakage Restrictions on Source Containers [9]

Brachytherapy Source Storage and In-house Transport Containers	Leakage Dose Rate in Air Kerma Rate (μGy/h)	
	5 cm from Surface	1 m from Source
Portable	500	20
Mobile	1000	520

The patient (and HDR unit) survey needs to be done in the case of a RAL HDR patient as well to ensure that the source has returned to the unit and not left behind in the patient. This can happen if the source welding gives way and the inbuilt detector fails to detect that only the cable retracted into the unit, without the source.

The survey report along with the details of survey (instrument used, surveying person's name, date of survey, etc.) must be recorded and retained for a period of time as per the national or regional regulations.

11.18.4 Ambient Survey (with Wall-Mounted Area Monitor)

In the case of isotopic machines (i.e., telecobalt or HDR machines), there is a small chance that the source may not return to the parking position, or return only partially. In such cases, the zone monitor (the wall-mounted area monitor) may indicate higher radiation levels indicating an emergency situation. Before entering the room after source retraction, the monitor reading must always be observed to see if there is an emergency situation.

The room survey is conducted the same way it was described for a linac room.

11.19 Brachytherapy Suite

Unlike linac treatment, brachytherapy treatment involves few additional steps:

1. An operation theater for placing the applicators in the patient
2. An imaging room for applicator localization
3. A source preparation room
4. A treatment room for afterloading the sources into the patient
5. A recovery room (for a MAL patient)
6. A rest room
7. Nurse station with patient intercom

These form a brachytherapy suite.

In the case of RAL HDR treatment, a source handling room or a nurse station is not necessary. Often, a C-arm in a nearby room is used for localization. The minimum facilities required in an HDR treatment facility are shown in Figure 11.21.

On entering the facility, the patient preparation room is on the right and the HDR treatment room is on the left, so that the patient can be moved from the operation room to the treatment room without coming into contact with the public in the hospital. A C-arm in the patient preparation room would be very useful for localization purposes without moving

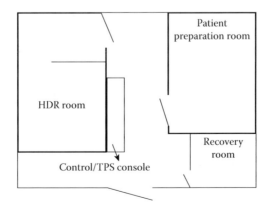

FIGURE 11.21
A typical HDR treatment facility.

the patient. The control console of the RAL HDR unit is placed just outside the HDR treatment room. Access to the enclosed facility is restricted since the whole area comes under a controlled area.

For some general references on brachytherapy physics the reader should consult the books by Thomadsen et al., and Baltas et al. [10,11].

References

1. ICRU 38, *Dose and Volume Specification for Reporting Intracavitary Therapy in Gynecology*, Betheda, MD, Oxford University Press, 1985.
2. ICRP 98, *Radiation Safety Aspects of Brachytherapy for Prostate Cancer Using Permanently Implanted Sources*, Elsevier, Amsterdam, 2005.
3. ARPANSA Radiation Protection Series (RPS) No. 14, *Practice for Radiation Protection in the Medical Applications of Ionizing Radiation*, Australian Radiation Protection and Nuclear Safety Agency (ARPANSA), Yallambie, Victoria, Australia.
4. NCRP 40, *Protection Against Radiation for Brachytherapy Sources*, NCRP, 7910, Woodmont Avenue, Bethesda, MD, USA, 1972.
5. NCRP 49, *NCRP 49: Structural Shielding Design and Evaluation for Medical Use of X-Rays and Gamma-Rays up to 10 MeV*, NCRP reports are published by NCRP, 7910, Woodmont Avenue, Bethesda, MD, USA, 1976.
6. IAEA Safety Report No. 17: *Lessons Learnt from Accidental Exposures in Radiotherapy*, IAEA, Vienna, Austria, 2000.
7. ICRP 97, *Prevention of High-dose-rate Brachytherapy Accidents*, Elsevier, Amsterdam, 2005.
8. AAPM 41, Remote afterloading technology, *Medical Physics* (20)6, 33–41, 1993.
9. Atomic Energy Regulatory Board, India Code (AERB/RF-MED/SC-1 (Rev. 1), 'Radiation Therapy Sources, Equipment and Installations'. http://www.aerb.gov.in/AERBPortal/pages/English/t/publications/CODESGUIDES/SC-MED-01R.pdf
10. B.R. Thomadsen et al., *Brachytherapy Physics*, Medical Physics Publishing, Madison, WI, 2005.
11. D. Baltas et al., *The Physics of Modern Brachytherapy for Oncology*, CRC Press, Boca Raton, FL, 2006.

Review Questions

1. ^{192}Ir wires used in breast cancer treatment is known as _____ brachytherapy (LDR/MDR/HDR).

2. What sources are normally used in permanent implants?

3. What are the advantages of RAL brachytherapy over MAL and direct lading brachytherapy practices, from a radiation safety point of view?

4. Given the source effective activity, A in GBq and the air kerma rate constant, Γ_{Ka} in Gy·m^2/h GBq, how will you calculate the air kerma rate of the source at distance d meters from the source?

5. How is a source leak tested, and what is the criterion for a leaking source?

6. What are the causes of radioactive contamination occurring in a brachytherapy room?

7. Why is the security of brachytherapy sources important?

8. What must a brachytherapy nurse be aware of?

9. Who may have access to keys to the HDR unit?
10. Should the physicist be present during HDR treatment?
11. What are the radiation survey requirements for ^{192}Ir HDR treatments?
12. State three checks that must be carried out daily in HDR brachytherapy.
13. What is the typical activity of a new ^{192}Ir HDR source, in curies?
14. What sources are used in HDR brachytherapy treatments?
15. In the case of an emergency where the ^{192}Ir HDR source is not returning to the safe after completion of treatment, what actions may be necessary?

12

Prevention of Accidents in Radiation Oncology

12.1 Introduction

Accidents are serious issues in radiation therapy for the following reasons:

- Dose rates are high and "permitted doses" to staff are very low in comparison, with a potential for overexposure to the patient and staff.
- They can adversely affect the treatment outcome even causing patient death.
- They can cause great suffering to patients.
- The reputation of the institution is affected.
- They can affect equipment manufacturers financially.
- Litigation can arise, with financial losses to the institution.

Overexposure causes clinical complications leading to death and underexposure causes recurrence of cancer making it difficult to treat the patient again. We can refer to incidents as those of relatively minor consequences and accidents as those of severe clinical complications. In this chapter, we will list both types of events that have occurred in radiation oncology. The accident may affect a single patient or a large number of patients, over a long period of time, until it is detected and rectified. For instance, a patient positioning error is a random event which may affect a single treatment session while a wrong calibration of the beam output is a systematic error which will affect all the patients treated until the error is detected.

Treatment of cancer involves several steps from prescription to treatment delivery (imaging for target localization, treatment aids fabrication, simulation, treatment planning, dose delivery with treatment verification). Errors can occur at any of these stages which will affect the treatment outcome. In discussing accidents and incidents, two issues must be kept in mind. Major accidents must of course be prevented from occurring from the patient safety point of view but even incidents like dosage errors beyond 5% from the prescribed dose must be prevented, as far as possible, since it can affect the treatment outcome leading to treatment failure or recurrence of cancer. Many safety features in the equipment and regulatory requirements in the practice of radiation oncology over the years have led to relatively few major accidents. However, a large number of over and underexposures have occurred in radiation oncology practice, compromising the treatment outcome. Overdose errors of the order of 10–25% are difficult to detect unless there is a meticulous patient follow-up, which may not be possible in less-developed countries. It is relatively easier to detect severe overdose cases from acute tissue reactions like tissue necrosis. Concentration needs to be on incidents that may go unnoticed but will have a negative impact on the treatment outcome. With the increasing complexity of the treatment technologies, over the years, the chances of committing errors in delivering the planned treatments have also increased proportionately and elaborate patient pretreatment quality audit (QA) procedures are required to avoid treatment errors in modern radiation oncology treatments.

A study of the history of accidents in radiation oncology reveals that a comprehensive QA program could have avoided many accidents. However, the number one culprit is not the QA program. An analysis of the reported accidents shows that more than 90% of the accidents had occurred not because of lack of resources or lack of elaborate QA programs, but because of complacency and lack of redundancy. Some of the prominent accidents and incidents that have occurred over the years will be briefly described in this chapter, so lessons can be learned from the previous mistakes to avoid them in the future. For more comprehensive information on accidents and prevention, the reader should refer to the publications of the International Atomic Energy Agency (IAEA), and ICRP [1,2].

12.2 Some Accidents in External Beam Therapy

12.2.1 Beam Calibration Very Much Off

Initial calibration of a ^{60}Co beam was correct, but the decay curve for ^{60}Co was wrongly drawn, underestimating the dose rate. The "defense in depth" principle of safety was not followed. Treatment times based on this were longer than appropriate, leading to overdoses, which increased with time reaching up to 50% when the error was discovered. There were no beam measurements in 22 months and a total of 426 patients were affected. Of the 183 patients who survived 1 year, 34% had severe complications.

In another incident relating to ^{60}Co teletherapy, a ^{60}Co miscalibration occurred in Costa Rica in 1996 leading to 60% underestimation of the dose rate. Longer than usual treatment times, because of the lower dose rate, went unnoticed for 35 days affecting 115 patients, 17 of them dying of overexposure, within 2 years of the incident. In a third ^{60}Co teletherapy incident that occurred in the United States in 1987–1988, following a source change, all the treatment planning system (TPS) files were updated except one file which was not deleted. The use of this old file led to 33 people receiving a 75% overdose.

Many such incidents involving incorrect beam calibration have been reported in the literature. Some of the causes of wrong calibration were:

1. Application of incorrect decay data for telecobalt source (for 2 years without making an actual check)
2. Wrong interpretation of calibration certificate
3. N_{Dw} taken as "dose in air," N_{Dair}
4. Incorrect use of the chamber (e.g., wrong orientation)
5. Improper use of correction factors
6. Incorrect calibration procedure
7. Human error (e.g., 0.3 minutes taken as 30 seconds instead of 18 seconds for beam output determination causing 66% underestimation of dose rate increasing the treatment time to the same extent causing an overdose, Costa Rica incident).

12.2.2 Accelerator Software Problems

1. Software from an older accelerator design was used for a new, substantially different, design leading to six accidental overexposures and the death of three patients.
2. A very serious accident occurred with a scanning beam linac (Therac 25 machine), built by Canada and France in collaboration, and based on the earlier models of the machine. The software was never thoroughly tested. In the X-mode, the machine withdrew the target and treated in the e-mode but without the scanning mechanism and with 100 times higher current because the machine was in X-mode. This resulted in a high-intensity pencil beam of electrons penetrating the patient. About six accidents occurred between 1985 and 1987 resulting in several deaths and serious injuries.

12.2.3 Incorrect Accelerator Repair and Communication Problems (Spain, 1990)

An attempt to repair an accelerator fault restored the electron beam but it was always delivering a 36 MeV electron beam regardless of the energy selected. A beam check by the physicist would have revealed the beam energy, but treatment was resumed without notifying physicists for a beam check (which is a normal QA practice). Also, the energy selected did not match the energy displayed but this was attributed to a faulty indicator and was not investigated and resolved. A total of 27 patients suffered from massive overexposures leading to 17 deaths.

12.2.4 Power Failure Leading to an Accident (Poland, 2001)

Power failure interrupted an electron mode treatment and the treatment was resumed when the power retuned but without checking the beam characteristics (which is a QA requirement). Two patients complained of burning. A dose rate check revealed that it was 10 times higher than intended. Power failure had also led to the damage of a safety interlock and improper functioning of the dose monitoring system. Five patients were affected by the overdose and one patient died. Following any unusual occurrence it is important to check the output and energy. However, this was overlooked because power failure was a frequent occurrence and the machine had not changed its beam characteristics on the previous occasions.

12.3 Some Accidents in Treatment Planning

12.3.1 Incomplete Understanding and Testing of a TPS (UK, 1982–1990)

In a hospital, most of the treatments were with a source–surface distance (SSD) of 100 cm and for non-standard SSDs, the SSD correction was done manually by the technicians. When a new TPS was acquired, the technician continued to apply the correction manually without realizing that the TPS algorithm already accounted for the SSD correction. This application of distance correction twice led to 30% under-dosage to the patient. A written procedure would have prevented this incident from occurring. The problem remained undiscovered for 8 years affecting 1045 patients. Of these, 492 patients developed recurrence perhaps due to an underdose.

TABLE 12.1

A Few Accidents in EBT in India

	External Beam Therapy		
Cause of Accident/Incident	Source	Nature of Accident/Incident	No. of Accidents/Incidents
Failure of source to return to OFF position	^{60}Co, ^{137}Cs	Overexposure to technologist and patient undergoing treatment	12
Falling down of source drawer during source replacement/repair	^{60}Co	Overexposure to workers involved in the procedure	2

Note: EBT:External Beam Therapy.

12.3.2 Untested Change of Procedure for Data Entry into TPS (Panama, 2000)

A TPS allowed data entry for four shielding blocks, one block at a time. A need for five blocks led to a deviation from this procedure, entering data for several blocks in a single step. There was also some ambiguity in the user instructions. The TPS computed treatment time was double the correct one leading to 100% overdose. A written procedure for data entry did not exist which would have prevented this error from occurring. Computer output was not checked against manual calculations which would have uncovered the error. The error affected 28 patients and 1 year after the incident, at least five patients had died of overexposure.

Apart from the above prominent accidents, other causes of the reported accidents are:

- Lack of agreement between chambers
- Simulation error
- Treatment planning errors
- Lack of knowledge on TPS algorithms
- Incorrect basic data in TPS
- Incorrect data input into the TPS
- Incorrect patient dose calculations
- Incorrect treatment setup
- Patient mix-up
- Improper setting up of accessories
- Lack of communication (human error)

Table 12.1 shows a few accidents that have occurred in India in electronic brachytherapy (EBT).

12.4 Some Accidents in Brachytherapy

12.4.1 Malfunction of High-Dose-Rate Brachytherapy Equipment (USA, 1992)

A high-dose rate (HDR) ^{192}Ir source became detached from the drive cable and remained in the patient, and only the cable was retracted despite the console display indicating source retraction. The zone monitor indicated a radiation level in the room which was ignored. The source remained in the patient for several days and the patient died of overexposure.

A Geiger–Muller detector in the HDR unit would have indicated whether the source returned or not. This provision was introduced in the later versions of the HDR equipment. A survey of the patient before leaving the room also would have indicated the presence of source in the patient. This is a mandatory requirement these days to rule out the possibility of source left in the patient.

12.4.2 Treatment of Wrong Site

In ^{192}Ir seed treatment of the lung, a kink in the catheter used to insert the seeds, resulted in the seeds treating a site 26 cm away from the target site. The error was detected only at the end of the treatment.

A study of brachytherapy accidents reveals the following as the causes of accidents:

- Equipment failure/malfunction
- Source order and delivery
- Source specification/calibration
- Source preparation for treatment
- Incorrect use of source strength in TPS
- Inconsistent use of source specification
- Selection of wrong source strength
- Incorrect supplier data/no verification at the hospital
- Wrong isotope data for treatment time
- Source implantation errors
- Treatment planning and dose calculation errors
- Loss or dislodging of sources
- Improper handling of sources
- Removal time error in calculation
- Source removal errors

Table 12.2 shows details of a few accidents that have occurred in India in brachytherapy.

TABLE 12.2

Some Brachytherapy Accidents in India

Brachytherapy			
Cause of Accident/Incident	Source	Nature of Accident/Incident	No. of Accidents/Incidents
Loss of source	^{60}Co, ^{137}Cs	Loss of brachytherapy tubes and needles	23
Improper maintenance	^{60}Co	Dislodge of source pencil from a RAL unit causing overexposure to workers	1
Breakage of a sealed source	^{60}Co	Breakage of a sealed source caused contamination	1
Dislodge of a source from the catheter	^{192}Ir	Dislodge of an interstitial source from the open end of the applicator resulted in excessive exposure to the patient	1

Note: RAL, remote afterloading.

12.5 Clinical Consequences of Radiation Accidents

The side effects of radiation therapy treatments are usually minor (e.g., skin reddening) and transient. The normal tissue complications, however, are more severe and long lasting. It is the complications that limit the dose that can be safely delivered for curative treatment.

The normal tissue complications are of two types: (1) acute (usually found in fast-dividing cells) and (2) chronic (late, usually found in slowly proliferating cells). Acute effects manifest early (within days to weeks) and the recovery is fast and complete. Late effects manifest after a considerable time interval and the damage is often permanent.

One of the objectives of radiation therapy is to achieve local tumor control without causing much damage to normal tissues. It is the unacceptable normal tissue complications that determine the amount of dose that can be delivered to the patient for the purposes of a cure. All modern treatment technologies attempt to optimize tumor dose to normal tissue dose (complications). The tissue complications are kept at an acceptable level so the normal tissues retain the structure and function of the organ. This very much depends on the type of organ tissues. This dose is known as the tissue tolerance dose (TD).

Figure 12.1 shows the dose–response curves for the tumor, and the normal tissues that limit the tumor dose. A 5% complication rate is generally taken as the acceptable level of complications (i.e., complications must be manifested only in 5% of the cases). This decides the dose that can be delivered to the normal tissues which in turn decides the deliverable tumor, as shown in Figure 12.1.

A simple model is used to explain the TD of normal tissues. The normal tissues are organized as functional subunits (FSUs)—the smallest grouping of cells within which a

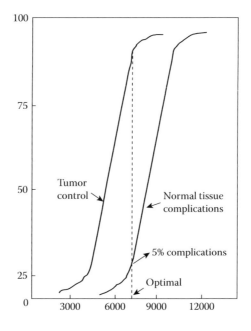

FIGURE 12.1
Dose–response curves for tumor and normal tissues.

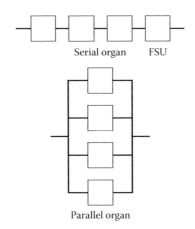

FIGURE 12.2
Arrangement of FSUs in serial and parallel organs.

single cell survival is sufficient for the regeneration of the entire FSU. Some tissues behave as though the FSUs are arranged in series and some tissues behave as though the FSUs are arranged in parallel. So, the organs are generally divided into serial organs and parallel organs, as shown in Figure 12.2. More complex organs exhibiting a combination of both also exist.

In serial organs, damage to one FSU is enough to incapacitate the whole organ. For example, rectal mucosa is a serial organ. The spinal cord and intestines are other examples of serial organs. Here, an FSU or a minimum volume exceeding the TD can cause major complications (e.g., paralysis). In a parallel organ, if one FSU is damaged other FSUs can retain the organ function. These organs exhibit a graded response depending on the volume of tissue irradiated. The lung is an example of parallel organ.

Late effects may take months to years to develop and are permanent. Some complications reported in the literature are (1) extensive fibrosis of the left groin with limitation of hip motion as a result of accidental overexposure, (2) a young woman becoming quadriplegic as a result of accidental overexposure to the spinal cord, and (3) a child affected by overdoses to brain and spinal cord and losing the ability to speak and walk.

12.6 Prevention of Accidents in Radiation Oncology

A single error or fault of equipment or human should not develop into an accident. This can be achieved by employing defense in depth in the design of the equipment and independent checks of any physicist's measurement or calculations. Employment of a primary monitor, secondary monitor, and a timer to control the treatment beam delivery is an example of the defense in depth concept. If the primary monitor fails, the secondary monitor terminates the treatment and if both fail, the timer will terminate the treatment. The secondary monitor and the timer are suitably set for this purpose. Similarly, an independent check would have detected the errors and prevented many of the incidents reported here.

A comprehensive QA program is a must to prevent many accidents. The majority of accidents occurred in hospitals that had no comprehensive QA program, or did not properly implement the program. Some important steps which could have prevented many of the accidents reported in the literature are:

- Training
- In-house comprehensive QA program for all the equipment (including the TPS) and dosimeters, with everyone's role clearly defined
- Regular beam calibration using an accepted protocol and independent verification
- External dose QA program
- Record keeping/written procedures (documentation)
- In vivo dosimetry program
- Education and training
- Careful observation of patients for unusual or unexpected symptoms
- Allocation of resources
- Patient identification and patient chart

Effective patient identification (through photographs) and treatment charts (double check of chart data at the beginning of treatment, before changes in the course of treatment [e.g., a new field], and at least once a week).

Some additional QA programs are required for the radioactive sources in brachytherapy:

- Source leakage tests
- Source air kerma strength check and its use in the TPS
- Source uniformity check with autoradiograph or by other means
- Provisions for checking source activity and source identification before use
- Automatic retraction of HDR remote afterloading sources in case of obstruction or no transfer tube connected to the indexer
- TPS dose calculation verification
- Source positioning and source removal
  - Provisions to verify source position
  - Provisions to ensure that sources do not remain in the patient (including monitoring patients and clothes)

To summarize, a comprehensive QA program/defense in depth is the key element in prevention of accidents.

12.7 Security of Radioactive Sources

Safe keeping of radioactive sources is a serious issue that is being widely discussed these days. In countries where the regulatory mechanisms are not strong, or in countries troubled by violence, security of sources becomes particularly important. Table 12.3 lists some major accidents that have occurred in the past due to abandoned sources.

TABLE 12.3

Accidents Due to Abandoned Sources

Country	Year	Consequences on Humans
Mexico	1984	4000 people exposed (5 of them from 2–7 Gy)
Brazil	1988	4 persons died due to radiation exposure; 112 persons monitored for contamination
Georgia	1994	1 person died due to radiation exposure
Turkey	1999	11 persons severely exposed to radiation
Thailand	2000	3 persons died due to radiation exposure

References

1. *Lessons Learnt from Accidental Exposures in Radiotherapy*, IAEA Safety Report No. 17, IAEA, Vienna, Austria, 2000.
2. ICRP 86, *Prevention of Accidental Exposures to Patients undergoing Radiation Therapy*, ICRP, Pergamon Press, UK, 2000.

Review Questions

1. How do accidents affect the patient, staff, or the institution?
2. What is the important consequence of overexposure or underexposure?
3. At what stage of patient treatment can errors occur?
4. List some causes of accidents in EBT.
5. List some causes of accidents in brachytherapy.
6. Give two examples of serial and parallel organs.
7. What are acute and late effects in radiotherapy?
8. Mention three important steps which can help to avoid many accidents that have been reported in the literature.
9. Why is security of sources of great importance?

Index